Communications
in Computer and Information Science   31

T0100365

Sushil K. Prasad   Susmi Routray
Reema Khurana   Sartaj Sahni (Eds.)

Information Systems, Technology and Management

Third International Conference, ICISTM 2009
Ghaziabad, India, March 12-13, 2009
Proceedings

 Springer

Volume Editors

Sushil K. Prasad
Deptartment of Computer Science
Georgia State University
Atlanta, GA, USA
E-mail: sprasad@gsu.edu

Susmi Routray
Institute of Management Technology
Ghaziabad, India
E-mail: sroutray@imt.edu

Reema Khurana
Institute of Management Technology
Ghaziabad, India
E-mail: rkhurana@imt.edu

Sartaj Sahni
CISE Department
University of Florida, Gainesville, FL, USA
E-mail: sahni@cise.ufl.edu

Library of Congress Control Number: 2009921607

CR Subject Classification (1998): K.6, D.2, H.2.8, D.4.6, C.2

ISSN   1865-0929
ISBN-10   3-642-00404-0 Springer Berlin Heidelberg New York
ISBN-13   978-3-642-00404-9 Springer Berlin Heidelberg New York

springer.com

© Springer-Verlag Berlin Heidelberg 2009
Printed in Germany

Typesetting: Camera-ready by author, data conversion by Scientific Publishing Services, Chennai, India
Printed on acid-free paper   SPIN: 12624317   06/3180   5 4 3 2 1 0

Message from the Program Co-chairs

Welcome to the Third International Conference on Information Systems, Technology and Management (ICISTM 2009), held at IMT Ghaziabad, India.

The main program consisted of 30 regular papers selected from 79 submissions. The submissions represented all the six inhabited continents and 15 different countries, with top six being India, Iran, USA, Tunisia, Brazil, and Sweden. These submissions were distributed among four track areas of Information Systems (IS), Information Technology (IT), Information Management (IM), and Applications. The four Track Chairs, Shamkant Navathe (IS), Indranil Sengupta (IT), Subhajyoti Bandyopadhyay (IM), and Mahmoud Daneshmand (Applications), ably assisted us. The submissions went through a two-phase review. First, in coordination with the Track Chairs, the manuscript went through a vetting process for relevance and reassignment to proper tracks. Next, the Track Chairs assigned each paper to four Program Committee (PC) members for peer review. A total of 274 reviews were collected. A few external reviewers also helped in this process. Finally, 30 papers were selected to be presented during ten sessions.

The conference proceedings are published in Springer's CCIS series. This required a review by Springer's editorial board, and represents a significant improvement in the publication outlet and overall dissemination—we thank Stephan Goeller and Alfred Hofmann, both from Springer, for their assistance. The 30 regular papers were allocated 12 pages each, with two additional pages on payment of excess page charges. Additionally, some authors were invited to present posters of their work in a separate poster session—two-page extended abstracts of these posters also appear in the conference proceedings.

For the best paper selection, we invited PC members to send their preferences and each track chair to nominate one or more papers from their respective areas. The best paper selected with unanimity was "A Reduced Lattice Greedy Algorithm for Selecting Materialized Views" by T.V. Vijay Kumar and Aloke Ghoshal.

For a newly-created conference, it has been our goal to ensure improvement in quality as the primary basis to build upon. In this endeavor, we were immensely helped by our Program Committee members and the external reviewers. They provided rigorous and timely reviews and, more often than not, elaborate feedback to all authors to help improve the quality and presentation of their work. Our sincere thanks go to the 56 PC members, 21 external reviewers, and the four Track Chairs. Another important goal has been to ensure participation by the international as well as India-based research communities in the organizational structure and in authorship.

Throughout this process, we received the sage advice from Sartaj Sahni, our conference Co-chair, and help from Reema Khurana, also a conference Co-chair; we are indeed thankful to them. We thank the director of IMT Ghaziabad,

B. S. Sahay for providing constant support during the conference and for providing the premises and facilities of IMT for hosting the conference. We are also thankful to Rajshekhar Sunderraman for taking up the important and time-consuming job of Publications Chair, interfacing with all the authors, ensuring that Springer's formatting requirements were adhered to, and that the requirements of copyrights, author registrations and excess page charges were fulfilled, all in a timely and professional manner. Finally, we thank all the authors for their interest in ICISTM 2009 and for their contributions in making this year's technical program particularly impressive. We wish all the attendees and authors a very informative and engaging conference.

January 2009   Sushil K. Prasad
                                                                  Susmi Routray

Organization

General Co-chairs

Reema Khurana   Institute of Management Technology, Ghaziabad, India

Sartaj Sahni   University of Florida, USA

Program Co-chairs

Sushil K. Prasad   Georgia State University, USA

Susmi Routray   Institute of Management Technology, Ghaziabad, India

Track Chairs

Information Systems   Shamkant Navathe, Georgia Tech., USA

Applications   Mahmoud Daneshmand, AT&T Labs - Research, USA

Information Technology   Indranil Sengupta, IIT, Kharagpur, India

Information Management   Subhajyoti Bandyopadhyay, Univ. of Florida, USA

Workshop and Tutorial Chair

Rajat Moona   Indian Institute of Technology, Kanpur, India

Publications Chair

Rajshekhar Sunderraman   Georgia State University, USA

Publicity Chairs

S. Balasundaram   Jawaharlal Nehru University, Delhi, India

Paolo Bellavista   University of Bologna, Italy

Rajkumar Buyya   The University of Melbourne, Australia

Guihai Chen   Nanjing University, China

Yookun Cho   Seoul National University, South Korea

Mario Dantas   Federal University of Santa Catarina, Brazil
Suthep Madarasmi   King Mongkut's University of Technology,
                          Thonburi
Milind S. Mali   University of Pune, India
Jayanti Ranjan   Institute of Management Technology, Ghaziabad,
                          India

Finance Chair

Prateechi Agarwal   Institute of Management Technology, Ghaziabad,
                          India

Local Arrangements Chair

Roop Singh   Institute of Management Technology, Ghaziabad,
                          India

ICISTM 2009 Program Committee

Information Systems

Rafi Ahmed   Oracle Corporation, USA
Angela Bonifati   ICAR CNR, Italy
David Butler   Lawrence Livermore National Lab, USA
Susumu Date   Osaka University, Japan
Amol Ghoting   IBM T.J. Watson Research Center, USA
Minyi Guo   University of Aizu, Japan
Song Guo   University of Aizu, Japan
Ruoming Jin   Kent State University, USA
Kamal Karlapalem   IIIT, India
Vijay Kumar   University of Missouri, Kansas City, USA
Jason Liu   Florida International University, USA
Rajshekhar
   Sunderraman   Georgia State University, USA
Feilong Tang   Shanghai Jiao Tong University, China
Wai Gen Yee   Illinois Institute of Technology, USA

Information Technology

Rekha Agarwal   Amity Delhi, India
Paolo Bellavista   University of Bologna, Italy
Yeh-Ching Chung   National Tsing Hua University, Taiwan
Poonam Garg   IMT Ghaziabad, India

Kohei Ichikawa	Kansa University, Japan
Ravindra K. Jena	IMT Nagpur, India
Bjorn Landfeldt	Sydney University, Australia
Young Lee	Sydney University, Australia
Xiaolin (Andy) Li	Oklahoma State University, USA
Manish Parasher	Rutgers University, USA
Sanjay Ranka	University of Florida, USA
Mrinalini Shah	IMT Ghaziabad, India
Arun M. Sherry	CDIL, India
Javid Taheri	Sydney University, Australia
Shirish Tatikonda	Ohio State University, USA
David Walker	Cardiff University, UK
Albert Zomaya	Sydney University, Australia

Information Management

Anurag Agarwal	University of Florida, USA
Manish Agarwal	University of Florida, USA
Haldun Aytug	University of Florida, USA
Seema Bandhopadhyay	University of Florida, USA
Hsing Kenneth Cheng	University of Florida, USA
Jeevan Jaisingh	UST Hong Kong, China
Karthik Kannan	Purdue University, USA
Akhil Kumar	Penn State, USA
Chetan Kumar	California State University, San Marcos, USA
Subodha Kumar	University of Washington, USA
H. R. Rao	University of Buffalo, USA
Jackie Rees	Purdue University, USA
Tarun Sen	Virginia Tech., USA
Chandrasekar Subramanian	UNC Charlotte, USA
Surya Yadav	Texas Tech. University, USA
Wei T. Yue	University of Texas, Dallas, USA

Applications

Janaka Balasooriya	University of Missouri, Rolla, USA
Akshaye Dhawan	Georgia State University, USA
Siamak Khorram	North Carolina State University, Raleigh, USA
Kamna Malik	U21, India
Sita Misra	IMT Ghaziabad, India
Fahad Najam	AT&T Labs Research, USA
Eric Noel	AT&T Labs Research, USA
Prabin K. Panigrahi	IIM Indore, India
M.L. Singla	India

External Reviewers

Muad Abu-Ata
Rasanjalee Dissanayaka
Shaoyi He
Jia Jia
Saurav Karmakar
Sarah Khan
Nanda Kumar
Fan-Lei Liao
Shizhu Liu
Srilaxmi Malladi
Chibuike Muoh
Linh Nguyen
Neal Parker
Raghav Rao
Fadi Tashtoush
Orcun Temizkan
Navin Viswanath
Xiaoran Wu
Zai-Chuan Ye

Table of Contents

Session 4. Scheduling and Distributed Systems

Session 5. Advances in Software Engineering

Session 6. Case Studies in Information Management

Session 7. Algorithms and Workflows

Session 8. Authentication and Detection Systems

Session 9. Recommendation and Negotiation

Session 10. Secure and Multimedia Systems

Extended Abstracts of Posters

Embedded Sensor Networks

Sitharama S. Iyengar

Roy Paul Daniels Professor and Chair
Sensor Network Laboratory - Dept. of Computer Science
Louisiana State University, USA

Abstract. Embedded sensor networks are distributed systems for sensing and in situ processing of spatially and temporally dense data from resource-limited and harsh environments such as seismic zones, ecological contamination sites are battle fields. From an application point of view, many interesting questions arise from sensor network technology that go far beyond the networking/computing aspects of the embedded system. This talk presents an overview of various open problems that are both of mathematical and engineering interests. These problems include sensor-centric quality of routing/energy optimization among other graph theoretic problems.

Bio: Dr. S. S. Iyengar is the Chairman and Roy Paul Daniels Distinguished Professor of Computer Science at Louisiana State University and is also Satish Dhawan Chaired Professor at Indian Institute of Science. He has been involved with research in high-performance algorithms, data structures, sensor fusion, data mining, and intelligent systems since receiving his Ph.D. degree in 1974. He has directed over 40 Ph.D. students, many of whom are faculty at major universities worldwide or scientists or engineers at national labs/industry around the world. His publications include 15 books (authored or coauthored, edited; Prentice-Hall, CRC Press, IEEE Computer Society Press, John Wiley & Sons, etc.) and over 350 research papers in refereed journals and conferences. He has won many best paper awards from various conferences. His research has been funded by NSF, DARPA, DOE-ORNL, ONR among others. He is a Fellow of ACM, Fellow of the IEEE, Fellow of AAAS. Dr. Iyengar is the winner of the many IEEE Computer Society Awards. Dr. Iyengar was awarded the LSU Distinguished Faculty Award for Excellence in Research, the Hub Cotton Award for Faculty Excellence, and many other awards at LSU. He received the Prestigious Distinguished Alumnus Award from Indian Institute of Science, Bangalore in 2003. Also, Elected Member of European Academy of Sciences (2002). He is a member of the New York Academy of Sciences. He has been the Program Chairman for many national/international conferences. He has given over 100 plenary talks and keynote lectures at numerous national and international conferences.

S.K. Prasad et al. (Eds.): ICISTM 2009, CCIS 31, p. 1, 2009.
© Springer-Verlag Berlin Heidelberg 2009

Future of Software Engineering

Pankaj Jalote

Director, Indraprastha Institute of Information Technology (IIIT), Delhi,
Professor, Department of Computer Science and Engineering
Indian Institute of Technology, Delhi, India

Abstract. Software and software business are undergoing a fundamental change, being driven by the limitations of current software models, ubiquity of computing, and increased penetration of wireless devices. At the same time, software has historically also shown a longevity and stickiness which other forms of technology have not shown. As software and software business evolve to respond to the new scenario, software engineering will have to change to deliver. Though any prediction is hard, and predicting in technology is even harder, this talk will look at how some aspects of software engineering might be in Future.

Bio: Pankaj Jalote is currently the Director of Indraprastha Institute of Information Technology (IIIT) Delhi, an autonomous Institute created through an act by Delhi Govt and with a State University status. Before joining IIIT, he was the Microsoft Chair Professor at Dept of Computer Science and Engineering at IIT Delhi. Earlier he was a Professor in the Department of Computer Science and Engineering at the Indian Institute of Technology Kanpur, India, where he was also the Head of the CSE Department from 1998 to 2002. From 85-89, he was an Assistant Professor at University of Maryland at college Park. From 1996 to 1998, he was Vice President at Infosys Technologies Ltd., a large Bangalore-based software house, and from 2003 to 2004, was a Visiting Researcher at Microsoft Corporation, Redmond, USA. He has a B.Tech. from IIT Kanpur, MS from Pennsylvania State University, and Ph.D. from University of Illinois at Urbana-Champaign. He is the author of four books including the highly acclaimed CMM in Practice, which has been translated in Chinese, Japanese, Korean etc, and the best selling textbook An Integrated Approach to Software Engineering. He is an advisor to many companies and is on the Technical Advisor Board of Microsoft Research India (Bangalore). His main area of interest is Software Engineering. He was on the editorial board of IEEE Transactions on Software Engineering and is currently on editorial board of Intl. Journal on Empirical Software Engineering and IEEE Trans. on Services Computing. He is a Fellow of the IEEE.

S.K. Prasad et al. (Eds.): ICISTM 2009, CCIS 31, p. 2, 2009.
© Springer-Verlag Berlin Heidelberg 2009

Seeing beyond Computer Science and Software Engineering

Kesav Vithal Nori

Executive Director and Executive vice President
TCS Innovation Lab - Business Systems, Hyderabad, India
`kesav.nori@tcs.com`

Abstract. The boundaries of computer science are defined by what *symbolic computation* can accomplish. Software Engineering is concerned with effective use of computing technology to support automatic computation on a large scale so as to construct desirable solutions to worthwhile problems. Both focus on what happens within the machine. In contrast, most practical applications of computing support end-users in realizing (often unsaid) objectives. It is often said that such objectives cannot be even specified, e.g., what is the specification of MS Word, or for that matter, any flavour of UNIX? This situation points to the need for architecting what people do with computers. Based on Systems Thinking and Cybernetics, we present such a viewpoint which hinges on Human Responsibility and means of living up to it.

Bio: Professor Kesav Vithal Nori is executive Director and executive Vice President of Tata Consultancy Services (TCS) and is one of the pioneers of TCS's computer-based adult literacy programme which has been successful all over India and in South Africa. Nori's research interests include: programming languages and compliers, meta tools and derivation of tools for software processes, modeling, simulation and systems engineering applied to enterprise systems. Nori has a distinguished career that has alternated between academics and industrial research within TCS. Previously, Nori was with the Tata Research Development and Design Centre (TRDDC), the R&D wing of TCS. Nori received his MTech from IIT Kanpur where he continued to work as a senior research associate from 1968 to 1970. Nori then became a research scientist at the Computer Group in the Tata Institute of Fundamental Research (TIFR). He then returned to teach at IIT Kanpur. In addition, Nori was Visiting Faculty at Carnegie Mellon University (CMU) in Pittsburgh, USA and finally returned to TRDDC and TCS.

S.K. Prasad et al. (Eds.): ICISTM 2009, CCIS 31, p. 3, 2009.
© Springer-Verlag Berlin Heidelberg 2009

System on Mobile Devices Middleware:
Thinking beyond Basic Phones and PDAs

Sushil K. Prasad

Director, GEDC-GSU Distributed and Mobile Systems (DiMoS) Laboratory
Chair, IEEE Computer Society Technical Committee on Parallel Processing (TCPP)
Professor of Computer Science
Georgia State University, USA
sprasad@gsu.edu

Abstract. Several classes of emerging applications, spanning domains
such as medical informatics, homeland security, mobile commerce, and
scientific applications, are collaborative, and a significant portion of these
will harness the capabilities of both the stable and mobile infrastructures
(the "mobile grid"). Currently, it is possible to develop a collaborative
application running on a collection of heterogeneous, possibly mobile,
devices, each potentially hosting data stores, using existing middleware
technologies such as JXTA, BREW, Compact .NET and J2ME. However, they require too many ad-hoc techniques as well as cumbersome
and time-consuming programming. Our System on Mobile Devices (SyD)
middleware, on the other hand, has a modular architecture that makes
such application development very systematic and streamlined. The architecture supports transactions over mobile data stores, with a range of
remote group invocation options and embedded interdependencies among
such data store objects. The architecture further provides a persistent
uniform object view, group transaction with Quality of Service (QoS)
specifications, and XML vocabulary for inter-device communication. I
will present the basic SyD concepts, introduce the architecture and the
design of the SyD middleware and its components. We will discuss the basic performance figures of SyD components and a few SyD applications on
PDAs. SyD platform has led to developments in distributed web service
coordination and workflow technologies, which we will briefly discuss.
There is a vital need to develop methodologies and systems to empower
common users, such as computational scientists, for rapid development
of such applications. Our BondFlow system enables rapid configuration
and execution of workflows over web services. The small footprint of the
system enables them to reside on Java-enabled handheld devices.

Bio: Sushil K. Prasad (BTech'85 IIT Kharagpur, MS'86 Washington State, Pull-
man; PhD'90 Central Florida, Orlando - all in Computer Science/Engineering)
is a Professor of Computer Science at Georgia State University (GSU) and Director of GSU-GEDC Distributed and Mobile Systems (DiMoS) Lab hosted at
Georgia Institute of Technology, Atlanta. He has carried out theoretical as well as
experimental research in parallel, distributed, and networked computing, resulting in 85+ refereed publications, several patent applications, and about $1M in

S.K. Prasad et al. (Eds.): ICISTM 2009, CCIS 31, pp. 4–5, 2009.
© Springer-Verlag Berlin Heidelberg 2009

external research funds as PI and over \$4M overall (NSF/NIH/GRA/Industry). Recently, Sushil successfully led a multi-year, Georgia Research Alliance (GRA) funded interdisciplinary research project with seven GSU faculty, three Georgia Tech faculty, and over two dozen students on developing SyD middleware for collaborative distributed computing over heterogeneous mobile devices, resulting in several patents, dissertations, and publications. Sushil has been very active in the professional community, serving on the organizations of top conferences, on NSF and other review panels, on advisory committees of conferences and State of Georgia funding agency Yamacraw, and carrying out editorial activities of conference proceedings and journal special issues. Sushil has received invitations for funded research visits from a variety of organizations nationally and internationally (Oak Ridge National Lab, 2008; Indian Inst. of Science, 2007; University of Melbourne and NICTA, Australia, 2006; University of New Brunswick, Canada, 2005). In May 2007, he was conferred an Honorary Adjunct Professorship at University of New Brunswick, Canada, for his collaborative research on ACENET project to establish high performance computing infrastructures in Atlantic Canada. Sushil has been elected as the chair of IEEE Technical Committee on Parallel Processing (2007-09).

A Reduced Lattice Greedy Algorithm for Selecting Materialized Views

T.V. Vijay Kumar and Aloke Ghoshal

School of Computer and Systems Sciences,
Jawaharlal Nehru University,
New Delhi-110067

Abstract. View selection generally deals with selecting an optimal set of beneficial views for materialization subject to constraints like space, response time, etc. The problem of view selection has been shown to be in NP. Several greedy view selection algorithms exist in literature, most of which are focused around algorithm HRU, which uses a multidimensional lattice framework to determine a good set of views to materialize. Algorithm HRU exhibits a high run time complexity. One reason for it may be the high number of re-computations of benefit values needed for selecting views for materialization. This problem has been addressed by the algorithm Reduced Lattice Greedy Algorithm (RLGA) proposed in this paper. Algorithm RLGA selects beneficial views greedily over a reduced lattice, instead of the complete lattice as in the case of HRU algorithm. The use of the reduced lattice, containing a reduced number of dependencies among views, would lead to overall reduction in the number of re-computations required for selecting materialized views. Further, it was also experimentally found that RLGA, in comparison to HRU, was able to select fairly good quality views with fewer re-computations and an improved execution time.

Keywords: Materialized View Selection, Greedy Algorithm, Reduced Lattice.

1 Introduction

A materialized view selection problem generally deals with selecting an optimal set of views for materialization subject to constraints like space, response time, etc. [17]. An optimal selection of views usually refers to the set of views that are most beneficial in answering user queries having least operational costs [17]. As per [20], there are three possible ways of selecting views to materialize, namely materialize all, materialize none and materialize a select few. Each of these ways trades, in different amounts, the space requirement for storing views, against the response time for answering user queries. Although materializing all views (2^d views, for d dimensions) may result in the least possible response time, storing these views would entail a high space overhead. At the other extreme, with no views materialized, the view's results would have to be recomputed each time, increasing the response time. Thus, the only option

S.K. Prasad et al. (Eds.): ICISTM 2009, CCIS 31, pp. 6–18, 2009.
© Springer-Verlag Berlin Heidelberg 2009

available is to selectively materialize a subset of views statically that are likely to be most beneficial in answering user queries, while keeping the remaining views dynamic [15, 20]. However, due to the exponentially high number of possible views existing in a system, the search space is very large, so the problem of selecting the optimal subset from it for materialization has been shown to be in NP [4, 8, 15]. Thus, for all practical purposes, the selection of the subset of views to materialize is done by pruning of the search space either empirically or heuristically[16]. The empirical selection is based on the past user querying patterns[1, 2]. On the other hand, the heuristics based selection identify the best possible subset of views using algorithms such as Greedy [3, 5, 7, 8, 9, 11, 14, 20], A*[6, 7], Genetic [12, 13], etc.

This paper focuses on the greedy algorithms for selecting views for materialization. These algorithms at each step selects an additional view, that has the maximum benefit per unit space and that fits within the space available for view materialization. Several papers discuss the greedy heuristic for selecting views for materialization [3, 5, 7, 8, 9, 11, 14, 20]. One of the fundamental greedy based algorithms was proposed in [11]. This algorithm, referred to as HRU hereafter in this paper, uses a multidimensional lattice framework [11, 15], expressing dependencies among views, to determine a good set of views to materialize.

Algorithm HRU requires benefit values of the all the views in the lattice to be computed other than those that are already materialized. Selection of a beneficial view may affect the benefit values of other views in the lattice. The affected views' benefit values are re-computed for selecting the next beneficial view. That is, with each selection, the benefit values of other views in the lattice may get affected and require re-computation. Since all possible dependencies are depicted by the complete lattice, the number of re-computations with every selection would be high. This number of re-computations would grow with an increase in the number of dimensions of the data set. This in turn may lead to high run time complexity[14, 15, 18, 19]. An attempt has been made in this paper to address this problem by reducing the number of re-computations of benefit values.

In this paper, a Reduced Lattice Greedy Algorithm (RLGA) is proposed that selects beneficial views greedily over a reduced lattice, instead of the complete lattice as in HRU. RLGA first arrives at a reduced lattice, based on a lattice reduction heuristic, followed by greedy selection of beneficial views over it. The reduced lattice, containing fewer dependencies, may result in lesser number of re-computations of benefit values. This may improve the performance of the system.

The paper is organized as follows: Section 2 discusses greedy based algorithms with emphasis on HRU. The proposed algorithm RLGA is given in section 3. Section 4 gives an example illustrating selection of views using HRU and RLGA followed by their comparisons in section 5. Section 6 is the conclusion.

2 Greedy Based Algorithms

As discussed earlier, greedy based algorithms are widely used for selecting views to materialize. The problem of selection of the set of views to materialize for data cubes is NP-complete[11]. HRU addressed this problem by selecting top-k views, in decreasing order of their benefit per unit space values over a multidimensional lattice.

This model was extended in [8] by considering factors like probability of a view being queried and space constraints for storing the materialized views with respect to a hypercube lattice. [8, 9] presented a polynomial time greedy algorithm for selecting views for materialization using AND graphs and OR graphs, and an exponential time algorithm for AND-OR graphs. However, only the cost of materialization, subject to maximum space available for materialization of views, was considered. The cost of maintenance was later included in [7]. A scalable solution to HRU was presented as PGA in [14], which selects views in polynomial time. An algorithm is given in [5] that select optimum set of views for materialization using the number of dependent relationships and the frequency of update to the base relations. An adapted greedy algorithm is presented in [3], which uses the hybrid approach to select views for materialization. The hybrid approach only selects those views for materialization that offer higher benefit, while the rest of the views are kept virtual. Most of these algorithms are focused around algorithm HRU, which is discussed next.

2.1 Algorithm HRU

HRU is a heuristic based greedy algorithm that uses a multidimensional lattice framework to determine a good set of views to materialize. HRU works with a linear cost model, where cost is computed in terms of the size of the view that needs to be known in advance. The basis for this model is that there is almost a linear relationship between the size of the view on which a particular query is executed and the execution time of the query, as all tuples of the view may be needed to answer a query. In HRU, the benefit of a view is computed using the cost (size) associated with the view[11]. The benefit value is a function of the number of dependent views and the difference in size of a view and that of its nearest materialized ancestor view. The benefit value of view V i.e. B_V is defined as

$$B_V = (SizeNMA_V - Size_V) \times D_V$$

where V- View, $Size_V$ – Size of V, $SizeNMA_V$ – Size of the Nearest Materialized Ancestor View of V, D_V – Number of Dependents of V and B_V – Benefit of V.

Initially the root node (view) of the lattice is assumed as materialized and is used to compute all benefit values. In each of the iterations, the most beneficial view is selected for materialization. With each view selection, the benefit values of other views in the lattice change and are recomputed with respect to their nearest materialized ancestor view. The algorithm continues to select the best view for materialization till a predefined number of views have been selected.

In HRU, at each of the iterations, the benefit value of views other than the views that were already selected for materialization is recomputed. This may result in a high number of re-computations, which could become phenomenally high for higher dimensional data sets. This in turn would lead to a high run time complexity[14, 15, 18, 19]. This problem has been addressed by the proposed algorithm RLGA. Algorithm RLGA aims to reduce the total number of re-computations by selecting beneficial views greedily from a reduced lattice, instead of a complete lattice. The algorithm RLGA is discussed next.

3 Algorithm RLGA

As discussed above, algorithm HRU exhibits high run time complexity. One reason for this is the high number of re-computations of benefit values required while selecting views for materialization. This can be improved upon by decreasing the number of re-computations. The proposed algorithm RLGA attempts to reduce these re-computations by selecting views from a reduced set of dependencies among views. RLGA is based on a reduced lattice framework where, instead of considering the complete lattice, as in HRU, a reduced lattice is arrived at from the given set of dependencies among the views. This reduction of the lattice is based on a heuristic, using which only those views are retained that are likely to be beneficial and have high likelihood of being selected for materialization. The heuristic is defined as:

For each view, its dependency with its smallest sized parent view is retained

The basis of this heuristic is that in HRU algorithm the view having the smallest size, among all views at a level of the lattice, has the maximum benefit at that level vis-à-vis all the views on which it is dependent. This implies that selecting the smallest sized view can maximize the benefit with respect to views above it in a lattice. In other words, retaining dependency of each view to its smallest sized parent view is most beneficial, as the smallest sized parent will in turn obtain maximum benefit with respect to the view above it and thus have a higher likelihood of being selected. To understand this better, consider the lattice shown in Fig. 1 (a). Now as per the heuristic the dependent view V3 retains its direct dependency only with its smallest parent view, i.e. view V1, resulting in the reduced lattice shown in Fig. 1 (b).

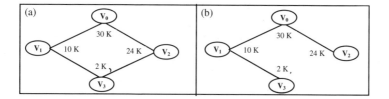

Fig. 1. Complete Lattice and Reduced Lattice

Again as per the heuristic, the view V_1, by virtue of being smaller in size, has a higher likelihood of being materialized as compared to view V_2 at the same level in the lattice. This assumption behind the heuristic is proven to be right, as the first view to be selected for materialization from this lattice is indeed the view V_1, having the highest benefit. As a result, in the next iteration, its dependent view V_3 will consider this view V_1 as its nearest materialized ancestor view. Thus, by adjusting dependencies as per the reduced lattice heuristic the same decision can be made, in terms of selecting the most beneficial views for materialization, without having to evaluate all the other dependencies from the lattice, like the dependency between views V_3 and its other parent view V_2.

The above heuristic is used to arrive at a reduced lattice from the given dependencies among the views. The method to arrive at a reduced lattice is given in Fig. 2. The resultant reduced lattice comprises of nodes, corresponding to views to be materialized, and edges, expressing dependencies, from each node to its smallest sized parent node.

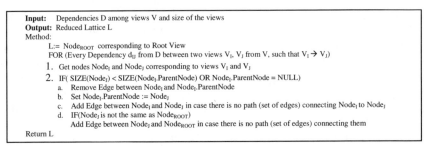

Fig. 2. Method ConstructReducedLattice

RLGA selects beneficial views greedily over the reduced lattice. The algorithm RLGA is given in Fig. 3. RLGA takes a set of dependencies among views as input and produces the top T beneficial views for materialization as output.

```
Input:    Dependencies D among Views V
Output:   Top T views
Method:
Step 1: //Arrive at a reduced lattice L
              L=ConstructReducedLattice(D)
Step 2: //Select Top T Views
              Count :=0; MatViewList:=NULL;
              // Select Views for Materialization
         1.  For every view N from the reduced lattice L, compute:
              a.  NoDependents_N := COUNT (Number of views appearing at a lower level in the lattice and having View_N as an ancestor)
              b.  BenefitValView_N := (Size_ROOT - Size_N) * NoDependents_N
              c.  Prepare OrderedListBeneficialViews_L = ORDER_DESC (BenefitValView_N * NoDependents_N)
         2.  WHILE (Count < T) DO
              a.  Select top view (V_TOP) from OrderedListBeneficialViews and add V_TOP to MatViewsList
              b.  // Update benefit value and no of dependents
                   i.  For every view V_J from a higher level in the lattice than V_TOP, s.t. V_TOP → V_J
                        Adjust NoDependents_J := NoDepedents_J - NoDependents_TOP
                   ii.  For every view V_M from a lower level in lattice than V_TOP, s.t. V_M → V_TOP
                        Adjust BenefitValView_M = (Size_TOP - Size_M) * NoDependents_M
              c.  Remove V_TOP from OrderedListBeneficialViews
              d.  Re-compute OrderedListBeneficialViews values
              e.  Count++
         3.  Return MatViewList
```

Fig. 3. Algorithm RLGA

The algorithm RLGA comprises two stages. In the first stage, the set of dependencies among views, given as input, are evaluated one by one to arrive at a reduced lattice using the heuristic defined above. Next top T beneficial views are selected greedily over the reduced lattice. The benefit value of each view is computed in the same way as computed by the HRU algorithm i.e. by taking the product of number of its dependents with the cost (size) difference from its nearest materialized ancestor view. In the initial iteration, the benefit value for all views is computed with respect to the root view, which is assumed to be materialized. The view having the highest benefit value among them is then selected for materialization. The benefit values of all views that were dependent on the view selected for materialization are accordingly recomputed. However, unlike the HRU algorithm that operates over a complete lattice, the RLGA operates over a reduced lattice thereby substantially reducing the number of nodes dependent upon the view selected for materialization. This results in fewer nodes requiring re-computation of their benefit values thereby resulting in an overall reduction in the number of re-computations. The complexity analysis of algorithm RLGA is given next.

3.1 Complexity Analysis – RLGA

As discussed above, the RLGA heuristic is based upon the reduced lattice. The reduced lattice retains only the dependencies between a node and its smallest parent node. As a result each node of the reduced lattice, except the root node, contains only one edge to its immediate parent. Thus, for a total of N nodes (views) in the reduced lattice, there are N-1 dependencies between each node and its smallest immediate parent node. Conversely, looking at the dependencies from parent to its child nodes, in the worst case, the maximum possible dependents on a node of the reduced lattice is N/2-1, for a node at a level below the root node. Similarly, in the worst case the maximum number of dependents of a node at two levels below the root node is N/4-1. This feature of the reduced lattice can be clearly observed in the reduced lattice for a 4 dimensional dataset, containing 16 views (i.e. N=16), shown in Fig. 4.

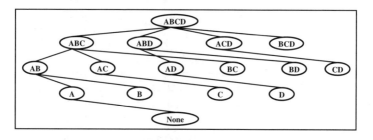

Fig. 4. Reduced Lattice

In the reduced lattice of Fig. 4, view ABC, at a level below the root node of the lattice has (N/2-1) i.e. 7 dependents. Similarly the view AB, at two levels below the root node, has (N/4-1) i.e. 3 dependents. In this way, the maximum number of dependents on a node can be computed as:

$$(N/2 - 1) + (N/4 - 1) + \dots (N/N - 1) = N - \log_2 N$$

Further, in each of the iteration of the algorithm RLGA, a view is selected for materialization. As a result of this selection, the benefit value for all dependent views needs to be recomputed. Therefore, in the worst case, at the most the benefit value for $(N - \log_2 N)$ nodes would have to be recomputed in a single iteration of the algorithm. Therefore, the complexity of the RLGA algorithm for selecting top K views, by executing K iterations of the algorithm, is $O(K) * O(N - \log_2 N) = O(KN)$.

An example illustrating selection of views using algorithms HRU and RLGA is given next.

4 An Example

For the following dependencies D, among Views V, of Sizes S, select the top 4 beneficial views using both HRU and RLGA (X→Y implies, X is dependent on Y).

V = {ABC, AB, AC, BC, A, B, C}
S = {5K, 5K, 0.9K, 3K, 0.09K, 1K, 0.5K}
D = {AB→ ABC, AC→ ABC, BC → ABC, A→ ABC, A→AB, A→AC,
    B→ABC, B → AB, B →BC, C → ABC, C → AC, C → BC}

The selection of top 4 views, using HRU and RLGA, are shown in Fig. 5 and Fig. 6 respectively.

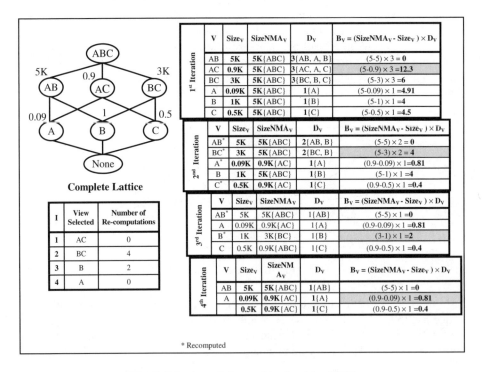

Fig. 5. Selection of views using algorithm HRU

The top 4 beneficial views selected by the HRU algorithm are AC, BC, B and A. RLGA also selects the same top 4 beneficial views, AC, BC, B and A, for materialization. It is observed that the RLGA requires fewer re-computations i.e. 3 as compared to that required by HRU algorithm i.e. 6. This difference is attributed to the fewer dependencies in the reduced lattice leading in effect to fewer views requiring re-computation. Although the top 4 views selected by both algorithms are same, the number of re-computations required in case of RLGA is less than that of HRU. This difference would increase with increase in the number of dimensions of the data set. Thus, for higher dimensional data sets, RLGA is likely to have a better execution time as compared to HRU.

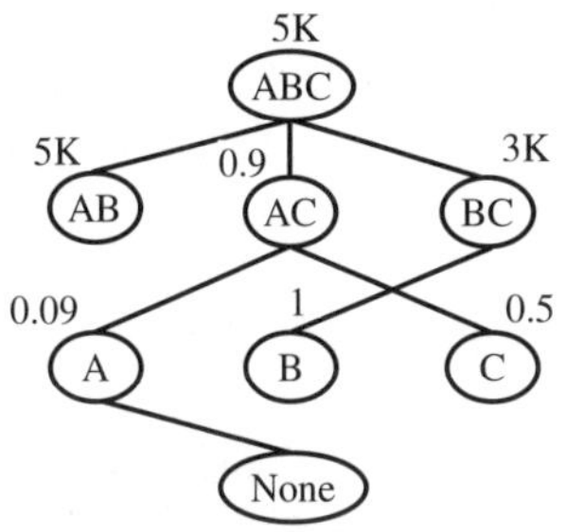

Reduced Lattice

5K ABC — 5K AB, 0.9 AC, 3K BC — 0.09 A, 1 B, 0.5 C — None

I	View Selected	Number of Re-computations
1	AC	0
2	BC	2
3	B	1
4	A	0

1st Iteration

V	Size$_V$	SizeNMA$_V$	D$_V$	B$_V$ = (SizeNMA$_V$ - Size$_V$) × D$_V$
AB	5K	5K{ABC}	1{AB}	(5-5) × 1 = 0
AC	0.9K	5K{ABC}	3{AC, A, C}	(5-0.9) × 3 =12.3
BC	3K	5K{ABC}	2{BC, B}	(5-3) × 2 =4
A	0.09K	5K{ABC}	1{A}	(5-0.09) × 1 =4.91
B	1K	5K{ABC}	1{B}	(5-1) × 1 =4
C	0.5K	5K{ABC}	1{C}	(5-0.5) × 1 =4.5

2nd Iteration

V	Size$_V$	SizeNMA$_V$	D$_V$	B$_V$ = (SizeNMA$_V$ - Size$_V$) × D$_V$
AB	5K	5K{ABC}	1{AB}	(5-5) × 1 =0
BC	3K	5K{ABC}	2{BC, B}	(5-3) × 2 = 4
A*	0.09K	0.9K{AC}	1{A}	(0.9-0.09) × 1=0.81
B	1K	5K{ABC}	1{B}	(5-1) × 1 =4
C*	0.5K	0.9K{AC}	1{C}	(0.9-0.5) × 1 =0.4

3rd Iteration

V	Size$_V$	SizeNMA$_V$	D$_V$	B$_V$ = (SizeNMA$_V$ - Size$_V$) × D$_V$
AB	5K	5K{ABC}	1{AB}	(5-5) × 1 =0
A	0.09K	0.9K{AC}	1{A}	(0.9-0.09) × 1 =0.81
B*	1K	3K{BC}	1{B}	(3-1) × 1 =2
C	0.5K	0.9K{ABC}	1{C}	(0.9-0.5) × 1 =0.4

4th Iteration

V	Size$_V$	SizeNMA$_V$	D$_V$	B$_V$ = (SizeNMA$_V$ - Size$_V$) × D$_V$
AB	5K	5K{ABC}	1{AB}	(5-5) × 1 =0
A	0.09K	0.9K{AC}	1{A}	(0.9-0.09) × 1 =0.81
C	0.5K	0.9K{AC}	1{C}	(0.9-0.5) × 1 =0.4

* Recomputed

Fig. 6. Selection of views using algorithm RLGA

In order to compare the performance of RLGA with respect to HRU, both the algorithms were implemented and run on the data sets with varying dimensions. The performance based comparisons of HRU and RLGA is given next.

5 Comparison – RLGA vs. HRU

The RLGA and HRU algorithms were implemented using Sun Java 5 on an Intel based 2.3 GHz PC having 2 GB RAM. The two algorithms were compared on parameters like number of re-computations and execution time, for selecting the top 40 views for materialization, by varying the number of dimensions in the dataset from 6 to 14. First the total number of re-computations needed was captured against the number of dimensions. The graph is shown in Fig. 7.

It is observed from the graph that the number of re-computations required for RLGA is lower than that for HRU. As the number of dimensions increase, this difference becomes significant. This reduction in re-computations in the case of RLGA algorithm can be attributed to the lesser number of dependencies that exists within the reduced lattice. To better understand the difference, the number of

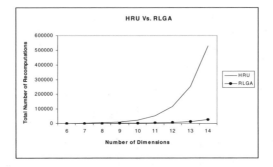

Fig. 7. HRU Vs. RLGA: Total Number of Re-computations Vs. Number of Dimensions

re-computations was plotted individually for the dimensions 8, 10, 12 and 14 against the iteration of the algorithms. These graphs are shown in Fig. 8. It is evident from the graphs that the number of re-computations for RLGA is far lower than that of HRU. These graphs firmly establish that use of reduced lattice substantially lowers the number of re-computations.

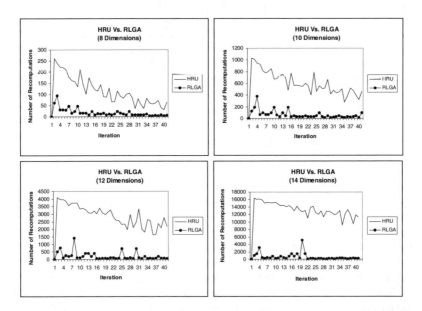

Fig. 8. HRU Vs. RLGA: Number of Re-computations Vs. Iterations: 8 - 14 Dimensions

Next, graphs were plotted to ascertain improvement in execution time with reduction in the number of re-computations. The graph showing the execution time (in milliseconds) against the number of dimensions is shown in Fig. 9.

As can be observed from the graph, the execution time for RLGA is lower than that for HRU. This may be as a consequence of lower number of re-computations in the

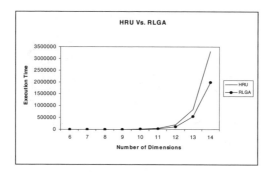

Fig. 9. HRU Vs. RLGA: Execution Time Vs. Number of Dimensions

case of RLGA. Further this improvement can also be seen in the Execution Time versus Iteration graphs plotted for the individual dimensions. These graphs are shown in Fig. 10. In each of these graphs, the execution time of RLGA is found to be lower than that of HRU. These graphs validate the basis of the heuristic used by RLGA that fewer re-computations would be required for selecting beneficial views. This in turn would lead to improvement in execution time.

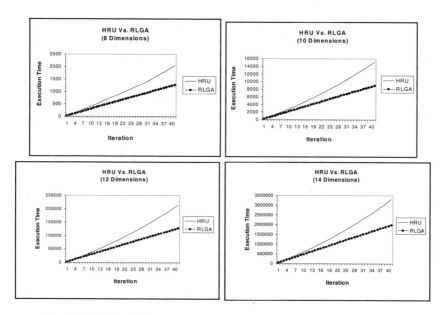

Fig. 10. HRU Vs. RLGA: Execution Time Vs. Iteration: 8 - 14 Dimensions

Next, in order to ascertain the quality of the views being selected by RLGA, in comparison with those being selected by HRU, two graphs were plotted capturing the percentage of views common to both algorithms. The first of these graphs, shown in Fig. 11, was plotted by varying the Percentage of Views Selected (PVS) for

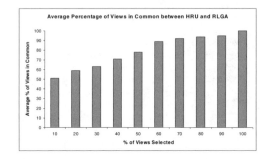

Fig. 11. Average PVC between HRU and RLGA Vs. PVS

Materialization from 10% to 100%, against the Average Percentage of Views in Common (PVC) between RLGA and HRU algorithm. Here, PVS for Materialization is a percentage of the maximum number of views selected per dimension (i.e. percentage of 2^d for a d-dimensional dataset). For each dimension, the PVC value is captured for each of the PVS units from 10% to 100%. The averaged PVC values are then plotted on the graph. The tests were restricted to dimensions 8 due to constraints in the existing system configuration.

From the above graph, it is observed that as the PVS value increases, the average PVC value also increases. The average PVC value is lowest at the 10% mark and touches 100% at the top. This is due to the fact that, as a larger percentage of views are selected for materialization, there is a higher likelihood of there being views in common selected by the two algorithms. At the midpoint, when PVS is 50 %, the PVC value is around 80%, which is an acceptable value and is further explained by the graph shown in Fig. 12. This graph captures PVC between the two algorithms HRU and RLGA for dimensions 3 to 8 when 50 % views are selected i.e. PVS is 50%.

It can be seen from the graph that on an average 78 percent of the times RLGA selects views that are as good as those selected by HRU algorithm. This establishes that RLGA selects fairly good quality views for materialization.

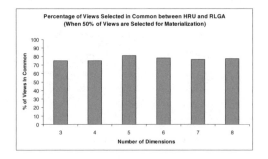

Fig. 12. PVC between HRU and RLGA when 50 % PVS for dimensions 3 to 8

The above graphs show that RLGA not only selects views efficiently i.e. with a better execution time due to lower number of re-computations, but on an average also selects views that are fairly good in comparison to those selected by HRU. While the two algorithms may at times select some percentage of different views, due to the different heuristics used by them, it is found that when higher percentage of views is selected for materialization this difference becomes low.

6 Conclusion

This paper focuses on greedy based selection of views for materialization, with emphasis on the fundamental greedy algorithm HRU. It has been found that one reason for the high run time complexity of algorithm HRU is the high number of re-computations of benefit values required for selecting views for materialization. This problem has been addressed by the proposed algorithm RLGA. RLGA uses a reduced lattice, instead of the complete lattice, for selecting beneficial views for materialization. The use of reduced lattice, which depicts the reduced set of dependencies among views, reduces the number of re-computations after each selection of beneficial views thereby leading to a reduction in the overall number of re-computations vis-à-vis the HRU algorithm. Further, to validate the claim of RLGA, experiment based comparison of RLGA and HRU algorithms was carried out on parameters like the number of re-computations, execution time and the percentage of views selected in common by both algorithms. The results show that the algorithm RLGA requires a lower number of re-computations, when compared with algorithm HRU, resulting in improved execution time. Also a high percentage of views selected by RLGA were in common with those selected by HRU. This shows that RLGA is effective and efficient in selecting beneficial views for materialization.

References

1. Agrawal, S., Chaudhuri, S., Narasayya, V.: Automated Selection of Materialized Views and Indexes for SQL Databases. In: 26th VLDB Conference, Cairo, Egypt, pp. 496–505 (2000)
2. Baralis, E., Paraboshi, S., Teniente, E.: Materialized view selection in a multidimensional database. In: 23rd VLDB Conference, pp. 156–165 (1997)
3. Chan, G.K.Y., Li, Q., Feng, L.: Optimized Design of Materialized Views in a Real-Life Data Warehousing Environment. International Journal of Information Technology 7(1), 30–54 (2001)
4. Chirkova, R., Halevy, A.Y., Suciu, D.: A formal perspective on the view selection problem. In: 27 VLDB Conference, Italy, pp. 59–68 (2001)
5. Encinas, M.T.S., Montano, J.A.H.: Algorithms for selection of materialized views: based on a costs model. In: Eighth Mexican International Conference on Current Trends in Computer Science, pp. 18–24. IEEE, Los Alamitos (2007)
6. Gou, G., Yu, J.X., Choi, C.H., Lu, H.: An Efficient and Interactive A*- Algorithm with Pruning Power- Materialized View Selection Revisited. In: DASFAA, p. 231 (2003)
7. Gupta, H., Mumick, I.S.: Selection of Views to Materialize in a Data Warehouse. IEEE Transactions on Knowledge and Data Engineering 17(1), 24–43 (2005)
8. Gupta, H., Harinarayan, V., Rajaraman, A., Ullman, J.D.: Index Selection for OLAP. In: 13th ICDE Conference, pp. 208–219 (1997)

9. Gupta, H.: Selection of Views to Materialize in a Data Warehouse. In: Afrati, F.N., Kolaitis, P.G. (eds.) ICDT 1997. LNCS, vol. 1186, pp. 98–112. Springer, Heidelberg (1996)
10. Hanson, E.R.: A Performance Analysis of View Materialization Strategies. In: ACM SIGMOD Management of Data, pp. 440–453 (1987)
11. Harinarayan, V., Rajaraman, A., Ullman, J.: Implementing Data Cubes Efficiently. In: Proc. ACM SIGMOD, Montreal, Canada, pp. 205–216 (1996)
12. Horng, J.T., Chang, Y.J., Liu, B.J., Kao, C.Y.: Materialized View Selection Using Genetic Algorithms in a Data Warehouse System. In: IEEE CEC, vol. 2, p. 2227 (1999)
13. Lawrence, M.: Multiobjective Genetic Algorithms for Materialized View Selection in OLAP Data Warehouses. In: ACM GECCO 2006, US, pp. 699–706 (2006)
14. Nadeua, T.P., Teorey, T.J.: Achieving Scalability in OLAP Materialized View Selection. In: DOLAP 2002, pp. 28–34. ACM, New York (2002)
15. Shah, B., Ramachandaran, K., Raghavan, V.: A Hybrid Approach for Data Warehouse View Selection. International Journal of Data Warehousing and Mining 2(2), 1–37 (2006)
16. Teschke, M., Ulbrich, A.: Using Materialized Views to Speed Up Data Warehousing (1997)
17. Theodoratos, D., Bouzeghoub, M.: A General Framework for the View Selection Problem for Data Warehouse Design and Evolution. In: 3rd ACM Intl. Workshop on Data Warehousing and On-Line Analytical Processing (DOLAP 2000), Washington, DC., U.S.A., pp. 1–8. ACM Press, New York (2000)
18. Uchiyama, H., Runapongsa, K., Teorey, T.J.: A Progressive View Materialization Algorithm. In: ACM DOLAP, Kansas city, USA, pp. 36–41 (1999)
19. Valluri, R., Vadapalli, S., Karlapalem, K.: View Relevance Driven Materialized View Selection in Data Warehousing Environment. In: 13th Autralasian Database Conference (ADC 2002), Melbourne, Autralia, vol. 24(2), pp. 187–196 (2002)
20. Yin, G., Yu, X., Lin, L.: Strategy of Selecting Materialized Views Based on Cache-updating. In: IEEE International Conference on Integration Technology, ICIT 2007 (2007)

Improving Expression Power in Modeling OLAP Hierarchies

Elzbieta Malinowski

Department of Computer and Information Sciences
University of Costa Rica
emalinow@cariari.ucr.ac.cr

Abstract. Data warehouses and OLAP systems form an integral part of modern decision support systems. In order to exploit both systems to their full capabilities hierarchies must be clearly defined. Hierarchies are important in analytical applications, since they provide users with the possibility to represent data at different abstraction levels. However, even though there are different kinds of hierarchies in real-world applications and some are already implemented in commercial tools, there is still a lack of a well-accepted conceptual model that allows decision-making users express their analysis needs. In this paper, we show how the conceptual multidimensional model can be used to facilitate the representation of complex hierarchies in comparison to their representation in the relational model and commercial OLAP tool, using as an example Microsoft Analysis Services.

1 Introduction

Organizations today are facing increasingly complex challenges in terms of management and problem solving in order to achieve their operational goals. This situation compels managers to utilize analysis tools that will better support their decisions. **Decision support systems** (DSSs) provide assistance to managers at various organizational levels for analyzing strategic information. Since the early 1990s, **data warehouses** (DWs) have been developed as an integral part of modern DDSs. A DW provides an infrastructure that enables users to obtain efficient and accurate responses to complex queries. Various systems and tools can be used for accessing and analyzing the data contained in DWs, e.g., **online analytical processing** (**OLAP**) systems allow users to interactively query and automatically aggregate the data using the **roll-up** and **drill-down** operations. The former, transforms detailed data into summarized ones, e.g., daily sales into monthly sales; the latter does the contrary.

The data for DW and OLAP systems is usually organized into fact tables linked to several dimension tables. A **fact table** (FactResellerSales in Fig.1) represents the focus of analysis (e.g., analysis of sales) and typically includes attributes called **measures**; they are usually numeric values (e.g., amount) that allow a quantitative evaluation of various aspects of an organization. **Dimensions** (DimTime in Fig.1) are used to see the measures from different perspectives,

S.K. Prasad et al. (Eds.): ICISTM 2009, CCIS 31, pp. 19–30, 2009.
© Springer-Verlag Berlin Heidelberg 2009

e.g., according different periods of time. Dimensions typically include attributes that form **hierarchies**. Hierarchies are important in analytical applications, since they provide the users with the possibility to represent data at different abstraction levels and to automatically aggregate measures, e.g., moving in a hierarchy from a month to a year will yield aggregated values of sales for the various years. Hierarchies can be included in a flat table (e.g., City-StateProvince in the DimGeography table in Fig.1) forming the so-called **star schema** or using a normalized structure (e.g., DimProduct, DimProductSubcategory, and DimProductCategory in the figure), called the **snowflake schema**. However, in real-world situations, users must deal with different kinds and also complex hierarchies that either cannot be represented using the current DW and OLAP systems or are represented as star or snowflake schemas without possibility to capture the essential semantics of multidimensional applications.

In this paper we refer to different kinds of hierarchies already classified in [7] that exist in real-world applications and that are required during the decision-making process. Many of these hierarchies can already be implemented in commercial tools, e.g., in Microsoft SQL Server Analysis Services (SSAS). However, these hierarchies cannot be distinguished either at the logical level (i.e., star of snowflake schemas) or in the OLAP cube designer. We will show the importance of using a conceptual model, such as the MultiDim model [7], to facilitate the process of understanding users' requirements by distinguishing different kinds of hierarchies. This paper does not focus on details as described in [7] for representing different kinds of hierarchies. The main objective is to show how many concepts already implemented in commercial tools and accepted by the practitioners can be better understood and correctly specified if the practice for DW design changes and includes a representation at the conceptual level. We use the MultiDim model as an example of conceptual model without pretending that this model is the only one that responds to analytical need. At the contrary, we leave to the designer the decision of using a conceptual multidimensional model among several already existing models, e.g., [1,6].

We have chosen as an example of a commercial tool SSAS, since it provides different kinds of hierarchies that can be included in the cube without incurring to any programming effort, i.e., using a wizard or the click-and-drag mechanism. We will compare the representation of different kinds of hierarchies in the MultiDim model, in relational model, and in OLAP cube designer.

Section 2 surveys works related to DW and OLAP hierarchies. Section 3 introduces a motivating example that is used throughout this paper. Section 4 briefly presents the main features of the MultiDim model. Section 5 refers to the conceptual representation and implementation of different kinds of hierarchies. Finally, the conclusions are given in Section 6.

2   Related Work

The advantages of conceptual modeling for database design have been acknowledged for several decades and have been studied in many publications. However, the analysis presented in [11] shows the small interest of the research community in conceptual multidimensional modeling. Some proposals provide graphical representations based on the ER model (e.g., [12]), on UML (e.g., [1,6]), or propose new notations (e.g., [2,5]), while other proposals do not refer to a graphical representation (e.g., [9,10]).

Very few models include a graphical representation for the different kinds of hierarchies that facilitates their distinction at the schema and instance levels (e.g., [6,12]). Other models (e.g., [2,12]) support only simple hierarchies. This situation is considered as a shortcoming of existing models for DWs [3].

Current commercial OLAP tools do not allow conceptual modeling of hierarchies. They usually provide a logical-level representation limited to star or snowflake schemas. Some commercial products, such as SSAS, Oracle OLAP, or IBM Alphablox Analytics, can cope with some complex hierarchies.

3   Motivating Example

In this section we briefly describe an example that we use throughout this paper in order to show the necessity of having a conceptual model for representing different kinds of hierarchies for DW and OLAP applications.

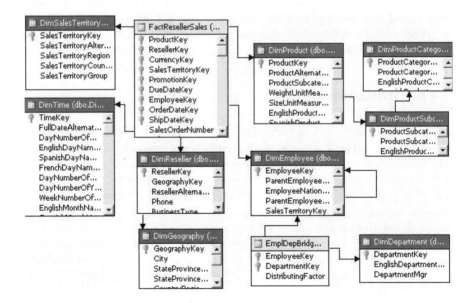

Fig. 1. An extract of the AdventureWorksDW schema

The schema in Fig.1 shows an extract of the AdventureWorksDW database issued by Microsoft[1]. The DW schema in the figure is used for analysis of sales by

---

[1] We do not refer to the correctness of the AdventureWorksDW schema.

resellers (the fact table FactResellerSales). These sales are analyzed from different perspectives, i.e., dimensions. The Product dimension includes a hierarchy using the snowflake structure representing products, subcategories and categories. The Time dimension include attributes that allow users to analyze data considering calendar and fiscal periods of time. Another perspective of analysis is represented by the DimSalesTerritory table which allows decision-making users to analyze measures considering geographical distribution of a sales organization. The DimReseller table in Fig.1 includes stores that resale products and has attached a table (the DimGeography) indicating geographical distribution of these stores. In addition, this schema contains an employee dimension (the DimEmployee table in the figure) with an organizational hierarchy of supervisors and subordinates. We modified slightly the DimEmployee table and deleted the attribute DepartmentName. Instead, we created a new table that represents different departments. Since we assigned some employees to two different departments, we had to create an additional table (the EmplDepBridge table in Fig.1). This table represents all assignments of employees to their corresponding departments and in addition, it includes an attribute called DistributingFactor that indicates how to distribute measures between different departments for employees that work in more than one department, e.g., assign 70% of sales to the department 10 and 30% of sales to the department 14.

As can be seen in Fig.1, even though there are several hierarchies that users are interested in exploring, only the hierarchy represented as snowflake schema (e.g., Product-Subcategory-Category) can be distinguished. We will see in the next section, how this situation can be changed using a conceptual model.

4   The MultiDim Model

The MultiDim model [7] is a multidimensional model that allows designers to represent at the conceptual level all elements required in data warehouse and OLAP applications, i.e., dimensions, hierarchies, and facts with associated measures. In order to present a brief overview of the model[2], we use the example in Fig.2. The schema in this figure corresponds to the logical schema in Fig.1. We include in the schema only those hierarchies that are relevant for the article and we omit the attributes since they are the same as in Fig.1.

A **schema** is composed of a set of dimensions and a set of fact relationships. A **dimension** is an abstract concept that groups data sharing a common semantic meaning within the domain being modeled. A dimension is composed of a level or a set of hierarchies.

A **level** corresponds to an entity type in the ER model. It describes a set of real-world concepts that have similar characteristics, e.g., the Product level in Fig.2. Instances of a level are called **members**. A level has a set of **attributes** that describe the characteristics of their members and one or several **keys** that identify uniquely the members of a level. These attributes can be seen in Fig.3a.

---

[2] The detailed model description and formalization can be found in [7].

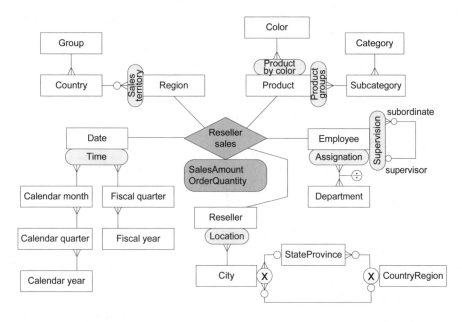

Fig. 2. Conceptual representation of hierarchies using the MultiDim model

A hierarchy comprises several related levels. Given two related levels of a hierarchy, the lower level is called the **child** and the higher level is called the **parent**. The relationships between parent and child levels are characterized by **cardinalities**, indicating the minimum and the maximum number of members in one level that can be related to a member in another level. We use different symbols for indicating cardinalities: ——○ (0,1), —— (1,1), ——○< (0,n), and ——< (1,n). Different cardinalities may exist between parent and child levels leading to different kinds of hierarchies, to which we refer in more detail in the next sections.

The level in a hierarchy that contains the most detailed data is called the **leaf level**; its name is used for defining the dimension's name. The last level in a hierarchy, representing the most general data, is called the **root level**.

The hierarchies in a dimension may express various structures used for analysis purposes; thus, we include an **analysis criterion** to differentiate them. For example, the Product dimension in Fig.2 includes two hierarchies: Product groups and Product by color. The former hierarchy comprises the levels Product, Subcategory, and Category, while the latter hierarchy includes the levels Product and Color.

A **fact relationship** expresses a focus of analysis and represents an n-ary relationship between leaf levels, e.g., the Reseller sales fact relationship relates the Product, Region, Employee, Reseller, and Date levels in Fig.2. A fact relationship may contain attributes commonly called **measures** that usually contain numeric data, e.g., SalesAmount and OrderQuantity in Fig.2.

5   Hierarchies: Their Representation and Implementation

In this section, we present various kinds of hierarchies using the MultiDim model that provides clear distinction at the schema and instance levels. We also show that even though, some commercial tools, such as SSAS, allow designers to include and manipulate different kinds of hierarchies, the distinction between them is difficult to make.

5.1   Balanced Hierarchies

A **balanced hierarchy** has only one path at the schema level, e.g., Product groups hierarchy in Fig.2 composed by the Product, Subcategory, and Category levels. At the instance level, the members form a tree where all the branches have the same length, since all parent members have at least one child member, and a child member belongs to only one parent member, e.g., all subcategories have assigned at least one product and a product belongs to only one subcategory. Notice that in Fig.2 we have another balanced hierarchy Product by color that could not be distinguished in the logical level in Fig.1 since it is included as an attribute in the DimProduct table.

Balanced hierarchies are the most common kind of hierarchies. They are usually implemented as a star or a snowflake schema as can be seen in Fig.1. On the other hand, SSAS uses the same representation for all kinds of hierarchies (except recursive as we will see later) as shown for the Sales Territory hierarchy in Fig.4a.

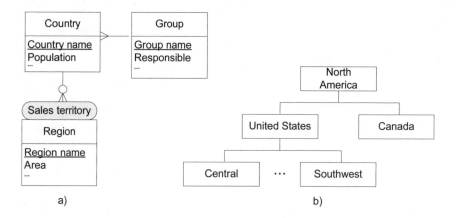

Fig. 3. Unbalanced hierarchy: a) schema and b) examples of instances

5.2   Unbalanced Hierarchies

An **unbalanced hierarchy**[3] has only one path at the schema level and, as implied by the cardinalities, at the instance level some parent members may not

---

[3] These hierarchies are also called heterogeneous [4] and non-onto [9].

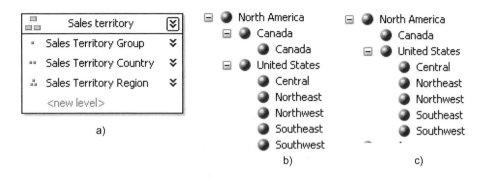

Fig. 4. Unbalanced hierarchy in SSAS: a) schema and b), c) instances

have associated child members. Fig.3a shows a Sales territory hierarchy composed of Region, Country, and Group levels. However, the division in some countries does not include region ((e.g., Canada in Fig.3b).

At the logical level this kind of hierarchy is represented as a star or a snowflake schema (the DimSalesTerritory table in Fig.1). At the instance levels, the missing levels can include placeholders, i.e., the parent member name (e.g., Canada for the region name) or null values.

SSAS represents unbalanced hierarchy as shown in Fig.4a. For displaying instances, the designers can choose between two options: to display the repeated member (Fig.4b) or not to include this member at all (Fig.4c). To select one of these options, designers should modify the HideMemberIf property by indicating one of the following options: (1) OnlyChildWithParentName: when a level member is the only child of its parent and its name is the same as the name of its parent, (2) OnlyChildWithNoName: when a level member is the only child of its parent and its name is null or an empty string, (3) ParentName: when a level member has one or more child members and its name is the same as its parent's name, or (4) NoName: when a level member has one or more child members and its name is a null value.

Notice that this is an incorrect assignment, since for unbalanced hierarchies, only the first or second option should be applied, i.e., the parent member will have at most one child member with the same name, e.g., the name Canada in Fig.3b will be repeated in the missing levels until the tree representing the instances will be balanced.

5.3   Recursive Hierarchies

Unbalanced hierarchies include a special case that we call **recursive hierarchies**[4]. In this kind of hierarchy the same level is linked by the two roles of a parent-child relationship. An example is given in Fig.2 for the Employee dimension where the Supervision recursive hierarchy represents the employee-supervisor relationship. The subordinate and supervisor roles of the parent-child relationship

---

[4] These are also called parent-child hierarchies [8].

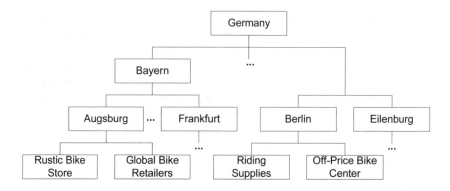

Fig. 5. Some instances of non-covering hierarchy

are linked to the **Employee** level. Recursive hierarchies are mostly used when all hierarchy levels express the same semantics, e.g., where an employee has a supervisor who is also an employee.

At the logical levels this kind of hierarchy is represented by the inclusion of a foreign key in the same table that contains a primary key as can be seen in Fig.1 for the **DimEmployee** table. This kind of hierarchy is not represented as a hierarchy in SSAS; only a hierarchy symbol 🗂 is attached to the attribute that represents a parent key.

5.4 Non-covering Hierarchies

A **non-covering** or **ragged** hierarchy contains multiple exclusive paths sharing at least the leaf level. Alternative paths are obtained by skipping one or several intermediate levels of other paths. All these paths represent one hierarchy and account for the same analysis criterion. At the instance level, each member of the hierarchy belongs to only one path. We use the symbol $\otimes$ to indicate that the paths are exclusive for every member.

Fig.2 includes a **Location** non-covering hierarchy composed of the **Reseller**, **City**, **StateProvince**, and **CountryRegion** levels. However, as can be seen by the straight lower line and the cardinalities, some countries do not have division in states. Fig.5 shows some hypothetical instances that we use for this hierarchy[5]. Notice that the cities of Berlin and Eilenburg do not have assigned any members for the **StateProvince** level.

This hierarchy is represented in the logical schema as a flat table, e.g., the **DimGeography** table in Fig.1, with corresponding attributes. At the instance level, similar to the unbalanced hierarchies, placeholders or null values can be included in the missing members.

SSAS, for representing these hierarchies, uses a similar display as shown for the **Sales Territory** hierarchy in Fig.4a. For displaying the instances, SSAS

---

[5] We modify the instance of the AdventureWorksDW to represent this kind of hierarchy.

provides the same four options as described in Sec. 5.2. However, for non-covering hierarchies, the third or fourth option should be applied since, for our example in Fig.5, two children roll-up to the Germany member included as a placeholder for the missing StateProvince level.

Even though the unbalanced and non-covering hierarchies represent different situations and can be clearly distinguished using a conceptual model (Fig.2), SSAS considers implementation details that are very similar for both hierarchies and states that "it may be impossible for end users to distinguish between unbalanced and ragged hierarchies" [8]. These two hierarchies also differ in the process of measure aggregation. For an unbalanced hierarchy, the measure values are repeated from the parent member to the missing child members and cannot be aggregated during the roll-up operations. For the non-covering hierarchies, the measures should be aggregated for every placeholder represented at the parent level.

5.5   Non-strict Hierarchies

For the hierarchies presented before, we assumed that each parent-child relationship has many-to-one cardinalities, i.e., a child member is related to at most one parent member and a parent member may be related to several child members. However, many-to-many relationships between parent and child levels are very common in real-life applications, e.g., an employee can work in several departments, a mobile phone can be classified in different product categories. We call a hierarchy **non-strict** if at the schema level it has at least one many-to-many relationships. Fig.2 shows the Assignation hierarchy where an employee can belong to several departments. Since at the instance level a child member may have more than one parent member, the members form an acyclic graph.

Non-strict hierarchies induce the problem of double-counting measures when a roll-up operation reaches a many-to-many relationship, e.g., if an employee belongs to the two departments, his sales will be aggregated to both these departments, giving incorrect results. To avoid this problem one of the solutions[6] is to indicate that measures should be distributed between several parent members. For that, we include an additional symbol $\oplus$ called a **distributing factor**.

The mapping to the relational model [7] will provide the same solution as presented in Fig.1: the DimEmployee, DimDepartment, and EmplDepBridge tables for representing the Employee, Department levels and many-to-many cardinalities with distributing factor attribute, respectively. However, having the bridge table we loose the meaning of a hierarchy that can be used for the roll-up and drill-down operations. This is not the case when using a conceptual model.

The SSAS requires several different steps in order to use this hierarchy and to have correct results. First, the bridge table is considered as another fact table and Employee and Department dimensions are handled as separate dimensions (Fig.6a) that later on can be combined to form a hierarchy in a cube data browser. In the next step, designers must define in the Dimension Usage that in order to aggregated the measure from Fact Reseller Sales table,

---

[6] Several solutions can be used as explained in [7].

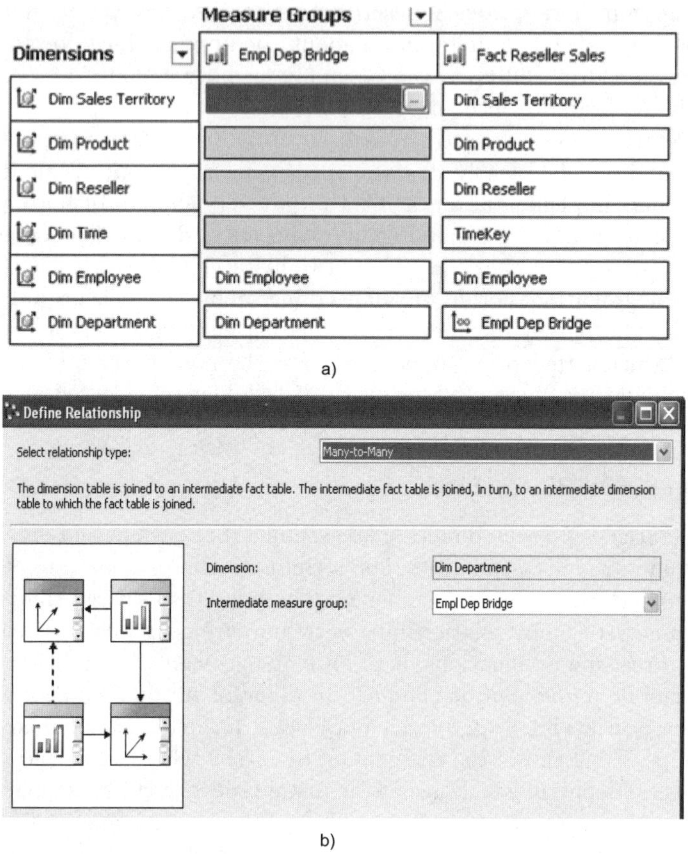

a)

b)

Fig. 6. Non-strict hierarchy in SSAS: a) dimension usage and b) and representation of many-to-many relationship

many-to-many cardinalities must be considered (Fig.6a). Notice the SSAS representation of this cardinality in Fig.6b. Finally, in order to use a distributing factor from the bridge table, for every measure of the fact table the Measure Expression property must be modified, e.g., for SalesAmount measure we include [SalesAmount]*[DistributingFactor].

5.6 Alternative Hierarchies

Alternative hierarchies represent the situation where at the schema level there are several non-exclusive simple hierarchies sharing at least the leaf level and accounting for the same analysis criterion. The Time hierarchy in Fig.2 is an example of alternative hierarchies, where the Date dimension includes two hierarchies corresponding to the usual Gregorian calendar and to the fiscal calendar of an organization.

Alternative hierarchies are needed when the user requires analyzing measures from a unique perspective (e.g., time) using alternative aggregation paths. Since the measures from the fact relationship will participate totally in each composing hierarchy, measure aggregations can be performed as for simple hierarchies. However, in alternative hierarchies it is not semantically correct to simultaneously combine the different component hierarchies to avoid meaningless intersections, such as Fiscal 2003 and Calendar 2001. The user must choose only one of the alternative aggregation paths for his analysis and switch to the other one if required.

The logical schema does not represent clearly this hierarchy since all attributes forming both paths of alternative hierarchies are included in the flat DimTime table (Fig.1).

The current version of SSAS does not include this kind of hierarchy and the designers should define two different hierarchies, one corresponding to calendar and another to fiscal time periods, allowing combinations between the alternative paths and creating meaningless intersections with null values for measures.

5.7 Parallel Hierarchies

Parallel hierarchies arise when a dimension has associated several hierarchies accounting for different analysis criteria, e.g., the Product dimension in Fig.2 with Product by color and Product groups parallel hierarchies. Such hierarchies can be independent where composed hierarchies do not share levels or dependent, otherwise.

Notice that even though both multiple and parallel hierarchies may share some levels and may include several simple hierarchies, they represent different situations and should be clearly distinguishable. This is done by including only one (for alternative hierarchies) or several (for parallel dependent hierarchies) analysis criteria. In this way the user is aware that in alternative hierarchies it is not meaningful to combine levels from different composing hierarchies, while this can be done for parallel hierarchies, e.g., for the Product dimension in Fig.2, the user can issue the query "what are the sales figures for products that belong to the bike category and are black".

6   Conclusions

DWs are defined using a multidimensional view of data, which is based on the concepts of facts, measures, dimensions, and hierarchies. OLAP systems allow users to interactively query DW data using operations such as drill-down and roll-up, and these operations require the definition of hierarchies for aggregating measures.

A hierarchy represents some organizational, geographic, or other type of structure that is important for analysis purposes. However, there is still a lack of a well-accepted conceptual multidimensional model that is able to represent different kinds of hierarchies existing in real-world applications. As a consequence, even though some commercial tools are able to implement and manage different kinds of hierarchies, users and designers have difficulties in distinguishing them. Therefore, users cannot express clearly their analysis requirements and designers as well as implementers cannot satisfy users' needs.

References

1. Abelló, A., Samos, J., Saltor, F.: YAM^2 (yet another multidimensional model): An extension of UML. Information Systems 32(6), 541–567 (2006)
2. Golfarelli, M., Rizzi, S.: A methodological framework for data warehouse design. In: Proc. of the 1st ACM Int. Workshop on Data Warehousing and OLAP, pp. 3–9 (1998)
3. Hümmer, W., Lehner, W., Bauer, A., Schlesinger, L.: A decathlon in multidimensional modeling: Open issues and some solutions. In: Kambayashi, Y., Winiwarter, W., Arikawa, M. (eds.) DaWaK 2002. LNCS, vol. 2454, pp. 275–285. Springer, Heidelberg (2002)
4. Hurtado, C., Gutierrez, C.: Handling structural heterogeneity in OLAP. In: Wrembel, R., Koncilia, C. (eds.) Data Warehouses and OLAP: Concepts, Architectures and Solutions, ch. 2, pp. 27–57. IRM Press (2007)
5. Hüsemann, B., Lechtenbörger, J., Vossen, G.: Conceptual data warehouse design. In: Proc. of the Int. Workshop on Design and Management of Data Warehouses, p. 6 (2000)
6. Luján-Mora, S., Trujillo, J., Song, I.: A UML profile for multidimensional modeling in data warehouses. Data & Knowledge Engineering 59(3), 725–769 (2006)
7. Malinowski, E., Zimányi, E.: Advanced Datawarehouse Design: From Conventional to Spatial and Temporal Applications. Springer, Heidelberg (2008)
8. Microsoft Corporation. SQL Server 2005. Books Online (2003),
   http://technet.microsoft.com/en-us/sqlserver/bb895969.aspx
9. Pedersen, T., Jensen, C., Dyreson, C.: A foundation for capturing and querying complex multidimensional data. Information Systems 26(5), 383–423 (2001)
10. Pourabbas, E., Rafanelli, M.: Hierarchies. In: Rafanelli, M. (ed.) Multidimensional Databases: Problems and Solutions, pp. 91–115. Idea Group Publishing (2003)
11. Rizzi, S.: Open problems in data warehousing: 8 years later. In: Proc. of the 5th Int. Workshop on Design and Management of Data Warehouses (2003)
12. Sapia, C., Blaschka, M., Höfling, G., Dinter, B.: Extending the E/R model for multidimensional paradigm. In: Proc. of the 17th Int. Conf. on Conceptual Modeling, pp. 105–116 (1998)

A Hybrid Information Retrieval System for Medical Field Using MeSH Ontology

Vahid Jalali and Mohammad Reza Matash Borujerdi

Amirkabir University of Technology,
Tehran, Iran
{vjalali,borujerm}@aut.ac.ir

Abstract. Using semantic relations between different terms beside their syntactical similarities in a search engine would result in systems with better overall precision. One major problem in achieving such systems is to find an appropriate way of calculating semantic similarity scores and combining them with those of classic methods. In this paper, we propose a hybrid approach for information retrieval in medical field using MeSH ontology. Our approach contains proposing a new semantic similarity measure and eliminating records with semantic score less than a specific threshold from syntactic results. Proposed approach in this paper outperforms VSM, graph comparison, neural network, Bayesian network and latent semantic indexing based approaches in terms of precision vs. recall.

Keywords: Medical field, MeSH ontology, Semantic information retrieval, Semantic similarity measure.

1 Introduction

If we consider information retrieval roots in Vannevar Bush's article titled "As We May Think" published in 1945, it can be said that information retrieval has a history of about 60 years. Although during the course of its history, information retrieval has experienced several trends and directions, but its importance has always been growing. Especially after the emergence of World Wide Web and intensive growth of the size of human knowledge stored over it in digital format, this branch of computer science became more critical. At that time new mechanisms should be introduced to index and search large amounts of knowledge in appropriate ways, so that everyone could get access to required information from this mess of digital documents.

Early approaches of information retrieval had statistical basis and narrowed their attention to syntactic similarities between words in queries and documents. Well known methods like Boolean, probabilistic and vector space models are among these techniques. These approaches suffered from a set of drawbacks which in turn leaded to introduction of semantic approaches for information retrieval.

The main idea of semantic information retrieval is that the lack of common terms in two documents does not necessarily means that the documents are not related.

S.K. Prasad et al. (Eds.): ICISTM 2009, CCIS 31, pp. 31–40, 2009.
© Springer-Verlag Berlin Heidelberg 2009

There are cases in which semantically related terms are totally irrelevant from the syntactic view. Consider situations like the presence of synonyms or hyponyms (word that indicates a more specific meaning than the other related word) with different spelling in two documents which would be ignored by classic information retrieval systems. Yet the question which remains unanswered is that how can we determine if two terms are semantically related. There are different strategies to approach this question. Some methods like latent semantic indexing distinguish a kind of semantic similarity by statistical analysis of co-occurrences of the terms over multiple documents while other approaches make use of thesauruses or ontologies for this purpose.

In this paper we use preexisting MeSH (Medical Subject Headings) ontology for measuring semantic similarity between concepts in queries and documents. MeSH is a huge controlled vocabulary for the purpose of indexing journal articles and books in the life sciences, Created and updated by the United States National Library of Medicine (NLM). This vocabulary at the time contains 24767 subject headings, also known as descriptors. Most of these are accompanied by a short definition, links to related descriptors, and a list of synonyms or very similar terms (known as entry terms). The test collection which is used for evaluating our system is Medline dataset created in Glasgow University and contains 1033 documents, 30 queries and relevant judgments for these queries.

The paper is organized as follows. Section 2 talks about the related work in field of semantic information retrieval. Section 3 describes our proposed approach for calculating semantic relatedness and combining it with classic information retrieval. Experiment and results are reported in Section 4, further related discussions are covered in Section 5 and finally Section 6 concludes the paper.

2 Related Work

Conceptual search, i.e., search based on meaning rather than just character strings, has been the motivation of a large body of research in the IR field long before the Semantic Web vision emerged [1] [2]. This drive can be found in popular and widely explored areas such as Latent Semantic Indexing [3] [4], linguistic conceptualization approaches [5] [6], or the use of thesaurus and taxonomies to improve retrieval [7] [8]. In continue we introduce most eminent works in the field of semantic information retrieval which exploit a kind of auxiliary vocabulary or knowledgebase in performing the retrieval task.

OWLIR [9] indexes RDF triples together with document content. OWLIR treats distinct RDF triples as indexing terms. RDF triples are generated by natural language processing techniques based on textual content. Search can be performed based on words and RDF triples with wildcards. Shah [9] reports an increase of Average Precision using an approach taking semantic information into account opposed to a text-only approach.

Mayfield and Finin combine ontology-based techniques and text-based retrieval in sequence and in a cyclic way, in a blind relevance feedback iteration [10]. Inference over class hierarchies and rules is used for query expansion and extension of semantic annotations of documents. Documents are annotated with RDF triples, and ontology-based queries are reduced to Boolean string search, based on matching RDF statements with wildcards, at the cost of losing expressive power for queries.

KIM [11] relies on information extraction and relates words in documents with concepts from an ontology. Before indexing, documents are enriched with identifiers for the ontological concepts the words in the document represent. These identifiers are directly inserted into the indexed text. For homonyms the same identifier is used. Queries are formed using concepts and relations from the ontology. The KIM platform provides infrastructure and services for automatic semantic annotation, indexing, and retrieval of documents. It allows scalable and customizable ontology-based information extraction (IE) as well as annotation and document management, based on GATE natural language processing framework. In order to provide a basic level of performance and to allow the easy bootstrapping of applications, KIM is equipped with an upper-level ontology and a massive knowledge base, providing extensive coverage of entities of general importance.

In 2003 Guha introduced TAP which aims at enhancing search results from the WWW with data from the Semantic Web [12]. It performs a graph based search on a RDF graph from the web. It starts at one or more anchor nodes in the RDF graph, which have to be mapped to query terms. A breath first search is performed in the RDF graph, collecting a predefined amount of triples. Optionally only links of a certain type are followed in traversing the RDF graph.

QuizRDF combines keyword-based search with search and navigation through RDF(S) based annotations [13]. Indexing in QuizRDF is based on content descriptors, which can be terms from the documents and literals from RDF statements. A query is formulated in using terms. In addition, the type of the resource that should be returned can be specified. Search results can be filtered by property values of resources.

Rocha presents a Hybrid Approach for Searching in the Semantic Web which combines full-text search with spreading activation search in an ontology [14]. Search starts with a keyword based query. Results to the full text search are instances from the ontology. Those instances are used to initiate a spreading activation search in the ontology to find additional instances.

CORESE is an ontology-based search engine which operates on conceptual graphs internally [15]. COROSE is queried using one or a combination of triples. The query language is similar to SPARQL, SeRQL or RDQL but allows for approximate search. Approximate search is based on semantic distance of two classes in a common hierarchy and the rdfs:seeAlso property. The relevance of a result is measured by the similarity to the query.

Castells combines SPARQL based search with full text search [16]. For ranking results of a SPARQL query he weights annotations of documents with concepts from an ontology using a tf*idf-like measure. Then he combines the results of a full-text-search with the ranked list obtained via the SPARQL query using the CombSUM strategy. For performing the full-text search he extracts certain parts of the SPARQL query and uses them as query terms.

3 Proposed Approach

Although in general, semantic similarity measures are used to detect relevancy of documents, they also can be used in the reverse fashion. This paper advocates the use of semantic similarity for determining non-relevancy of results. Further details about calculating and using semantic measures are described in this section.

3.1 Applying Classic Information Retrieval Techniques to the Corpus

For performing classic retrieval, the system exploits Lucene which is a free information retrieval library, originally created in Java and is supported by Apache Software Foundation. Lucene scoring uses a combination of the Vector Space Model (VSM) and the Boolean model to determine how relevant a Document is to the User's query.

3.2 Semantic Annotation of Documents

Identifying MeSH concepts starts with extracting noun phrases of the queries and documents. In this experiment GATE is used as our natural language processing framework. GATE provides a noun phrase chunker which can be used for extracting noun phrases of a text document. The NP chunker application is a Java implementation of the Ramshaw and Marcus BaseNP chunker, which attempts to insert brackets marking noun phrases in text which has been marked with POS tags in the same format as the output of Eric Brill's transformational tagger.

Having extracted noun phrases, they are splitted from coordinating conjunctions (CC). For example if a query contains "head injuries and infectious" CC splitter will extract a couple of noun phrases from it which would be "head injuries" and "head infectious". Noun phrases after being CC splitted are passed to another module, which checks whether they exist in MeSH ontology or not. We try to find concepts with more specific meaning at this stage; it means that finding larger phrases is preferred to shorter ones. If a noun phrase has a relating concept in MeSH ontology, then that concept will be stored, else the noun phrase will be further splitted to its sub-noun items. Then each noun item again is checked to find whether there is an appropriate concept for it in the ontology. The output of this level will determine the concepts each document contains and their related frequencies. Furthermore extra statistical facts about concepts in this level would be determined like number of documents containing a specific concept.

3.3 Semantic Retrieval Algorithm

The core of a conceptual retriever is the semantic similarity measure it uses. In recent years, a large volume of research and publication is dedicated to find innovative algorithms for measuring semantic relatedness between concepts in an ontology. A complete report on semantic similarity measures and different approaches used in this field is reported in [17], and a more specific report which attacks directly the semantic similarity measure problem in MeSH ontology can be found in [18].

The semantic similarity measure used in our algorithm is based on Wu and Palmer measure [21], which considers the position of concepts and in the taxonomy relatively to the position of the most specific common concept C.

$$Sim_{w\&p}(c_1, c_2) = \frac{2H}{N_1 + N_2 + 2H}$$

(1)

Where and is the number of IS-A links from and respectively to the most specific common concept C, and H is the number of IS-A links from C to the root of the taxonomy.

The reason for using this similarity measure in our system is its appropriateness for MeSH hierarchical structure and taking into consideration both the shortest path between concepts and their distance from the root of ontology. In addition to Wu and Palmer similarity measure, we have examined Leacock [20] and Li [19] similarity measures which are based on edge counting like Wu and Palmer approach. Although Leacock and especially Li similarity measures have slight difference with selected measure, using Wu and Palmer similarity measure yields to best results in our algorithm. In Equation our similarity measure based on Wu and Palmer approach is introduced.

$$Sim_J(c_1, c_2) = Sim_{w\&p}(c_1, c_2) > \alpha\,?$$
$$Sim_{w\&p}(c_1, c_2) \times (Sim_{w\&p}(c_1, c_2) - \alpha) \times \beta;$$
$$Sim_{w\&p}(c_1, c_2)$$

(2)

Where and are positive integers that can be determined regarding the ontology for which similarity measure is being calculated. In our experiment best results are achieved for = 0.45 and = 1000.

For calculating the semantic relatedness of queries and documents in Medline corpus, extra facts other than semantic similarity between their concepts should be considered. For example we should distinguish between semantic similarity measure gained for a couple of concepts which are frequently repeated over the corpus and a couple of concepts which have lower frequency. The reason for using auxiliary frequency factor in our algorithm is because of the fact that the distance of a concept from the root is not always in direct relation with its frequency over the corpus. For example "human" concept in MeSH ontology is located at the depth of 8 from the root which should indicate that it is fairly a specific concept (considering the fact that maximum depth of MeSH ontology is 11) and should have repeated scarcely over the corpus, but in contrast to this supposition, "human" concept has repeated 98 times in the corpus which is a high frequency for Medline test collection. In proposed algorithm for conceptual ranking of documents and queries, 3 auxiliary factors accompany Wu and palmer semantic similarity measure. These factors include concept frequency in a text, inverse concept frequency over the corpus and the whole number of concepts of a document (or query). Equation 3 shows our proposed formula for calculating conceptual relatedness score between a query and a document.

$$CScore(Q_i, D_j) = \frac{\sum_{k=1}^{qc\#}\sum_{l=1}^{dc\#} \dfrac{Sim_l(c_1, c_2) \times qcf_k \times dcf_l}{\ln(60 \times (qcdf_k + 1)) \times \ln(60 \times (dcdf_l + 1))}}{\ln(0.3 \times qc\#+1) \times \ln(0.3 \times dc\#+1)} \qquad (3)$$

In this equation qc# and dc# accordingly denote the number of concepts in ith query and jth document. qcf_k is the frequency of kth concept in ith query and dcf_l is the frequency of lth concept in jth document. $qcdf_k$ is equal to the number of documents in the corpus which contain kth concept of ith query and $dcdf_l$ is equal to the number of documents in the corpus containing lth concept of jth document. It should be mentioned that before applying this formula to the corpus, a mechanism similar to stop word elimination in classic information retrieval should be performed. This mechanism decides to use which concept in calculating semantic score, by eliminating does concepts from both queries and documents which their frequencies over the corpus exceeds a specific threshold. The threshold used for checking concepts frequencies in this experiment is 30, which is 10% of the most frequent concept over the corpus.

3.4 Combining Semantic and Syntactic Approaches

Although search engines which rely only on semantic relatedness of concepts, nearly always yield to worse results compared to classic approaches, they have unique characteristics which can be used to ameliorate results from syntactic approaches. One of these interesting aspects of semantic similarity measure is that, if the calculated semantic relatedness between a pair of query and document is below a specific threshold, there is a great possibility that this pair is indeed non-relevant. So not only it won't be harmful to omit all those pairs with low semantic score from classic result set, but also it can improve the precision of the system.

The threshold of semantic similarity measure, after normalizing scores calculated in equation 3 is 0.2 and best results are gained when records with syntactic score below 0.02 and semantic score less than 0.2 are eliminated from other pairs returned by Lucene search engine.

4 Experiment and Results

Proposed approach in this paper is tested on Medline collection containing 30990 query, document pairs, from these pairs, 696 pairs are judged as relevant pairs by Medline. Lucene algorithm distinguishes 10038 pairs from which only 597 pairs are among those related pairs determined by Medline. So at first glance it can be seen that classic approach suffers from overshooting. By using semantic scores as it is mentioned in section 3, these statistics become so much better, and without losing significance recall, higher precisions are achieved. In Fig. 1 precision vs. recall diagrams for results of Lucene, semantic and hybrid approaches is shown.

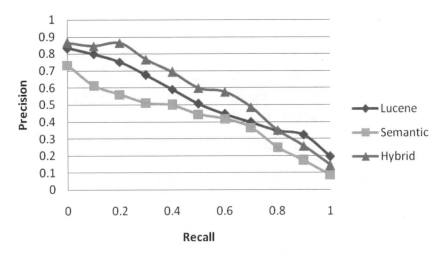

Fig. 1. Precision vs. recall diagram for Lucene, semantic and hybrid approaches

In Fig. 2 a comparison between proposed approach in this paper and several other approaches (with same test collection) is exhibited.

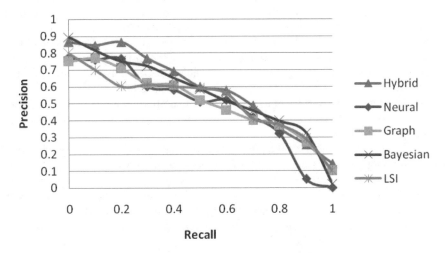

Fig. 2. Precision vs. recall diagram for proposed Hybrid approach and four other systems

The systems compared with our proposed approach are based on different techniques including using neural network [22], graph comparison [23], Bayesian network [24] and finally latent semantic indexing [25] in information retrieval field. For these systems 11 point average precision analysis is also done and you can see the results in Fig 3.

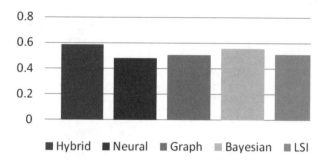

Fig. 3. Eleven-point average precision of different systems

5 Discussion

There are several issues about our algorithm that can be further discussed at this stage. Here, a couple of contentious choices taken in this experiment are clarified. First of all as it is mentioned in Section 2 of this paper, there are systems which make use of RDF and RDF query languages for representing user's queries in contrast to our system in which user provides queries in natural language format. Preferring queries in natural language to other formats in our opinion has two major reasons. The most prominent reason is that general users are more comfortable with natural language and hence we appreciate approaches that consider this fact. It is also possible to convert queries in natural language provided by user to other formats like RDF beneath the surface, so that the interface between user and the system is still natural language. However this conversion raises complexities which will hinder us from dealing directly with main problems in IR field. The other reason is that having queries in RDF format would be useful in cases that appropriate RDF infrastructures for corpus and auxiliary vocabularies preexist and apparently this premise does not hold in most situations.

The second point is that all the constants and thresholds in (2) and (3) are derived from a training set consisting of 20% of the documents in Medline corpus and their approximate values are those which results in a system with best average precision over 11 standard recall points. Also the main idea behind (3) is that the similarity between a query and a document has direct relation with their concepts similarity and their regarding frequencies, while it has reciprocal relation with generality of concepts forming a query or document.

The last point is about the reason we have used semantic techniques in our retrieval algorithm instead of working on statistical alternatives. There are several statistical techniques which can be used for improving the precision of the system. Pseudo relevance feedback and latent semantic indexing are among these techniques. Although these alternatives can be beneficial to some extent in IR field but we believe using semantic methods instead of them would be preferable. The reason for this preference is that pseudo relevance feedback and latent semantic indexing are both based on statistical and syntactic methods which have the same nature as classic VSM approach, in contrast semantic techniques are inherently different from classic approaches which can better cooperate with syntactic approaches. In our view, using semantic techniques in parallel with classic approaches is like examining a phenomenon from two different aspects which obviously yields to a more complete understanding of that phenomenon.

6 Conclusion and Future Work

In this paper we proposed a hybrid information retrieval system for medical field using MeSH as a domain specific ontology. In our approach both classic retrieval techniques and semantic measures are used together. Results show that semantic scores which are calculated in an appropriate way would be of great benefit in determining non-relevant results from those pairs rendered by syntactic approaches.

We strongly believe that exploiting existing domain specific ontologies or thesauruses would be an extremely appropriate choice for improving the quality of IR systems, and semantic approaches have still undiscovered potentialities in IR field which should be unraveled. Future work in this area would be using semantic scores more widely than what is done in this experiment. We are currently working on merging semantic and syntactic scores instead of using them exclusively in eliminating from syntactic results, and first results show that our proposed semantic similarity measure can be beneficial to this end. Also there are unused relations and attributes of MeSH ontology such as onlinenotes, seealso, annotation, pharmacologicalaction, and semantictype that are neglected in our approach. We have examined these useful sources of information and have found out they can be beneficial for our system, but extracting related and useful chunks of data from them is very crucial and needs large amount of effort and attention.

The last point is about constants and thresholds used in different parts of proposed approach. As a matter of fact, values for these parameters are determined based on experiment and in order to examine their rules in semantic information retrieval process, more studies with various test collections seems essential. One such test collection which authors of the paper have plan to work with, is OHSUMED data set, consisting of 348566 references from 270 medical journals.

References

1. Agosti, M., Crestani, F., Gradenigo, G., Mattiello, P.: An Approach to Conceptual Modelling of IR Auxiliary Data. In: Proc. IEEE Int'l. Conf. Computer and Comm. (1990)
2. Jorvelin, K., Kekalainen, J., Niemi, T.: ExpansionTool: Concept-Based Query Expansion and Construction. Information Retrieval 4(3-4), 231–255 (2001)
3. Deerwester, S., Dumais, D., Furnas, G., Landauer, T., Harshman, R.: Indexing by Latent Semantic Analysis. J. Am. Soc. Information Science 41(6), 391–407 (1990)
4. Letsche, T.A., Berry, M.W.: Large-Scale Information Retrieval with Latent Semantic Indexing. Information Sciences—Applications 100(1-4), 105–137 (1997)
5. Gonzalo, J., Verdejo, F., Chugur, I., Cigarran, J.: Indexing with WordNet Synsets Can Improve Text Retrieval. In: Proc. COLING/ ACL Workshop Usage of WordNet for Natural Language Processing (1998)
6. Madala, R., Takenobu, T., Hozumi, T.: The Use of WordNet in Information Retrieval. In: Proc. Conf. Use of WordNet in Natural Language Processing Systems, Montreal, pp. 31–37 (1998)

7. Christophides, V., Karvounarakis, G., Plexousakis, D., Tourtounis, S.: Optimizing Taxonomic Semantic Web Queries Using Labeling Schemes. J. Web Semantics 1(2), 207–228 (2003)
8. Gauch, S., Chaffee, J., Pretschner, A.: Ontology-Based Personalized Search and Browsing. Web Intelligence and Agent Systems 1(3-4), 219–234 (2003)
9. Shah, U., Finin, T., Joshi, A.: Information retrieval on the semantic web. In: CIKM 2002: Proc. of the 11th Int. Conf. on Information and Knowledge Management (2002)
10. Mayfield, J., Finin, T.: Information Retrieval on the Semantic Web: Integrating Inference and Retrieval. In: Proc. Workshop Semantic Web at the 26th Int'l. ACM SIGIR Conf. Research and Development in Information Retrieval (2003)
11. Kiryakov, A., Popov, B., Terziev, I., Manov, D., Ognyanoff, D.: Semantic annotation, indexing, and retrieval. Journal of Web Semantics: Science, Services and Agents on the World Wide Web 2, 49–79 (2004)
12. Guha, R., McCool, R., Miller, E.: Semantic search. In: WWW 2003: Proc. of the 12th Int. Conf. on World Wide Web (2003)
13. Davies, J., Krohn, U., Weeks, R.: Quizrdf: search technology for the semantic web. In: WWW 2002 Workshop on RDF & Semantic Web Applications, 11th Int. WWW Conf. (2002)
14. Rocha, C., Schwabe, D., Poggi de Aragao, M.: A hybrid approach for searching in the semantic web. In: Proc. of the 13th Int. Conf. on World Wide Web (WWW) (2004)
15. Corby, O., Dieng-Kuntz, R., Faron-Zucker, C., Gandon, F.: Searching the semantic web: Approximate query processing based on ontologies. IEEE Intelligent Systems 21(1), 20–27 (2006)
16. Castells, P., Fernndez, M., Vallet, D.: An adaptation of the vector-space model for ontologybased information retrieval. IEEE Trans Knowl. Data Eng. 19, 261–272 (2007)
17. Knappe, R.: Measures of Semantic Similarity and Relatedness for Use in Ontology-based Information Retrieval, Ph.D. dissertation (2006)
18. Hliaoutakis, A.: Semantic Similarity Measures in MeSH Ontology and their application to Information Retrieval on Medline, Diploma Thesis, Technical Univ. of Crete (TUC), Dept. of Electronic and Computer Engineering, Chania, Crete, Greece (2005)
19. Li, Y., Bandar, Z., McLean, D.: An Approach for Measuring Semantic Similarity between Words Using Multiple Information Sources. IEEE Transactions on Knowledge and Data Engineering 45 (2003)
20. Leacock, C., Chodorow, M.: Filling in a sparse training space for word sense identication. Ms (1994)
21. Wu, Z., Palmer, M.: Verb Semantics and Lexical Selection. In: Proceedings of the 32nd Annual Meeting of the Associations for Computational Linguistics (ACL 1994), Las Cruces, New Mexico, pp. 133–138 (1994)
22. Dominich, S.: Connectionist Interaction Information Retrieval. Information Processing and Management 39(2), 167–194 (2003)
23. Truong, D., Dkaki, T., Mothe, J., Charrel, J.: Information Retrieval Model based on Graph Comparison. In: International Days of Statistical Analysis of Textual Data (JADT 2008), Lyon, France (2008)
24. Indrawan, M.: A Framework for Information Retrieval Using Bayesian Networks, Monash University, PhD Thesis (1998)
25. Kumar, C.A., Srinivas, S.: Latent Semantic Indexing Using Eigenvalue Analysis for Efficient Information Retrieval. Int. J. Appl. Math. Comput. Sci. (2006)

Mining Rare Events Data for Assessing Customer Attrition Risk

Tom Au, Meei-Ling Ivy Chin, and Guangqin Ma

AT&T Labs, Inc.-Research, USA

Abstract. Customer attrition refers to the phenomenon whereby a customer leaves a service provider. As competition intensifies, preventing customers from leaving is a major challenge to many businesses such as telecom service providers. Research has shown that retaining existing customers is more profitable than acquiring new customers due primarily to savings on acquisition costs, the higher volume of service consumption, and customer referrals. For a large enterprise, its customer base consists of tens of millions service subscribers, more often the events, such as switching to competitors or canceling services are large in absolute number, but rare in percentage, far less than 5%. Based on a simple random sample, popular statistical procedures, such as logistic regression, tree-based method and neural network, can sharply underestimate the probability of rare events, and often result a null model (no significant predictors). To improve efficiency and accuracy for event probability estimation, a case-based data collection technique is then considered. A case-based sample is formed by taking all available events and a small, but representative fraction of nonevents from a dataset of interest. In this article we showed a consistent prior correction method for events probability estimation and demonstrated the performance of the above data collection techniques in predicting customer attrition with actual telecommunications data.

Keywords: Rare Events Data, Case-Based Sampling and ROC Curves.

1   Introduction

We study the problems in modeling the probability of a rare event. In statistics, the probability of rarity is often defined as less than 5%. Many of the events in real world are rare-binary dependent variables with hundreds of times fewer ones (events) than zeros ("nonevents"). Rare events are often difficult to predict due to the binary response is the relative frequency of events in the data, which, in addition to the number of observations, constitutes the information content of the data set. Most of the popular statistical procedures, such as logistic regression, tree-based method and neural network, will not only underestimate the event probabilities [1], but also more often result null models (no significant predictors).

In order to accurately estimate the probability of a rare event, researchers have been using disproportional sampling techniques, such as case-based sample, to do

S.K. Prasad et al. (Eds.): ICISTM 2009, CCIS 31, pp. 41–46, 2009.

the estimation and bias adjustment. For rare events study, case-based sampling is a more efficient data collection strategy than using random samples from population. With case-based sample, researchers can collect all (or all available) events and a small random sample of nonevents. In addition to gain accuracy and efficiency, researchers can focus on data collection efforts where they matter. For example, in survival analysis, statisticians have been using the case-based sample in proportional hazards analysis to evaluate hazards rate [2] ; in social sciences, researchers have been using the same sampling technique to evaluate the odds of rare events [1]. Resources available to researchers are different, in clinical trials, the number of subjects to study is limited, researchers relies on good quality of collection and sample design; in other disciplines, there may be a vast collection of information, researchers can be overwhelmed by the volume of information available, and the procedures used for assessing rare events can be very inefficient. To the later case, the case-based data sampling can become vital for a successful analysis.

Customer attrition refers to the phenomenon whereby a customer leaves a service provider. As competition intensifies, preventing customers from leaving is a major challenge to many businesses such as telecom service providers. Research has shown that retaining existing customers is more profitable than acquiring new customers due primarily to savings on acquisition costs, the higher volume of service consumption, and customer referrals [3] [4]. The importance of customer retention has been increasingly recognized by both marketing managers as well as business analysts [3] [5] [6] [4].

In this study, we emphasize the effectiveness of using case-based sample to modeling a rare event. We evaluated the performance of case-based sampling techniques in predicting customer attrition risk by using data from a major telecom service company and three popular statistical procedures, a logistic regression, a tree method and a neural network procedure.

2   Data Structure for Events Prediction

An event can occur at any time point to a subject of interest. To predict an event occurrence in a future period, we assume the event has not occurred to the subject at the time of prediction. We consider the following structure of events prediction data

$$\{X_{t-m}, X_{t-m+1}, X_{t-m+2}, \ldots, X_t | Y = 0/1 \; in \; (t, t+k)\}, \qquad (1)$$

where Xs are explanatory variables measured before the time of prediction t, Y is the event indicator. Y=1 if the event occurred during period (t, t+k); otherwise Y=0.

This data structure for events prediction is not only simple, but also bears real application scenarios. For example, when we score customer database with risk score for retention, we score all currently active customers who are yet to defect, disconnect, or cancel services, then we target those customers with high risk of leaving within a specified period. When we collect data for predictive modeling, we have to bear in mind its application scenario, that is, when we collect data, we need to define a reference time point that all collected individual subjects are 'active' (event not occurred yet), and a period after the time point that events are observed. The length of the window between start point to the reference point of 'active' can be determined by needs and resources available; the length of the period between the 'active' time point to the end of event observation window is specified by retention needs, say if you like to predict event probability for the next quarter or next six-month or next 12-month, etc.

3   Case-Based Sampling

As in most rare event history analyses, the rare event represents only a small percentage of the population (less than 5%). The low rate of event implies that if we draw a random sample to study the event, the sample will be characterized by an abundance of nonevents that far exceeds the number of events, we will need a very large sample to ensure the precision of the model estimation (or we would have to observe a long period of time to collect more cases of event.) Processing and analyzing the random sample will not only impact the efficiency of using resources, but also more importantly reduce the efficiency and accuracy of model estimation. To overcome this problem, practitioners used a type of sampling technique called the case-based sampling, or variations of the ideas like choice-based sampling (in econometrics), case-control designs (in clinical trials) [7], case-cohort designs (in survival analysis) [2]. Here we design a case-based sampling by taking a representative simple random sample of the data in (1) with response 0s, and all or all available subjects in (1) with 1s, such that, the ratio of 0s to 1s is about 3:1, 4:1, 5:1 or 6:1. The combined dataset is called a case-based sample of the data in (1).

The probability of event based on the case-based sample will be biased and different from population probability of event. We can show that this biasness can be corrected based on population prior information and the bias-corrected estimation is consistent.

Let N_0 be the number of nonevents in the population, and let n_0 be the number of nonevents from the random sample of the case-based sample. Denote event probability for individuals with $X = x$ in the case-base sample by $p(Y = 1|x, k)$, and the probability in the population by $P(Y = 1|x, k)$, then use similar arguments as in [1],

$$\frac{n_0}{N_0} \frac{p(Y = 1|x, k)}{(1 - p(Y = 1|x, k))} \xrightarrow{d} \frac{P(Y = 1|x, k)}{(1 - P(Y = 1|x, k))}, \tag{2}$$

in distribution as sample size increases, where $\xrightarrow{d}$ denote convergency in distribution. Similarly, the following bias-corrected event probability $p'(Y = 1|x, k)$ is consistent, that is,

$$p'(Y = 1|x, k) = \frac{n_0}{N_0} \frac{p(Y = 1|x, k)}{(1 - (1 - \frac{n_0}{N_0})p(Y = 1|x, k))} \xrightarrow{d} P(Y = 1|x, k), \tag{3}$$

in distribution as sample size increases. So the prior-adjusted event probability estimated based on a case-based sample is very similar to the population event probability.

A major task in customer retention is to rank customers by their attrition risks, and to target those most vulnerable customers as a retention effort. The consistency showed in (2) and (3) indicates that the ranking of customer attrition risk based on a case-based sample is consistent with the actual ranking, if there is one.

4   Predictive Modeling

There are several data mining methods that may be used to construct models to estimate event probability, such as the logistic regression method, the Cox regression method, the tree-based classification method, and, more recently, the artificial neural network method. Each may be more suitable for a particular application. In this exercise, we set out to use three different modeling methods that are most often used to predict customer attrition: (1) logistic regression - the regression-based method, (2) tree-based method, and (3) artificial neural network method.

The logistic regression is estimated by using the backward stepwise procedure with a level of significance at 10 percent. The tree-based method is developed by using a significant level for the Chi-square statistics less than 10 percent and with minimum number of observations in a node not less than 20. As for the artificial neural network method, we select a model with one hidden layer with three hidden neurons and a logistic activation function.

We choose to study a data set from a major telecom service company, which includes a segment of customers with a large number of service lines (in the order of millions) that are active at the beginning of the year. The unit of analysis is the individual subscriber line. For each active service line, information on service termination (yes/no) was collected over the next 3-month period (the first quarter of the chosen year). Customer usage and characteristics were collected over the 3-year period backward from the chosen year. This information includes type of service, service usage, marketing information, customer demographics, and service line transaction history.

We build the models by randomly splitting (50/50) a case-based sample into two data sets, one used for learning (training) and the other used for testing. We first develop the models using the learning data. We then test the models by applying it to the testing data to determine the extent to which the models may be generalized beyond the original learning data. We further validate the models by using a recent customer database which is called the validation data. For the validation data, the study window is shifted with event period not overlapping with the learning and testing data.

The performance of a predicting model may vary depending on a specific decision threshold used. Thus, an objective evaluation of a risk-predicting model should examine the overall performance of the model under all possible decision thresholds, not only one particular decision threshold. To achieve this, we adopt a useful tool, the receiver-operating characteristic (ROC) curve, to evaluate the performance of a risk predictive model. This curve is the locus of the relative frequencies of false event fraction (FEF) and true event fraction (TEF), which occupy different points on the curve corresponding to different decision thresholds. The area under this curve (on a unit square) is equal to the overall probability that the predictive model will correctly distinguish the "at risk" subjects from the "not at risk" subjects [8] [9].

5   Results

We repeated the analysis described in section 4 six times by using three different case-based samples and three different random samples. Figure 1 shows various performance characteristics of the three popular procedures based on the those samples. The performances of the three models estimated based on the random samples have deteriorated sharply on the learning and validation data. This deterioration is largely due to the uncertainty of model estimation based on a random sample.

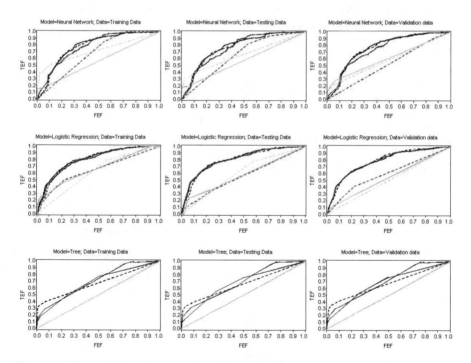

Fig. 1. ROC curves based on case-based samples (black and blue lines) and random samples (pink, red and orange lines)

When the customer attrition rate is low (less than 5%), we suggest to use the case-based sampling method for predictive modeling. The result in this paper will likely apply to some rare event data mining.

Since our intention here is to demonstrate the effectiveness of case-based sampling relative to random sampling in rare event modeling, models discrimination performance is not discussed here. We refer readers to our previous study on the applications applying and evaluating models to predict customer attrition [10].

References

1. King, G., Zeng, L.: Logistic Regression in Rare Events Data. Society for Political Methodology, 137–163 (February 2001)
2. Prentice, R.L.: A Case-cohort Design for Epidemiologic Cohort Studies and Disease Prevention Trials. Biometrika 73, 1–11 (1986)
3. Jacob, R.: Why Some Customers Are More Equal Than Others. Fortune, 200–201 (September 19, 1994)
4. Walker, O.C., Boyd, H.W., Larreche, J.C.: Marketing Strategy: Planning and Implementation, 3rd edn., Irwin, Boston (1999)
5. Li, S.: Applications of Demographic Techniques in Modeling Customer Retention. In: Rao, K.V., Wicks, J.W. (eds.) Applied Demography, pp. 183–197. Bowling Green State University, Bowling Green (1994)
6. Li, S.: Survival Analysis. Marketing Research, 17–23 (Fall, 1995)
7. Breslow, N.E.: Statistics in Epidemiology: The case-Control Study. Journal of the American Statistical Association 91, 14–28 (1996)
8. Hanley, J.A., McNeil, B.J.: The Meaning and Use of the Area under a ROC Curve. Radiology 143, 29–36 (1982)
9. Ma, G., Hall, W.J.: Confidence Bands for ROC Curves. Medical Decision Making 13, 191–197 (1993)
10. Au, T., Li, S., Ma, G.: Applications Applying and Evaluating Models to Predict Customer Attrition Using Data Mining Techniques. J. of Cmparative International Management 6, 10–22 (2003)

Extraction and Classification of Emotions for Business Research

Rajib Verma

Paris School of Economics, 48 BD Jourdan, 75014 Paris, France

Abstract. The commercial study of emotions has not embraced Internet / social mining yet, even though it has important applications in management. This is surprising since the emotional content is freeform, wide spread, can give a better indication of feelings (for instance with taboo subjects), and is inexpensive compared to other business research methods. A brief framework for applying text mining to this new research domain is shown and classification issues are discussed in an effort to quickly get businessman and researchers to adopt the mining methodology.

Keywords: Emotions, mining, methodology, business, marketing.

1 Introduction and Related Work

Text mining offers a way to collect dark data from the Internet for use in business research. While this has not been done significantly, it has widespread and important applications. Amassing this information will give a real time indication of the magnitude and polarity of sentiment, and it will provide us with data showing the history, direction, magnitude of changes, and the impact of events like product changes, price variation, policy changes, and the effect of marketing campaigns.

It provides an alternative to more traditional research methods like surveys and focus groups, so it is apt for areas like finance, CRM, and marketing. While this type of research has not been done extensively, it is a promising area since some work has been done in finance, politics, and other fields with positive results. The most significant application of this new line is seen in forecasting share price movements where sentiment indicators have made their way into commercial practice, although using a much smaller data set than the one advocated here. [10] and [2] show that online discussion can provide better forecasts than financial analysts, [9] demonstrates that the intensity in volume of after hours postings can predict the variation in stock market returns, and that the higher the amount of discussion the higher the volatility.

2 An Overview of Happiness Extraction

As the flowchart in Figure 1 shows, there are a series of steps on the path to extraction, which involve running automated programs, doing manual analysis, searching,

S.K. Prasad et al. (Eds.): ICISTM 2009, CCIS 31, pp. 47–53, 2009.

cleaning, and finally getting at the critical information. Since the information is spread throughout the Internet, every element of this global network acts as an input to the crawler, which is just a program that explores every node (e.g., web page) on the network and sifts out relevant data. This is then further refined (e.g. by cleaning up the hypertext, tagging key words, isolating negations, and clarifying abbreviations.) in order to get it ready for processing. After preparation, the raw data is scanned in order to see if there are any indications of sentiment. If there are, those parts are isolated and sent to a classifier and quantification module that determines if the sentiments are useful as happiness indicators; it then assigns an appropriate weight to them. The refined data is then organized in some meaningful way (for example by country), analyzed by both man and machine, and the results reported.

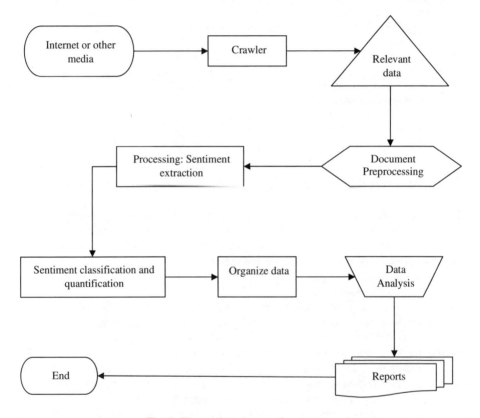

Fig. 1. The sentiment extraction process

2.1 Classification

There are a number of methods for sentiment classification but they can be broadly divided into statistical, informatic, linguistic, and qualitative approaches. In the first classification technique, the focus is on using metrics to select relevant terms or units. The second is based on computational methods, like machine learning. The third

relies on the grammatical construction of the text, the vocabulary used, word collocations, or prepared dictionary lists to select emotive passages. Finally, the last one uses
human subjectivity or non-technical methods (like visual inspection) to classify sentiments, this tends to be manual. While it is possible to use one or the other exclusively, the best results occur when one filter feeds into the other. For instance, initially
sieving the documents through a linguistic classifier increases the reliability of statistical measures since you are using them on a set of more likely candidates.

There are two things being classified, one is the text as a whole. The other is an in-
termediary step, the selection of individual sentiment laden terms or units. These are
used to build a lexicon or otherwise select the relevant texts for analysis [8]. While
this can be used as a first step in the classification process, it is not the best way to
calculate precision (the percentage of extracted terms that are also on the reference
list) since the system can extract the correct sentiment word even though it is not on
the list, in which case it would be rejected. Nevertheless, this is the most basic component of the classification system and many metrics rely on it.

2.1.1 Reference Lists

Experts can be used to classify a text directly or to validate the results of an automated system. They make it much easier to analyze complex texts at the expense of
time, but in the end may not give much better results for simple texts. More complicated ones still need people to either validate the algorithm's results or go through the
semantics entirely. Since terms tend to be somewhat vague, each expert needs to be
trained on their meaning; in any event, they tend to make their evaluation based on
their own intuition, external factors, and their experience with terms that they have
already evaluated. This creates inter-expert variation.

While the evaluation divergences can be reduced for a particular team, independent
teams continue to have their differences, suggesting that there is a structural difference in meaning—so by trying to homogenize the interpretations you are actually
destroying information and introducing errors. Assuming that some words are more
useful than others, disagreements in the interpretation of words among different people can be used as a way to filter out terms that are not central to the category. [1]
shows that 21.3% of the words found in different lists can obscure the classification
process, with the accuracy ranging from 52.6 percent when using lists with more
ambiguous words, to a high of 87.5 percent when more definite sentiment laden
words are used. Moreover, they find that the boundary between sentiment and neutral
words accounts for 93 percent of the misclassifications. This implies that large and
random lexicons may not be representative of the population of categories.

Nevertheless, through expert validation it is possible to calculate the algorithms
precision by comparing its extractions to those that have been kept by the experts.
However, there are some limitations when relying on experts. For instance, it is impossible to evaluate the system's recall ability (the percentage of terms in the lexicon
that have been selected by the algorithm.) when experts are used alone because there
is no reference list for comparison. As well, since expert based validation methods
apply directly to the results extracted by a system they are inherently non-scalable,
after all the results of a new algorithm classifying the same text would have to be
evaluated again, as would the same algorithm on a new text.

Some automation would help overcome these limitations. Perhaps the simplest relies on frequency / word count, which filters out messages or selects sentiment words by counting the number of relevant units. This can be easily implemented by developing a specialized lexicon and comparing it to the message's content. However, it may not accurately capture the meaning of the message, especially as vagueness of the message increases.

Fortunately, [1] shows that the gross and the net overlap (GOM / NOM) scores may overcome this to some extent; these lexicons tend to have the highest frequency classifications, hence, they tend to contain terms that have more sentiment content, which causes vagueness to diminish. While the GOM does provide more information than simple frequency, it can still be somewhat prone to the problems of vagueness when frequency is low and when the word is classified with both polarities. Fortunately the NOM adjusts for this, even though at low frequency it may not be effective.

Less ambiguity corresponds with higher classification accuracy since NOM scores of zero (i.e. for categories composed of neutral sentiments or those that mix positive and negative scores equally) give an accuracy of 20%, those with a zero score but only containing positive or negative terms have 57 percent accuracy, and since in general accuracy increases with higher absolute scores (exceeding 90 percent for NOM scores above seven) [1]. In order to use the overlap methods effectively over different corpora a list of high frequency sentiment words would need to be generated; which is a more computationally expensive and time consuming process than using simple frequencies. Even though simple word counts have some limitations, [7] finds that they make the best compromise for creating lexicon's for sentiment extraction, which is supported by [4] since true terms tend to have high frequency and because the metric is computationally very light. Not surprisingly, [6] finds that statistical measures do worse when applied to low frequency terms.

2.1.2 Linguistic Methods

Several techniques fall under this umbrella including information extraction (IE), semantic orientation (SO), and natural language processing (NLP). IE takes text from disparate sources and stores it in a homogenous structure, like a table. The idea is just to analyze the relevant portions of the original text, not necessarily the whole document. The second approach can be applied to individual words or the entire passage. A word's orientation depends on how it relates to other words in the text and on its relation to strength indicators, which includes words like great or horrible. By adding up the semantic orientation of each word, you can calculate the orientation of the entire text. The caveat is that this method is based on the assumption that the text pertains to only one topic. Consequently, passages containing many different subjects would give an incorrect semantic orientation score, and it could lead to them being assigned the wrong polarity.

NLP improves on the situation. It usually implies parsing, and mapping the results onto some kind of meaning representation. Since it allows sub-topic sentiment classification, you can pick out particular things that make someone happy or unhappy. Also, by using the information in the grammatical structure to determine what the referent is, it can reduce the errors made when associating the sentiment to the topic.

This can be done by: isolating phrases that contain definite articles, using preposi-tional phrases and relative clauses, and looking for particular structures of verb/noun-adjective phrases or their particular location in a sentence. By extracting only those portions of the text that can grammatically contain the concept, you automatically exclude words that would otherwise pass through your filter or be mistakenly recog-nized, so it makes the selection procedure execute faster (since there's a much smaller input), and it keeps errors lower.

[5] points out that text content can be grouped into two domains, unithood and termhood. Although this classification scheme allows you to search for good senti-ment candidates based on two linguistic dimensions instead of just one, [7] finds that measures for both of these seem to select similar terms, especially those that are ranked highest. Since they contain essentially the same information, it may be possi-ble to generate good classification lists using just one of the dimensions, which would reduce the computational overhead. However, which one is better is unclear; in [7] 2/3's of the effective measures are geared towards termhood, but they find that the log likelihood ratio does not effectively recognize sentiment in unithood, whereas [3] and others do. So more work needs to be done to see how effective the likelihood ratio is. If it turns out to be less effective, then algorithms could focus more on termhood. If it turns out to be an effective predictor, than perhaps both termhood and unithood should be retained.

Quantitative indicators of the unit or term's relevance can be broadly classified as heuristic or association, measures. The first tend to lack a strong theoretical basis, instead they are based on empirical observations or logical intuition. Consequently, they cannot guarantee a good solution, but like frequency and C-value, they tend to give reasonable answers. In fact, they can give better solutions then their theoretically sound counterparts, one explanation is that they correct for an underlying weakness in the theoretical measure. Take the mutual information metric, since it tends to select words that occur infrequently in texts, its power to discriminate meaningful sentiment words is reduced. However, by cubing the measure there is a tendency to overcome this weakness, at least in the highest-ranking terms.

Association measures tend to quantify the degree of collocation between words, which is based on the observed frequency values in the unithood contingency table. Unfortunately, estimating these frequencies is not a trivial matter since the overall probability distribution becomes more complex with greater variables. Thus, for cate-gorical variables it becomes impractical to estimate the individual cell counts. Luck-ily, the Bayesian network offers a solution. If you can summarize the distribution in terms of conditional probabilities, and if a small number of these are positive, then the number of parameters that need to be estimated declines immensely and the precision increases.

To extend these inferences to the population, a random sample model should be used to estimate the collocation values. [3] points out that if the collocations are mu-tually independent and stationary, then the elements for a 2 unit collocation can be taken from a Bernoulli distribution. In this case, each element of the table has its own probability of occurrence, each of which need to be estimated. Degrees of association metrics directly calculate these parameters using maximum likelihood estimation [7], but they are replete with estimation errors, especially when frequencies are low. Whereas significance of association measures ameliorate this by assuming that each of the table's cell probabilities are the product of the independent marginals, which are computed using maximum likelihood estimation. So the expected frequency of the cell is equal to the mean of the binomial distribution.

Empirically, measures based on frequency and significance of association outperform degree of association metrics in terms of precision. This occurs because discriminatory power lies with just the highest-ranking words, which tend to have the highest frequency scores too. This is similar to the notion of centrality. Since words that carry the most relevant emotion necessarily have high centrality, they tend to be used more often; so the extremely low precision values of the degree of association measures can be explained by the fact that they rely too much on low frequency terms. This is borne-out by the fact that the frequency metric (and those dependent on frequency) tends to consistently have the highest precision, mutual information (which tends to select low frequency terms) has the lowest, and its cube (which selects more high frequency terms) does much better than it at lower percentiles.

This suggests that, among these measures, frequency based scores will tend to extract sentiment most precisely, and that methods relying on the null hypothesis of independence (like the t-score or the log likelihood ratio) should be preferred to those that only try to approximate the probability parameters (like the dice factor). Computationally, the implication is that it may be best to use the frequency metric as the primary basis for term selection—it appears to be the most effective and needs the least computing power. Perhaps though, the other measures could be used to validate the frequency-based list in an attempt to improve precision further by isolating the remaining outliers, but this idea needs to be investigated further.

3 Conclusion and Future Work

A hybrid approach to mining and classification is recommended, and an overview of the mining process is given along with discussions about lexicon construction, linguistic methods, and the empirical basis of them.

Since mining is limited in emotions research, there are many avenues for interdisciplinary work. It could be used to study how influence, the connectedness / structure of sites, and how social contagion influences customer satisfaction. Results from political blogs suggest that influence via trust can drastically affect the sentiment index, suggesting that perhaps a similar effect could be seen in online revealed emotions data. Management scientists often use qualitative methods. Mining can aid in the detection of patterns in interview transcripts and the classification of textual data into novel categories. More generally, we could generate descriptive statistics, and use the data to make temporal and geographic comparisons.

On the technical end, MIS / ICT departments could develop a rich research agenda. Issues like the quantification of noise, determining which methods improve frequency-based lexicons, and developing algorithms for better searching, classification, cleaning, reducing, and analyzing mined data are all areas that need more work.

At a more applied level, emotions mining has immediate applications in areas like marketing and finance. Some applications have already hit the market, the most notable being the development of online sentiment indexes for financial markets. Mining can act as an early warning system to failing marketing campaigns, and allow us to better understand the impact of changes to the product or campaign. In short, it will improve competitiveness through better CRM. Certainly, in the near future this trend will continue and the application domains will become broader.

References

1. Andreevskaia, A., Bergler, S.: Mining Wordnet for Fuzzy Sentiment: Sentiment Tag Extraction from WordNet Glosses. In: Proceedings of the 5th International Conference on Language Resources and Evaluation (LREC) (2006)
2. Antweiler, W., Frank, M.: Is All That Talk Just Noise? The Information Content of Internet Stock Message Boards. Journal of Finance, American Finance Association 59(3) (2005)
3. Dunning, T.: Accurate Methods for the Statistics of Surprise and Coincidence. Computational Linguistics 19(1) (1994)
4. Evert, S., Krenn, B.: Methods for the qualitative evaluation of lexical association measures. In: Proceedings of the 39th Annual Meeting of the Association for Computational Linguistics, Toulouse, France (2001)
5. Kageura, K., Umino, B.: Methods of automatic term recognition. Terminology 3(2) (1996)
6. Krenn, B.: Empirical Implications on Lexical Association Measures. In: Proceedings of the Ninth EURALEX International Congress, Stuttgart, Germany (2000)
7. Pazienza, M.T., Pennacchiotti, M., Zanzotto, F.M.: Terminology Extraction. An Analysis of Linguistic & Statistical Approaches. In: Sirmakessis, S. (ed.) Knowledge Mining: Proceedings of the NEMIS 2004 Final Conference VIII (2005)
8. Verma, R.: Choosing Items for a Bifurcated Emotive-Intensive Lexicon using a Hybrid Procedure. In: Proceedings of SAMT 2008 Posters & Demos located at the 3rd International Conference on Semantic and Digital Media Technologies, Koblenz, Germany (December 3, 2008), http://resources.smile.deri.ie/conference/2008/samt/short/20_short.pdf
9. Wysocki, P.: Cheap Talk on the Web: The determinants of Stock Postings on Message Boards. WP 98025. University of Michigan Business School (1998)
10. Zaima, J.K., Harjoto, M.A.: Conflict in Whispers and Analyst Forecasts: Which One Should Be Your Guide? Financial Decisions, Article 6 (Fall, 2005)

Data Shrinking Based Feature Ranking for Protein Classification

Sumeet Dua and Sheetal Saini

Department of Computer Science, Louisiana Tech University,
Ruston, LA 71272, USA
sdua@coes.latech.edu, ssa017@latech.edu

Abstract. High throughput data domains such as proteomics and expression mining frequently challenge data mining algorithms with unprecedented dimensionality and size. High dimensionality has detrimental effects on the performance of data mining algorithms. Many dimensionality reduction techniques, including feature ranking and selection, have been used to diminish the curse of dimensionality. Protein classification and feature ranking are both classical problem domains, which have been explored extensively in the past. We propose a data mining based algorithm to address the problem of ranking and selecting feature descriptors for the physiochemical properties of a protein, which are generally used for discriminative method protein classification. We present a novel data shrinking-based method of ranking and feature descriptor selection for physiochemical properties. The proposed methodology is employed to demonstrate the discriminative power of top ranked features for protein structural classification. Our experimental study shows that our top ranked feature descriptors produce competitive and superior classification results.

Keywords: Protein Classification, Feature Ranking, Feature Selection, Data Shrinking.

1 Introduction

The rise of high throughput data disciplines, such as bioinformatics, biomedical informatics, and clinical informatics, in last few decades have resulted in the accumulation of enormous amounts of data. Many data preprocessing strategies have been proposed to sufficiently handle the multidimensional nature of the data and avoid the infamous 'curse of dimensionality.' Dimensionality reduction methods, including feature ranking, feature selection, and feature extraction, among other reduction strategies, have proven to be powerful in reducing this impediment. In feature selection, irrelevant features are eliminated from further consideration, thereby leaving only important features to be considered for further analysis. Feature extraction strategies may involve the construction of a new discriminative feature space for data representation. The underlying assumption of dimensionality reduction approaches is that not all dimensions are important, i.e. some dimensions may be irrelevant and detrimental to the efficacy of further data analysis, and hence can be eliminated.

S.K. Prasad et al. (Eds.): ICISTM 2009, CCIS 31, pp. 54–63, 2009.

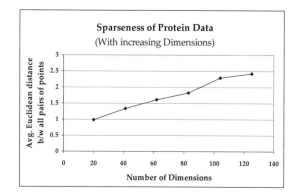

Fig. 1. The X-axis represents the number of dimensions and the Y-axis represents the average Euclidean distance between all pairs of points

Proteomics is high throughput data discipline, and multidimensionality is an inherent characteristic of the proteomic data. For example, the total number of feature descriptors generated for the physiochemical properties of proteins for the descriptive method of protein structural classification can be in the hundreds [1]. Proteomics data is characteristically high dimensional and exhibits sparseness (Fig. 1). Protein structural classification is an extensively researched area in proteomics. Finding protein structure helps researchers interpret protein functions and behavior and provides novel biological knowledge that helps scientists to design drugs, therapies, and other forms of treatment. Several computational methods such as eKISS [2], BAYESPROT [3], SVM and NN based methods [4], have been proposed for protein classification.

Furthermore, computational methods for protein structural classification can be divided into two major categories [3]: homology-based methods and discriminative methods. There are pros and cons to using both methods. The physiochemical property based classification scheme reported in [2], [3], and [4] utilizes the six physiochemical properties of amino acids and constructs 125 feature descriptors for structural classification. Our proposed work is focused on the discriminative method of protein structural classification which uses top ranked novel feature descriptors for physiochemical properties. The ranking of these feature descriptors is based on the shrinking profiles of these properties as explained in Section 3. Data shrinking is a data mining technique that simulates data movement. Such approaches often improve the performance and effectiveness of data analysis algorithms [5] and [6]. In this paper, we present a unique feature ranking and selection strategy based on a novel data adaptive shrinking approach (Section 3). The approach is applied to select unique feature descriptors from the 125 feature descriptors available. The ranked features are then used to structurally classify proteins.

The remainder of the paper is organized as follows. In Section 2, we present a brief history of related work. In Section 3, we discuss our methods and algorithms. In Section 4, we present our experiments and results, and, in Section 5, we discuss our conclusion.

2 Related Work

Many data mining methods have been proposed to cope with the protein structural and fold recognition problem and to improve performance for protein classification. There are two aspects of the related work, the first aspect pertains to data shrinking and its application for feature selection, and the other aspect pertains to protein structural classification and feature selection in protein structural classification.

The underlying hypothesis of our work is that feature ranking of feature descriptors for physiochemical properties can be derived by measuring feature sparseness in the feature domain. Work presented in [5] provides strong evidence that data shrinking can be used as feature ranking and selection technique. The efficacy of such ranking can be evaluated by classifying proteins in their respective structural or fold classes. In [5], the difference of the data distribution projected on each dimension through data-shrinking process is utilized for dimension reduction. The hypothesis is that alteration in the histogram variance through data shrinking is significant for relevant and discriminatory dimensions and good dimension candidates can be selected by evaluating the ratio of the histogram variances for each dimension before and after data-shrinking process. But, the existing work lacks experimental proofs.

Furthermore, a significant amount of research has been done to improve the methods for protein classification and feature selection application regarding this problem [2], [3], [4], [7], [8], and [9]. Some representative examples of work in this area are presented below; these works serve as partial motivation for our work. In [4], authors address issues related to the classification strategy one-vs-others and present studies pertaining to the unique one-vs-others and the all vs all methods for protein fold recognition. Similarly, in [2], authors attempt to improve classifier performance to cope with an imbalanced dataset and present an ensemble learning method 'eKISS' to solve problems related to protein classification. In [3], the authors present a framework based on Tree-Augmented Networks to address protein structure and fold recognition problems.

Interestingly, in the above mentioned works, the feature set derived for classifier building and testing has evolved little, and the focus has primarily been on classifier design and performance modeling. We believe that sparseness in these features has a detrimental affect on classifier performance and on attempts to rank features based on the degree of sparseness to boost the sensitivity and specificity of classifiers. Feature ranking methods can specifically improve classification performance when only the most relevant, top ranked features are used. Some related work in feature ranking for protein classification is as follows. In [8], n-grams descriptors based knowledge discovery framework involving feature ranking and selection is proposed to alleviate the protein classification problem. Furthermore, the authors of [9] propose a hybrid feature ranking method that combines the accuracy of the wrapper approach and the quickness of the filter approach in a two step process.

3 Methodology

In discriminative methods of structural classification, as presented in [2], [3], and [4], the physiochemical properties of amino acids (AA) are modeled as feature vectors and then deployed for classification. However, only a few descriptors contribute to most data discriminative power, and the rest are redundant or irrelevant [7]. Superior methods are needed to rank these descriptors and to use only those that are most relevant and informative.

Furthermore, hundreds of feature descriptors can be generated for amino acid (AA) properties [10]. This data is characteristically high dimensional in nature and exhibits sparseness. Sparseness, in this case, refers to an increase in the average Euclidean distance between all pairs of data points with an increase in the number of dimensions as shown in Fig. 1. Therefore, sparseness of high dimensional protein data presents a unique opportunity to apply a data shrinking strategy to diminish the sparseness. Research presented in [11] has inspired us to propose a novel method of ranking and selection for feature descriptors of protein physiochemical properties and to employ the method for supervised classification.

3.1 Data Shrinking

The data shrinking process utilizes the inherent characteristics of data distribution, and results in a more condensed and reorganized dataset. Data shrinking approaches can be broadly classified into gravitational law-based approaches and mean shift approaches. In gravitational law-based approaches, data points move along the direction of the density gradient. Points are attracted by their surrounding neighbors, and move toward the center of their natural cluster [5], [6], [12], [13], and [14]. Whereas, the mean shift approach originates from kernel density estimation, in this approach, data points are shrunk toward the cluster center using kernel functions and other criteria [15] and [16].

Furthermore, [5] and [6], propose a grid based shrinking method. However, this approach utilizes only fixed size grids. Fixed sized grids are "insensitive" to underlying data distribution and do not project the underlying distribution of the data. Consequently, they do not aggregate or shrink all data points into groups effectively. This problem is aggravated as the number of dimensions increases. We propose an "adaptive" grid generation method that captures the sparseness of the high dimensional data by utilizing the underlying data distribution in every dimension and exploiting it to generate grid boundaries. The grid boundaries are then further employed for data shrinking, and, finally, feature selection is performed for classification.

3.2 Proposed Shrinking Method

Let the dataset have N data points and n dimensions, and let the data be normalized to be within the unit hypercube $[0,1]^n$. Let dataset $X = \{x_1, x_2, x_3, \text{——} x_N\}$ be a set of data points and let $D = \{d_1, d_2, d_3, \text{——} d_n\}$ be a set of dimensions. In the following sections, we define our grid generation technique and present our shrinking algorithm.

3.2.1 Grid Generation

Grid structures are critical in grid based data shrinking. With that in mind, we have developed a grid generating algorithm that utilizes inherent data distribution characteristics and eventually generates adaptive grid boundaries for each dimension. The grids boundaries are determined by a wavelet transform-based coefficient aggregation approach to generate the data adaptive grid structure.

Algorithm:
1) Given a dimension d_i, sort the data values of the dimension in non-decreasing order.
2) Within the sorted values of a dimension, extract the windows of the desired size W and the overlap O_n.
3) Perform wavelet transform for each extracted window.
4) Once the wavelet transform has been performed on every window, choose a specified number of extracted coefficients $Coeff$ from every window.
5) Cluster transformed windows using hierarchical average linkage clustering.
6) Once transformed windows are clustered, accumulate corresponding data windows, i.e. the original windows in the same cluster.
7) Calculate the width of the partition or grid boundaries from the end points of the accumulated window.
8) Obtain partition width corresponding to other clusters as well.

Fig. 2. Grid generation algorithm to generate adaptive grid boundaries for every dimension

Initially, the data is normalized in unit hypercube $[0,1]^n$, assuming there are n dimensions in the data. (Fig. 2 shows the algorithm that is applied for grid generation). The following procedure is followed for the generation of a single dimension grid structure and is then applied for all the other dimensions independently.

3.2.2 Adaptive Data Shrinking Algorithm

Initially, using the data adaptive grid structure, we obtain the cell membership of all data points to find all distinct cells and to collect data points in those cells. We calculate the volume and density of the cells that are populated with data points. The density of a cell is defined as a fraction of the total number of data points in the cell over the cell volume. Cell volume is defined as a product of partition width (partition width corresponds to side length of a cell) over all dimensions. Density threshold is used to identify dense cells and to discard others. Next, we choose a dense cell, and all surrounding cells (that share an edge or a vertex with this cell) are captured in an adhoc cluster. The centroid of this cluster is computed. We then simulate movement of data points in the cell and move all points within the cell toward surrounding cells. Shrinking is performed on data grids that are variable both within and across the dimensions. We repeat this process for all dense cells. The algorithm terminates after a specified number of iterations, or, if termination criteria is satisfied. Fig. 3 below presents our data shrinking algorithm and a table of notation that describes all the notations used in the paper. Similarly, Fig. 4 demonstrates the data shrinking effect.

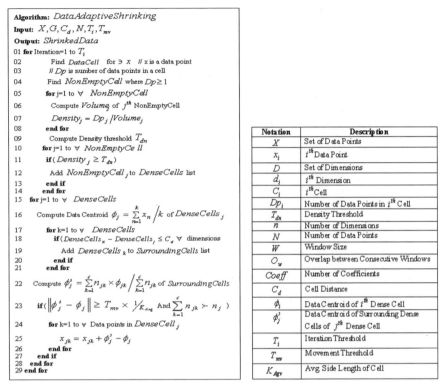

Algorithm: *DataAdaptiveShrinking*
Input: $X, G, C_d, N, T_i, T_{mv}$
Output: *ShrinkedData*
01 **for** Iteration=1 to T_i
02   Find *DataCell* for $\ni x$ // x is a data point
03   // Dp is number of data points in a cell
04   Find *NonEmptyCell* where $Dp \geq 1$
05   **for** j=1 to $\forall$ *NonEmptyCell*
06   Compute *Volume* of jth NonEmptyCell
07          $Density_j = Dp_j / Volume_j$
08   **end for**
09   Compute Density threshold T_{dn}
10   **for** j=1 to $\forall$ *NonEmptyCell*
11   **if**($Density_j \geq T_{dn}$)
12   Add *NonEmptyCell* $_j$ to *DenseCells* list
13   **end if**
14   **end for**
15 **for** j=1 to $\forall$ *DenseCells*
16   Compute Data Centroid $\phi_j = \sum_{n=1}^{k} x_n / k$ of *DenseCells* $_j$
17   **for** k=1 to $\forall$ *DenseCells*
18   **if**($DenseCells_k - DenseCells_j \leq C_d$ $\forall$ dimensions
19   Add *DenseCells* $_k$ to *SurroundingCells* list
20   **end if**
21   **end for**
22   Compute $\phi_j^s = \sum_{k=1}^{c} n_{jk} \times \phi_{jk} / \sum_{k=1}^{c} n_{jk}$ of *SurroundingCells*
23   **if**($\left\| \phi_j^s - \phi_j \right\| \geq T_{mv} \times 1/k_{Avg}$ And $\sum_{k=1}^{c} n_{jk} \succ n_j$)
24   **for** k=1 to $\forall$ Data points in *DenseCell* $_j$
25              $x_{jk} = x_{jk} + \phi_j^s - \phi_j$
26   **end for**
27   **end if**
28   **end for**
29 **end for**

Notation	Description
X	Set of Data Points
x_i	i^{th} Data Point
D	Set of Dimensions
d_i	i^{th} Dimension
C_i	i^{th} Cell
Dp_i	Number of Data Points in i^{th} Cell
T_{dn}	Density Threshold
n	Number of Dimensions
N	Number of Data Points
W	Window Size
O_w	Overlap between Consecutive Windows
$Coeff$	Number of Coefficients
C_d	Cell Distance
ϕ_i	Data Centroid of ith Dense Cell
ϕ_j^s	Data Centroid of Surrounding Dense Cells of jth Dense Cell
T_i	Iteration Threshold
T_{mv}	Movement Threshold
K_{Avg}	Avg. Side Length of Cell

(a)   (b)

Fig. 3. Part (a) represents pseudo code for adaptive data shrinking and part (b) represents the notational table

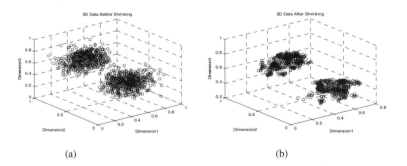

(a)   (b)

Fig. 4. Part (a) and (b) represents 3D data before shrinking and after shrinking. The X-axis, Y-axis and Z-axis represents dimension 1 dimension 2 and dimension 3 respectively.

3.3 Proposed Feature Selection Method

Fig. 5 below shows the algorithm that we apply for feature ranking and selection. The procedure below is followed.

Algorithm:
1) Calculate percentage of shrinking for every dimension:
   a. Calculate mean square distance between all pairs of data points for data 'before shrinking' and 'after shrinking'
   b. Calculate percentage change between mean square distances for data 'before shrinking' and 'after shrinking'.
2) Assign weight to dimensions for each protein class:
   a. Find dimension with maximum and minimum percentage of shrinking
   b. Perform min-max (range: 1-10) normalization of percentage of shrinking
   c. Repeat the process for all configuration of window size W and coefficients $Coeff$

   d. Sum all weights corresponding to every dimension across all window size W and coefficients $Coeff$ to obtain cumulative dimension weight for each protein class

   e. Repeat the process for every class
3) Perform ranking and selection of features
   a. Sort features in increasing order of their weights
   b. Select top ranked features containing 5% of overall energy of the weight signal (For every protein class we obtain a vector of accumulated weights, this is treated as weight signal) from every class
   c. Perform min-max (range: 1-10) normalization of the selected weights for every class
   d. Sum all the weights of features present across all classes to obtain overall cumulative weight of selected features
   e. Sort features in increasing order of their weights
   f. Select top ranked features containing 5% of overall energy of the final weight signal

---

Fig. 5. Feature Ranking and Selection Algorithm to rank feature descriptors for physiochemical properties of protein

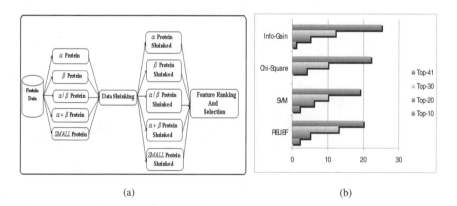

(a)   (b)

Fig. 6. Part (a) represents a block diagram of data shrinking-based feature ranking and selection methods and part (b) represents a comparative study of our top ranked features that are also reported by other classical feature selection methods

4   Experiments and Results

We conduct multiple experiments for the feature ranking and selection task. Data shrinking is first performed on individual protein classes, and then feature ranking and selection is performed. To further validate our feature ranking, a comparative study is conducted using different classifiers and different feature ranking methods. The dataset used for experiments contains both a training dataset and a test dataset and consists of five structural classes $(all-\alpha, all-\beta, \alpha/\beta, \alpha+\beta, SMALL$ and 125 feature descriptors. The training data has 408 training samples, and the test dataset has174 test samples from five structural classes. Feature descriptors are extracted from protein sequence information using the method discussed in [4]. This data is available at http://ranger.uta.edu/~chqding.

4.1   Comparative Study

We conduct experiments on the dataset to demonstrate that our proposed method is capable of effective feature ranking and selection. To demonstrate that our feature ranking method does work effectively, we compare classical feature ranking methods with our own method. For this purpose, we use the RELIEF algorithm, the Chi-Square filter, and the Information Gain based method and SVM [17], [18], and [19]. Table 1 displays the top ranked 41 features for the comparative methods and for our method. Table 1, shows approximately 45% - 60% of feature commonality between our top 41 ranked features (those indicated in bold) and the top ranked features in the comparative methods. We also conduct experiments on the dataset to validate our results and to showcase the effectiveness and superiority of our classification results. Our feature ranking and selection method results in the top ranked 41 feature descriptors for physiochemical properties. Thus, we also select the top ranked 41 features from other feature selection methods to conduct a comparative study.

Table 1. Common top ranked features

Feature Selection	Selected 41 Top ranked features arranged in decreasing order of their relevance
Info. Gain	84,**85**,88,**95**,**89**,100,86,92,**94**,**2**,96,**93**,**90**,**98**,97,**91**,**99**,105,110,**18**,**27**,**12** **1**,**79**,**111**,**32**,**69**,**116**,1,106,53,6,**37**,**48**,**58**,**74**,19,52,**87**,36,**10**,109
Chi-Square	84,**2**,86,100,88,**85**,**89**,**95**,96,92,**94**,**93**,**90**,**18**,97,**98**,110,**91**,**79**,**121**,105,1 06,**99**,**27**,6,**116**,**120**,118,**37**,**111**,1,36,**32**,**69**,109,**48**,53,117,**10**,19,119
SVM	84,**94**,**121**,**2**,**95**,118,109,**99**,**85**,**90**,43,86,**27**,1,**89**,105,6,106,**18**,**17**,**125**,**7** **9**,110,9,64,16,**78**,8,96,13,36,70,33,**62**,**37**,**93**,**32**,42,15,19,40
RELIEF	84,**94**,**99**,**93**,88,**98**,**89**,92,**85**,**2**,86,97,**95**,**91**,96,**90**,1,**18**,106,100,105,110, **32**,**27**,**87**,9,**58**,**37**,5,109,6,**48**,28,11,**121**,**79**,64,8,59, **10**,13
Proposed	104,**62**,**58**,31,**94**,**37**,**27**,**48**,**99**,**74**,**91**,**90**,**32**,**85**,17,102,**10**,83,125,**69**,103, **111**,**78**,73,**89**,115,49,**87**,**98**,**18**,**120**,80,122,**2**,**93**,**95**,**116**,57,**79**,**121**,**79**

A comparative study of F-score, average recall, precision, and accuracy, corresponding to logistic, neural network and (PART) rule-based classifiers for various feature selection methods is presented above in Tables 2 and 3, respectively. Our results demonstrate that our selected features give the best performance for all compared classifiers.

Table 2. Comparative study of F-score for feature selection methods using different classifiers

Classfier	Feature Selection Method	Class α	Class β	Class α/β	Class $\alpha+\beta$	Class SMALL
		F-score	F-score	F-score	F-score	F-score
Logistic Classifier	Chi-square	0.629	0.568	0.718	0	0.96
	Info Gain	0.527	0.444	0.661	0.061	0.96
	Proposed	**0.7**	**0.661**	**0.698**	**0.214**	**0.96**
PART Rule based Classifier	Chi-Square	0.753	0.727	0.741	0.211	1
	Info Gain	0.794	0.72	0.733	0.435	1
	Proposed	**0.844**	**0.678**	**0.73**	**0.545**	**1**
Neural Network Classifier	Chi-Square	0.789	0.582	0.748	0.353	1
	Info Gain	0.829	0.696	0.761	0.308	1
	Proposed	**0.848**	**0.741**	**0.761**	**0.444**	**1**

Table 3. Comparative study of Avg. recall, precision and accuracy for selection methods

Classifier	Logistic Based Classifier			NN Based Classifier			(PART) Rule Based Classifier		
Feature Selection	Avg. Recall	Avg. Precision	Accuracy	Avg. Recall	Avg. Precision	Accuracy	Avg. Recall	Avg. Precision	Accuracy
All Features	63.60	74.38	67.24	72.40	80.60	72.98	71.60	70.54	72.41
Chi-Square	60.20	57.70	64.37	68.80	84.80	71.84	69.80	70.86	72.99
Info Gain	53.80	55.16	53.45	72.20	72.40	74.14	73.00	75.58	74.14
Proposed	**65.40**	**65.50**	**66.67**	**76.00**	**76.40**	**76.44**	**74.20**	**80.04**	**74.14**

5 Conclusion

In this paper we have introduced a novel perspective of feature ranking and selection, based on data shrinking. Every dimension participates in the shrinking process, but every dimension shrinks differently. Some shrink a great deal; others shrink only a little. Thus, the way dimension shrinks decides its characteristics. These characteristics are used to find the most discriminative features. Our proposed approach and corresponding experimental study suggests that features that shrink less exhibit discriminative behavior. Our results confirm this hypothesis. Previous researchers have addressed the structural and fold recognition problem using the whole set of protein feature descriptors. Our proposed method and results have shown that a subset of informative features will get high accuracies in protein classification.

References

1. Ong, S., Lin, H., Chen, Y., Li, Z., Cao, Z.: Efficacy of Different Protein Descriptors in Predicting Protein Functional Families. BMC Bioinformatics 8 (2007)
2. Tan, A., Gilbert, D., Deville, Y.: Multi-Class Protein Fold Classification using a New Ensemble Machine Learning Approach. Genome Informatics 14, 206–217 (2003)
3. Chinnasamy, A., Sung, W., Mittal, A.: Protein Structural and Fold Prediction using Tree-Augmented Naïve Bayesian Classifier. In: Proceedings of 9th Pacific Symposium on Biocomputing, pp. 387–398. World Scientific Press, Hawaii (2004)

4. Ding, C., Dubchak, I.: Multi-Class Protein Fold Recognition using Support Vector Machines and Neural Networks. Bioinformatics Journal 17, 349–358 (2001)
5. Shi, Y., Song, Y., Zhang, A.: A Shrinking-Based Approach for Multi-Dimensional Data Analysis. In: Proceedings of 29th Very Large Data Bases Conference, pp. 440–451 (2003)
6. Shi, Y., Song, Y., Zhang, A.: A Shrinking-Based Clustering Approach for Multi-Dimensional Data. IEEE Transaction on Knowledge and Data Engineering 17, 1389–1403 (2005)
7. Lin, K., Lin, C.Y., Huang, C., Chang, H., Yang, C., Lin, C.T., Tang, C., Hsu, D.: Feature Selection and Combination Criteria for Improving Accuracy in Protein Structure Prediction. IEEE Transaction on NanoBioscience 6, 186–196 (2007)
8. Mhamdi, F., Rakotomalala, R., Elloumi, M.: Feature Ranking for Protein Classification. Computer Recognition Systems 30, 611–617 (2005)
9. Rakotomalala, R., Mhamdi, F., Elloumi, M.: Hybrid Feature Ranking for Proteins Classification. Advanced Data Mining and Applications 3584, 610–617 (2005)
10. Lin, C., Lin, K., Huang, C., Chang, H., Yang, C., Lin, C., Tang, C., Hsu, D.: Feature Selection and Combination Criteria for Improving Predictive Accuracy in Protein Structure Classification. In: Proceedings of 5th IEEE Symposium on Bioinformatics and Bioengineering, pp. 311–315 (2005)
11. Shi, Y., Song, Y., Zhang, A.: A Shrinking-Based Dimension Reduction Approach for Multi-Dimensional Data Analysis. In: Proceedings of 16th International Conference on Scientific and Statistical Database Management, Greece, pp. 427–428 (2004)
12. Kundu, S.: Gravitational Clustering: A New Approach Based on the Spatial Distribution of the Points. Pattern Recognition 32, 1149–1160 (1999)
13. Ravi, T., Gowda, K.: Clustering of Symbolic Objects using Gravitational Approach. IEEE Transactions on Systems, Man, and Cybernetics –Part B: Cybernetics 29, 888–894 (1999)
14. Gomez, J., Dasgupta, D., Nasraoui, O.: A New Gravitational Clustering Algorithm. In: Proceedings of 3rd SIAM International Conference on Data Mining, San Francisco (2003)
15. Georgescu, B., Shimshoni, I., Meer, P.: Mean Shift Based Clustering in High Dimensions: A Texture Classification Example. In: Proceedings of 9th IEEE International Conference on Computer Vision, vol. 1, pp. 456–464 (2003)
16. Wang, X., Qiu, W., Zamar, R.: CLUES: A Non-Parametric Clustering Method Based on Local Shrinking. Computational Statistics & Data Analysis 52, 286–298 (2007)
17. Duch, W., Wieczorek, T., Biesiada, J., Blachnik, M.: Comparison of Feature Ranking Methods Based on Information Entropy. In: Proceedings of IEEE International Joint Conference on Neural Networks, Budapest, pp. 1415–1419 (2004)
18. Kira, K., Rendell, L.: A Practical Approach to Feature Selection. In: Proceedings of 9th International Workshop on Machine Learning, pp. 249–256 (1992)
19. Liu, H., Setiono, R.: Chi2: Feature Selection and Descretization of Numeric Attributes. In: Proceedings of 7th International Conference on Tools with Artificial Intelligence, pp. 388–391 (1995)

Value-Based Risk Management for Web Information Goods

Tharaka Ilayperuma and Jelena Zdravkovic

Royal Institute of Technology
Department of Computer and Systems Sciences, Sweden
{si-tsi,jelena}@dsv.su.se

Abstract. Alongside with the growth of the World Wide Web, the exchange of information goods is tremendously increasing. Some information has entertainment value, and some has business value, but regardless of that, more and more people are willing to pay for information. Business models have become a common technique for representing the value exchanges among the actors involved in a business constellation. In this study, we utilize the notion of the business model to investigate and classify risks on the value of information goods and further, to propose adequate risk mitigation instruments.

Keywords: Risk Mitigation, Web Information Goods, Business Model, Value Model, Process Model.

1 Introduction

Owing to the World Wide Web, information goods, ranging from movies, music and books, to software code and business reports, have become the key drivers of world markets. The value of the Web lies in its capacity to provide immediate access to information goods.

The domain of the web-based business models is becoming an increasingly important area of research with a growing demand for offering economic resources over the Internet. A business model is used to describe who is responsible for creating certain goods (i.e. *economic resources*) and how the goods are offered. As such, business models (also named *value models*) help to reach a shared understanding among the involved parties about the core values exchanged in a business constellation. These models are furthermore operationalized in the form of business process models [1].

A number of studies [10], [3], [14] define risks and risk management as important forces of business model and business process model design. For instance, in a typical industrial goods delivery, the seller contracts a third-party warehouse for the stocking services as he would otherwise not be able to guarantee the delivery on the time agreed with the customer. As being a part of the value chain provided by the seller; the warehouse becomes a new party in the business model; the engagement of the warehouse enables the seller to minimize the risk of a late delivery.

S.K. Prasad et al. (Eds.): ICISTM 2009, CCIS 31, pp. 64–75, 2009.

With a growing market for web-based offering of information goods, it is even more demanding that businesses correctly identify the risks and the risk mitigations instruments related to offering those goods.

In this study we investigate the risks associated with the exchange of information goods, from their value perspective. Research questions related to this objective are: what shall be considered in a value-based risk analysis? How identification of risks affects a business model? And how can a business processes be identified effectively to deal with the risks identified in the risk analysis. We try to answer these questions by using the notion of the value model and its components. Based on these components, we then identify the risk mitigation instruments that will, by changing the core business scenario, contribute to ensuring the value of information goods as agreed between the involved parties. This approach has a practical relevance to business consultants and the system analysts. The former could use the approach in strategic analysis to justify the decision decisions made. The latter could use it to more deeply understand the organizations behavior and implement the processes accordingly.

The paper has the following structure. Section 2 introduces current research on business models and web-based information goods. Using the notion of business models, in Section 3 the value-based risks for information goods are identified and described. In Section 4 an approach for the risk mitigation is presented. Section 5 concludes the paper with a reflection on the results and future research directions.

2   Related Work

2.1   Business Models

There exist a number of efforts for business modelling in the literature, such as the business ontologies [7], [4], and [8]. For the purpose of this paper we will make use of a comprehensive and well established business model ontology, the e3-value [4], that is widely used for business modelling in e-commerce.

The e^3 *value* ontology [4] aims at identifying the exchanges of resources between actors in a business scenario. Major concepts in the e^3 *value* are *actors*, *value objects*, *value ports*, *value interfaces*, *value activities* and *value transfers* (see Figure 1). An actor is an economically independent entity. An actor is often, but not necessarily, a legal entity, such as an enterprise or end-consumer, or even a software agent. A value object (also called resource) is something that is of economic value for at least one actor, e.g., a car, Internet access, or a stream of music. A value port is used by an actor to provide or receive value objects to or from other actors. A value port has a direction: in (e.g., receive goods) or out (e.g., make a payment), indicating whether a value object flows into, or out of the actor. A value interface consists of in and out ports that belong to the same actor. Value interfaces are used to model economic reciprocity. A value transfer (also called value exchange) is a pair of value ports of opposite directions belonging to different actors. It represents one or more potential trades of resources between these value ports. A value activity is an operation that can be carried out in an economically profitable way for at least one actor.

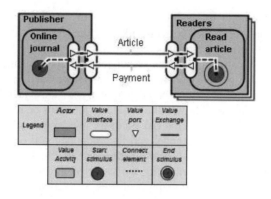

Fig. 1. Main concepts of e^3 *value*

In e^3 *value*, the following categories of value objects may be identified:

- Industrial goods, which are physical objects, like cars, refrigerators, or cell phones.
- Information goods, which are data in a certain context, like blueprints, referrals, contracts, customer profiles, and so forth.
- Services, which provide access to some value objects.
- Money, which is a medium for exchange

In this paper, we will focus on the information goods category of value objects.

A value object may have properties and associations to other objects, such as the representation media of an e-book, or the number of shops accepting a credit card. Such properties and associations are modeled by means of *value object features* [5].

In a transfer of value between actors, the three components of the value object may be exchanged: *right*, *custody*, and *evidence document* [2].

A right on a value object means that an actor is allowed to use that value object in some way. An example is the ownership of a book, which means that an actor is entitled to read the book, give it to someone else, or even destroy it. Another example of a right is borrowing a book, which gives the actor the right to read it, but not to give it away or destroy it.

The second component is the custody of the value object. An actor has the custody of a value object if he has immediate control of it, which typically implies a physical access to it. If an actor has the custody of a value object, this does not mean that she has all possible rights on it. For example, a reader may have the custody of an e-book, but he is not allowed to use the e-book in any way.

The third component is the evidence document. A transfer may include some evidence document that certifies that the buyer has certain rights on a value object. A typical example of an evidence document is a movie ticket that certifies that its owner has the right to watch a movie.

Regarding the relation between business models and process models, in [2] the authors have proposed a method for identifying the processes needed for operationalizing a given business model. Briefly, the method is based on associating the value transfers, that is, the transfers of the right and custody of the value object, or the document evidence, with processes. For instance, in Figure 1, we identify a process for transferring the right on the article, another process for the custody on the article, and the third process realizes the payment transfer.

2.2 Information Goods

Information goods hold special characteristics over the other goods both in themselves and in transferring them between different actors. For example, they can be easily reproduced and their transfers between actors can possibly be accessed by unauthorized actors.

Shapiro et.al. defines Information goods as anything that can be digitized - encoded as a stream of bits. Examples of those include books, movies, software programs, Web pages, song lyrics, television programs, newspaper columns, and so on [9].

Information goods are often categorized as experience products where evaluation needs direct exposure to the content [11], [6]. In [6], the authors further explains this as the information goods having a buying anomaly referring to them having trust and experience features and therefore making it difficult to evaluate before actually buying them. They further state that this difficult nature of the evaluation leads to an information asymmetry between the offering and the demanding actors. As such, they propose to enrich the information goods with two categories of metadata: *signaling of secondary attributes* such as quality ratings, reputation, etc. and *content projections* (static or dynamic) to mitigate the effects of it. Information goods also have a high cost of production and a low cost of reproduction. This makes them especially vulnerable to risks such as illegal redistribution.

3 Value-Based Risk Identification

When risks associated with value transfers are considered, different components of a value transfer should be taken into an account. Analysis of the components of a value transfer will also help us to easily identify the objects of value transferred between different stakeholders and the risks associated with them.

Among the three main components of a value transfer, i.e. right, custody and documentary evidence, the first two components are mandatory for an actor to exercise the control over the value object he has bought. Once the right and custody transfer is done, the value object will be in the possession of the buyer. Regardless of the type of rights and the conditions under which he has agreed to use the value object, once it is in his custody, he could use it without respecting to early agreements with the seller.

The third component, documentary evidence is optional in the sense that even without it, one can conclude that the value transfer is completed. This, of course, depends on the level of trust between the parties and also on the importance of having some document as a proof of having rights. In [13], we argue that there are two types of documentary evidences.

Some documentary evidences play the role of symbolic token of a transferred right. Example for such is a cinema ticket where the owner of the ticket is the holder of the right. This type of documentary evidences can be re-sold and therefore consider as a value object itself. The other type represents the documentary evidences, which are used just as a proof of buying the good, might carry a certain legal value but are not considered as value objects themselves.

The first type of documentary evidences qualify for being value objects by themselves and in our risk identification, we consider them along with the other value objects. Since the second type of documentary evidences carry no value in themselves, we consider them as message transfers related to intangible right transfers between actors. This type of documentary evidences is not explicitly modeled in value models and therefore we do not consider them in our value based risk analysis.

In risks identification and mitigation process, it is also important that features of value objects taken into consideration. In [6], the authors argue that in competitive markets, individuals base their buying decisions on restricted information which typically results in information asymmetry between buyers and sellers. Identification of risks associated with value object features will pave us the way to remedy such problems by offering information rich value objects.

In the list below, we consider different components attached to value transfers and also feature of value objects. Based on them we identify and classify risks involved in web-based information value objects transfer between different actors

In our approach we consider risks in three dimensions:

1. The risks present after transferring right and custody of a value object
2. The risks associated with transfer of custody between actors
3. The risks associated with value object features

3.1 Risks Present after Transferring Right and Custody

Distinguishing between the different rights a customer can obtain over a value object, is important not only to explain what kind of control he has over the object but also to identify different risks associated with the transfer. For example, when someone buys an information value object, he gets an ownership right. The ownership right the buyer gets in this case could be different from an ownership right he gets when he buys an industrial value object, for example, a car. Though he buys an information value object, the seller might not transfer him the right to share it with other actors but might rather transfer a restricted ownership right where only the buyer could use it.

Moreover, having a right over a resource will become meaningful only when the buyer gets its custody. That is, for the buyer to exercise his right over an information value object, it has to be in his possession and for that, the seller must enable him access the value object. When the buyer has both the right and the custody over the value object, he can use it to fulfill his needs. However, once he gets both right and the custody, it is not easy to restrict his power to use it in ways other than agreed before the value transfer. Therefore this becomes a cause of concern for the seller.

From the buyers' perspective, he has to compensate to the seller for the value object he bought. In a web-based value objects exchange this is generally done by means of transferring certain information where the seller can receive money using them (e.g. credit card information). In this case, the buyer transfers a use right over his credit card information, for example, to the seller and at the same time enables access to the information thereby giving the custody to the seller. However once the seller has the information in hid custody, there is a possible risk to the buyer that the information could be reused out of the original context.

Information value objects can be easily reproduced and reused if it is in ones possession. One way for some one to posses a value object is by having a right and custody over it. Below, we identify risks that could primarily caused by an actor having right and custody over an information value object.

Risks associated with rights and custody:

- The risk of reproduction
- The risk of reuse in unwarranted contexts

3.2 Risks Associated with the Transfer of Custody

The transfer of custody of a value object is one of the three main components in a value transfer. In a web-based information transfer, there is a potential risk that the data sent and received by actors may be accessible to unauthorized parties. Other than them being accessible to the third parties, there are other risks involved with a custody transfer over a network. The information transferred through network could get damaged due to loss of data during the transfer. There can also be service interruptions during which buyers are unable to access the value object. Further to that, there may be cases where the content is modified to accommodate certain information that is not part of the original content. In the following, we identify potential risks during the custody transfer.

Risks associated with the custody transfer of an information value object:

- The risk of damage during the transfer
- The risk of delay in delivery of information resource
- The risk of non-conformance to the original content
- The risk of accessible to unauthorized actors

3.3 Risks Associated with Value Object Features

The information value objects have properties like readability, file format, information content, etc. We call these properties, the features of that value object. These features describe the value object in a way that this information carries a certain value to its users. For example, the description about content may have a value to a user who looks for a particular type of information to fulfill his specific needs. Figure 2 depicts a business model where the value proposition offered to the buyer by the seller comprises selling online articles. This business model depicts an ideal situation in that it does not detail how readers asses the quality of the content. That is, it neglects the fact that the buyer needs to evaluate the value object to decide whether it fulfills his needs.

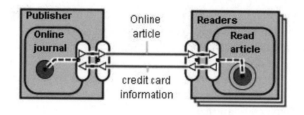

Fig. 2. e^3 *value* model with the risks associated with the value object features

Buying a value object without actually knowing what exactly it is would make it unusable. As such, identification of risks caused by lack of description of value objects plays an important role in classification of risks related to information goods and also to mitigate them.

Risks associated with Value Object Features:

- The risk of having unsupported formats
- The risk of having a poor product quality
- The risk of non-conformance to customer content requirements.

4   Risk Mitigation Instruments

Handling risks through risk mitigation instruments influences both the business and process models. For example, risk analysis may suggest delivery of goods should be insured against possible damages. This will result in adding new actor, say insurance company, in the business model with relevant value transfers. These changes to the business model will lead to identification of new processes in the process model. Identification of risks can be done both on the business level and the process level. The focus of this study is to identify risks from the value perspective. That is, we consider the different components of a value transfer explained in Section 2 along with the value objects transferred by them. Then, we identify risks and risk mitigation instruments based on these different components and value objects. As such, the solutions presented in this section will be mainly captured in terms of new business model components.

Our solution approach follows the following steps:

Step 1: For each risk category, determine possible causes of the identified risks and propose solutions.

Step 2: Identify new business model components such as actors, value transfers and value objects for each solution in step 1

Step 3: Identify new business processes according to step 2.

While we address the issues of step1 and step 2 in this section, the issues related to step 3 will be discussed in Section 5 where we identify process level changes based on the business level solutions proposed in this section.

In the following, we present mitigation instruments to handle the risks identified in Section 3.

4.1 Mitigating Risks Present after Transfer of Right and Custody

Mitigating risks associated with rights and custody over a value object involves identifying the causes for them. Illegal copying of an information object could happen for various reasons. For example, it may be due to buyer's perception of actual value of the value object vs. its price, or supply being lower than the demand, or due to the lack of legal means to protect intellectual property rights. While the outlined problems could be addressed by means of improving a current business model, the latter couldn't be addressed by a company alone. Since our focus is entirely on the value based risk mitigation, we do not consider the situation where lack of legal means lead to intellectual property rights violations. In fact, providing a value based solution for it is hardly possible since this could be solved on a political level rather than on a company level. Below, we capture the possible solutions for the risks related to illegal copying and reusing of information goods.

Solutions:

Increase supply to meet the demand

- This could be achieved by either outsourcing the service provisioning or increasing the current number of value transfers that offers the value object to the customers. If the outsourcing considered as a solution, introduce actor/market segment to the value model. Use either existing or add new value activity in the company to coordinate the outsourcing.
- If increasing current number of value transfers considered as a solution, then this change will not be explicitly visible in the value model. It will affect the structure of the current process that takes care of the value transfer between the company and the customers.

Introduce/Increase post-sale services

- If the focus is on introducing new services then add one value transfer for each new service. This may be through a new value interface or through the existing value interface which provides the information value object. If the new value transfer is added to the existing interface then it must be invoked with the other value transfers there, otherwise it is optional.
- Offer new value added services (local language supports, additional support for the registered users, etc)
- The value model components in this case are similar to the previous one. If these complementary value objects (local language supports, etc.) must be provided with the core value object, then new value transfers should be added to the existing value interface. Otherwise, new value interface are to be added, to offer new value objects.
- Add new market segment (e.g. expert reviewers) to sign the quality of the value object. Use existing or add new value activity in the company to coordinate with the new actor. Add necessary value transfers between the company and the new market segment. Also add new value transfer between the company and the customers to provide the new value object (e.g. expert reviews).

4.2 Mitigating Risks Associated with the Transfer of Custody

The transfer of custody of a value object involves the risks mainly related to the accessibility of unauthorized actors. Moreover, the information transferred over networks could be damaged and unusable when it is received by customers. Also, the content could be modified to include, for example, advertising objects or even harmful malicious information objects. Therefore, customers will be more comfortable if they could be given an assurance that the information they send and receive is error-free and protected. This will also convince them that their privacy needs and the confidentiality of the information is guaranteed. Here we do not go into detailed methods of protecting information against potential security threats, i.e. we set the focus on the business-level solutions to address the risks associated with the custody transfer.

Solutions:

A business model is extended with an additional, intermediate "actor" that foster the transfer as agreed (no more, no less) and also certifies the information transfer to a up-to-date technology

- Add new actor to the value model. Use existing value activity or use new value activity in the company to coordinate with the new actor. Also add relevant value transfers between them.

A business model that provides mirror sources and where the sources are protected by Confidentiality Integrity and Availability (CIA)

- Add new market segment. Use existing or add new value activity in the company to coordinate with the new market segment.

4.3 Mitigating Risks Associated with Value Object Features

To mitigate the risks in this category, we propose offering of the second order value in addition to the core (information) value object. Second order value can either be a particular way of providing the value object in concern, for example, one-click shopping facility provided by amazon.com or it could be a complementary object with value, for example, book reviews provided by amazon.com.

Since information goods have trust and experience features, it is hard to evaluate them. In [6], the authors propose to overcome this evaluation problem in two ways: by signaling secondary attributes, and by content projections. Both these solutions provide the ways to improve the search ability of the content. Offering more information-rich value objects will help to reduce information asymmetry. It will also lead to establish trust between company and customers and also better give company a competitive advantage over its competitors.

In the following, we propose possible mitigating instruments to the risks associated with the value object features.

Solutions:

Provide rich metadata with the information object (with applications needed to use the information resource, system requirements, etc.).

- Add new value transfers between the company and customers. Depending on the necessity to provide the new value object with the existing value objects, use existing or add new value interfaces.
- Add additional actors to sign the quality of resource.
- Change the business model to add the specification of the value object features in the form of the second order value.
- Add new value transfers between the company and the customers to offer complementary value objects. Depending on the necessity, offer new value object through an existing or a new value interface.

Figure 3 depicts how the business model in Figure 2 is changed after applying the risk mitigation instruments identified above. It models the various flows of value objects with second order values, such as static content browsing and expert reviews and certification.

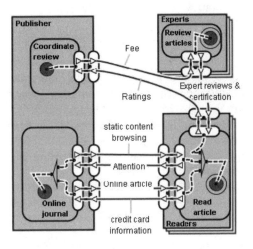

Fig. 3. e^3 *value* model after applying the risk mitigation instruments for the model in Figure 2

5 Identifying New Processes

The derivation of a process model from the business model is not straight forward. Since there are many ways to achieve a same goal, by considering costs and benefits, the designer should compare the alternatives and choose the best.

Identifying new processes from the business model with risk mitigation instruments is essentially about identifying new value activities, value transfers, value ports and value interfaces. In this section we intend to provide some guidelines to identify process model components based on the value model solutions that we have provided in Section 4. Among three components of a value transfer, the rights are not tangible and should be performed communicatively. In most cases rights and custody transfers happens simultaneously, for example during simple goods and money exchange. In cases like this, right transfer is implicit, and as a proof of the transfer of right, there may be documentary evidence transferred to the customer. Therefore, for each value transfer that includes a transfer of right, one process is introduced to handle message transfers between actors.

Transfer of custody of web-based information goods is generally similar to that of industrial goods. Although the transfers are physically different from each other, on both cases the focus is on giving access to the value object. The custody transfer can be performed by the seller him self or it could be outsourced to a third party. For example, the publisher may decide to direct delivery of the value object Expert reviews & certification from the expert's website to the reader (see Figure 3). In both cases, a process is introduced to handle the transfer of custody of the value object to the buyer.

In addition to the processes that we have identified above. there may be cases where there is a need for a coordination between various processes from different actors. For example, the publisher may decide to pay experts for their reviews on monthly basis or she may decide to pay reviewer for each review (see Figure 3). To handle this type of situation, we optionally introduce one process to coordinate internal logistics of the processes introduced in above.

Summarizing, the following processes can be identified after applying risk mitigation instruments:

• For each transfer where there is a transfer of right, one process is introduced for handling with communication aspects (e.g. transfer of documentary evidences in terms of messages for the transferred right) of it.
• For each value transfer, one process is introduced for dealing with transfer of custody between the actors.
• Optionally, one process is introduced for coordinating logistic aspects of the processes introduced above.

6 Conclusion

We have in this paper proposed a method for identification and management of the risks associated with the delivery and use of information goods. The method offers 1) a classification of the sources of risks using the notion of the business model, and its components, and 2) a set of heuristic guidelines for how to mitigate risks by changing or adding new components in the core business model, or in the process model.

The major benefit of the proposed approach is the elicitation of the mechanisms for enabling sustained values of information goods.

The method is used to, as first, systematically identify the risks associated with the transfer of information goods in certain business constellation; and as second, to choose one of the defined risk mitigation instruments to change the existing business model or the process model to enable the business with a minimal risk for the information values. However, in order to ensure mitigation of certain risks such as risks of reproduction, illegal reuse, etc, a legal framework should be used, which is beyond the scope of this paper. The solutions we have proposed in this work describe how a company can change its current business activities in order to mitigate the risks associated in a web-based offering of information goods.

Regarding the future work, it is our intention to investigate what types of risks are relevant for business models and what for process models; and also to identify the patterns for propagation of risks from business to process models.

References

1. Andersson, B., Bergholtz, M., Edirisuriya, A., Ilayperuma, T., Johannesson, P.: A declarative foundation of process models. In: Pastor, Ó., Falcão e Cunha, J. (eds.) CAiSE 2005. LNCS, vol. 3520, pp. 233–247. Springer, Heidelberg (2005)
2. Andersson, B., Bergholtz, M., Grgoire, B., Johannesson, P., Schmit, M., Zdravkovic, J.: From Business to Process Models a Chaining Methodology. In: Proc. of the 1st International Workshop on Business-IT Alignment (BUSITAL 2006) at 18th Int. Conf. on Advanced Information Systems Engineering (CAiSE 2006), pp. 233–247. Namur University Press (2006)
3. Bergholtz, M., Grégoire, B., Johannesson, P., Schmit, M., Wohed, P., Zdravkovic, J.: Integrated Methodology for Linking Business and Process Models with Risk Mitigation. In: Proceedings of the INTEROP REBNITA 2005 Workshop at the 13th Int. Conf. on Requirements Engineering (2005)
4. Gordijn, J., Akkermans, J.M., van Vliet, J.C.: Business modelling is not process modelling. In: Mayr, H.C., Liddle, S.W., Thalheim, B. (eds.) ER Workshops 2000. LNCS, vol. 1921, p. 40. Springer, Heidelberg (2000)
5. Hruby, P.: Model-Driven Design of Software Applications with Business Patterns. Springer, Heidelberg (2006)
6. Maass, W., Berhrendt, W., Gangemi, A.: Trading digital information goods based on semantic technologies. Journal of Theoretical and Applied Electronic Commerce Research, Universidad de Talca – Chile (2007)
7. McCarthy, W.E.: The REA Accounting Model: A Generalized Framework for Accounting Systems in a Shared Data Environment. The Accounting Review (1982)
8. Osterwalder, A.: The Business Model Ontology, Ph.D. thesis, HEC Lausanne (2004)
9. Shapiro, C., Varian, H.R.: Information rules - A Strategic Guide to the Network Economy. Harvard Business School Press (1999)
10. Schmitt, M., Grégoire, B., Dubois, E.: A risk based guide to business process design in inter-organizational business collaboration. In: Proceedings of the 1st International Workshop on Requirements Engineering for Business Need and IT Alignment, The Sorbonne, Paris (August 2005)
11. Varian, H.R.: Markets for Information Goods (accessed June 2008), http://people.ischool.berkeley.edu/~hal/Papers/japan/japan.pdf
12. Weigand, H., Johannesson, P., Andersson, B., Bergholtz, M., Edirisuriya, A., Ilayperuma, T.: Strategic Analysis Using Value Modeling, The c3-Value Approach. In: 40th Hawaii International Conference on Systems Science (HICSS-40 2007), Waikoloa, Big Island, HI, USA. IEEE Computer Society, Los Alamitos (2007) (CD-ROM/Abstracts Proceedings)
13. Weigand, H., Johannesson, P., Andersson, B., Bergholtz, M., Edirisuriya, A., Ilayperuma, T.: On the notion of value object. In: Dubois, E., Pohl, K. (eds.) CAiSE 2006. LNCS, vol. 4001, pp. 321–335. Springer, Heidelberg (2006)
14. Wieringa, R., Gordijn, J.: Value-oriented design of service coordination processes: correctness and trust. In: Proceedings of the 20th ACM Symposium on Applied Computing (SAC), Santa Fe, New Mexico, USA, pp. 1320–1327 (2005) ISBN 1-58113-964-0

Quality and Business Offer Driven Selection of Web Services for Compositions

Demian Antony D'Mello and V.S. Ananthanarayana

Department of Information Technology
National Institute of Technology Karnataka
Surathkal, Mangalore - 575 028, India
{demian,anvs}@nitk.ac.in

Abstract. The service composition makes use of the existing services to produce a new value added service to execute the complex business process. The service discovery finds the suitable services (candidates) for the various tasks of the composition based on the functionality. The service selection in composition assigns the best candidate for each tasks of the pre-structured composition plan based on the non-functional properties. In this paper, we propose the broker based architecture for the QoS and business offer aware Web service compositions. The broker architecture facilitates the registration of a new composite service into three different registries. The broker publishes service information into the service registry and QoS into the QoS registry. The business offers of the composite Web service are published into a separate repository called business offer (BO) registry. The broker employs the mechanism for the optimal assignment of the Web services to the individual tasks of the composition. The assignment is based on the composite service providers's (CSP) variety of requirements defined on the QoS and business offers. The broker also computes the QoS of resulting composition and provides the useful information for the CSP to publish thier business offers.

Keywords: Business Offers, QoS, Broker, Composition Graph.

1   Introduction

Web services are evolving rapidly as a mechanism to expose the computational and business capability to arbitrary devices over the Web. The widespread adoption of Web service technology makes the business to expand its services by building the new value added services. The building and execution of a value added service (composite service) can be carried out in *four* phases. In the first phase, the individual tasks (activity) of the composite service are identified and the control flow is defined for these tasks using a flow language (e.g. WSFL [2] or BPEL4WS [1]). In the second phase, the suitable Web services are discovered for each task based on the functionality. The third phase involves the optimal assignment of the discovered Web services for the tasks based on the local/global requirements. Finally, the execution phase executes the assigned services according to the predefined control flow. The selection of the most suitable (best in

S.K. Prasad et al. (Eds.): ICISTM 2009, CCIS 31, pp. 76–87, 2009.

terms of quality, compatibility and business offerings) Web services for the various tasks is a crucial issue in Web service composition. There have been efforts towards the optimal selection of services based on the QoS requirements defined on the multiple QoS properties [5], [6], [12]. In this paper, we propose the broker based architecture for QoS and business offer aware Web service composition which identifies the right services for the tasks based on the QoS and business offer requirements.

The organization of the paper is as follows. In the following sub-sections, we give a brief description about the related works, important QoS properties, QoS aggregation for the composition patterns and the motivattion. Section 2 presents the business offer model for the business Web services. The section 3 provides the representation scheme for the CSP's composition requirements. Section 4 proposes the broker based architecture for the QoS & business offer aware Web service composition. Section 5 describes the selection model for the service composition. Section 7 draws the conclusion.

1.1   Related Work

In literature, many researchers have addressed the issue of QoS-aware composition of the Web services [8], [9], [7], [6]. The work presented in this paper is closely related to the work proposed in [12], [11], [3], [10]. Zeng et-al [11] proposes a middleware which addresses the issue of the selection for compositions. The middleware employs a mechanism which assigns the best services to tasks based on the local and global constraints (global planning). The paper [3] proposes a model driven methodology for the QoS-optimized service composition based on the pattern-wise selection of services which are ranked using simple additive weight (SAW) method. Jaeger [10] explains the selection problem and points out the similarity to the KnapSack and resource constraint project scheduling problems. The authors propose heuristic algorithms such as greedy, discarding subsets etc to solve the selection problem. The paper [12] models the selection problem in two ways: the combinatorial model and graph model. The authors also define the selection problem as a multi-dimension, multi-choice 0-1 knapsack problem for the combinatorial model and multi-constraint optimal path problem for the graph model. They suggest the efficient heuristic algorithms for the composite processes involving different composition patterns with multiple constraints. So far there are no efforts from the researchers to assign the best candidates for the tasks of the composition plan based on the multiple requirements on QoS and business offers involving AND & OR operators with varied preferences.

1.2   QoS of Web Services

In this paper, we use the different QoS properties for the optimal service selection. The various Web service QoS properties have been introduced by the number of researchers [8], [11], [13]. We consider the following *six* QoS properties to describe the selection mechanism. They are: Execution Cost (EC), Execution Time (ET), Security (SC), Throughput (TP), Availability (AV) and Reputation

(RP). The definition for these QoS properties can be obtained from the paper
[13]. Note that, each QoS property is either *Beneficial* or *Lossy* in nature i.e. for
the beneficial QoS property, higher value indicates the better quality and vice
versa. The QoS properties like RP, TP, AV and SC are beneficial QoS properties
whereas EC and ET are lossy in nature.

1.3   QoS Aggregation for the Composition Patterns

We consider the following *seven* composition patterns described in [10], [5]
to represent the composite process (service). They are: Sequence, XOR-split
(XORs) followed by XOR-join (XORj), AND-split (ANDs) followed by AND-
join (ANDj), OR-split (ORs) followed by OR-join (ORj), AND-split (ANDns)
followed by m-out-of-n-join (ANDmj), OR-split (ORns) followed by m-out-of-n-
join (ORmj) and Loop. In order to estimate the QoS of the composition, aggregation functions are introduced for these composition patterns. We request the
reader to refer the QoS aggregation functions defined for the workflow patterns
in [9], [7].

1.4   Motivating Example

Nowadays, the research has been increased in various fields which makes the research communities and academic/industry organizations to arrange a forum for
the discussions through conferences/symposiums. There is a need for the single
service which arranges a conference involving various tasks. Here we present the
composite service scenario for the conference arrangement. Consider the conference arrangement in a city which requires booking of hall/hotel for presentations,
Illumination and Audio systems, Catering service for food on conference days,
Backdrop service, Conference bag supply service and Conference kit supply. Assume that, there exists many service providers for these atomic activities and
host Web services for their activities. We design a single conference arrangement
service involving these activities.

Figure 1 represents the execution flow of the conference arrangement service
using workflow patterns. The rectangles represent individual tasks and ovals
represent workflow patterns. Normally the conference organizing chair tries to

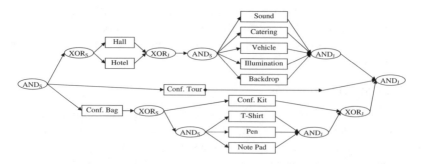

Fig. 1. Conference Arrangement Service Based on Composition Patterns

arrange the conference with low expenditure, expects good response from the reputed service providers and would like to have good amount of discounts on the service fee. As a motivating example, consider the conference chair's requirements as follows: "The service should be provided at the earliest with a few service discounts *OR* a popular (reputed) service provider is required who offers a service at the lowest cost". If the conference arrangement service provider gets such a request from the conference chair, with the requirements defined on the multiple QoS properties and business offers, then the composite service provider has to find and assign services to the tasks of composition plan in an optimal way meeting the requester's QoS and business offer demands.

2   Business Offer Model for Web Services

In today's e-business scenario, the business offers (offerings) have an inevitable importance in giving the buyer the most profitable deal. In order to attract the customers in good numbers, the service providers normally advertise a lot of attractive offers to improve their business thereby the profit. The business offer is an act of offering a reduction in the service fee or giving same/other service or an article as a gift for the service consumption. The business offers are classified as value based business offers, commodity based business offers, conditional business offers and probabilistic business offers.

2.1   Value Based Business Offers

The value based offers normally consist of unconditional discounts/cash gifts for the service consumption. These offers are given to the customers without any preconditions. The value based offers are further classified as, *cash offers (UC)* and *Discount offers (UD)*. In cash offers, the providers advertise a gift cheque/cash for the service consumption. The discount offers involve the discounts i.e. concessions in service fee for the service consumption.

2.2   Commodity Based Business Offers

The commodity based offers normally consist of gifts in the form of an item or a service. The *two* types of commodity based business offers are, *Article offer (UA)* and *Service offer (US)*. In article offer, the provider may offer an item (article) as a gift for the service consumption. The service offer delivers a free service for the service consumption. The service provider may offer a simple service (may be same or different) for the service consumption or a number of free services without any conditions.

2.3   Conditional and Probabilistic Business Offers

The business offers that are probabilistic in nature are called as probabilistic offers. The conditional service offers are defined based on the preconditions imposed by the service provider to enjoy the offer.

Lucky coupon offer (UL). A lucky coupon offer is a business offer where the lucky coupon of predefined value (amount) is given on the service consumption without any precondition.

Conditional service offer (CS). A conditional service offer is a service offer which provides a free service based on the precondition defined on the number of service executions.

Conditional discount offer (CD). A conditional discount offer is a discount offer which offers discount in service fee under the precondition defined on the number of service executions.

3   Composition Requirements on QoS and Business Offers

A CSP normally expects some requirements on the QoS and/or business offers to be satisfied by the primitive Web services in order to gain the profit. For example, in the conference arrangement scenario, the requirements on the execution time, cost and business offers are necessary to execute the composite service in a profitable manner. The CSP can impose different requirements on several QoS properties and/or business offers with varied preferences.

3.1   Composition Requirements

We classify CSP's composition requirements based on the structure as *Simple* and *Composite* requirements. A simple requirement (SR) is normally defined on a single QoS property or a single business offer. If the simple requirement is defined on QoS property then it is called as a *Simple Quality Requirement*. A simple requirement defined on a business offer is called as *Simple Offer Requirement*. A composite requirement (CR) is composed of two or more simple requirements using composition operators AND and/or OR. The CSP can enforce either simple or composite requirement with preferences to assign the optimal services to the composition tasks.

3.2   Composition Requirement Modeling

Consider the composition requirements to be satisfied on the QoS properties and/or business offers with preference to each simple and composite requirement. The CSP's composition requirement can be represented using a tree structure called *Composition Requirement Tree.*

Composition Requirement Tree (CRT). A composition requirement tree is a Weighted AND-OR tree [13] whose leaf node contains QoS property or business offer type. The internal node refers to composition operator AND/OR. The label W_{XY}, on the edge between any two nodes X and Y, represents the preference for the sub-tree rooted at the node Y while traversing from the root to leaf. i.e. the edge label represents the preference to either simple or composite requirement. The leaf node represents the simple requirement (SR) and any sub-tree rooted at internal node represents the composite requirement (CR).

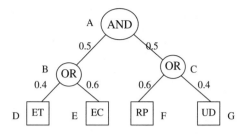

Fig. 2. Composition Requirement Tree (CRT) of CSP

As an illustration, consider the provider's composition requirements (to maximize the profit) as follows: (1) Execution time OR Execution cost (2) Reputation OR Discount. The CSP expects both the requirements to be equally satisfied for the composition with preferences 0.4, 06, 0.6 and 0.4 to Execution time, Execution cost, Reputation and Discount respectively. Fig. 2 presents the CRT for the composition requirements.

4   Broker Based Architecture for Compositions

We propose the broker based architecture for the QoS and business offer-aware Web service composition with an objective of selecting the best Web service for each task of the composition that satisfies requirements of the CSP. Fig. 3 depicts the various roles and operations involved in the architecture. The architecture uses the QoS and business offer vocabulary defined in section 1.2 and 2.

4.1   Roles and Operations

We define *three* additional roles to the conceptual Web service architecture [14] called *broker*, *QoS Registry* and *BO Registry*. The new operations are, *Select* and *Register*. The broker is defined between the registry services and the requester/provider. The QoS registry is a repository with the business sensitive,

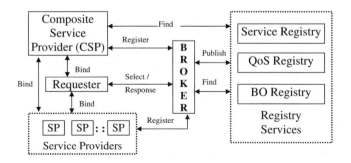

Fig. 3. The Broker Based Architecture for Web Service Compositions

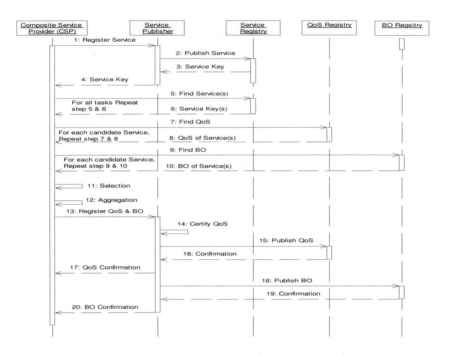

Fig. 4. Sequence Diagram for the Composite Service Publishing

performance sensitive and response sensitive QoS lookup and access support. The BO registry is a repository for the business offers of the services. The BO registry provides an business offer based and profit index (Refer section 5.1) based Web service search and retrieval. The select operation is defined between the broker and the requester, which facilitates the requester to select the best Web service based on the requirements [13]. The register operation is defined between the Web service providers and the broker to facilitate the QoS and business offer aware Web service publishing.

4.2   Component Interactions

We design the broker with *three* internal components namely *Service Selector*, *Service Publisher* and *Service Manager*. For each component, we define a set of functions to fulfill the requester's and service provider's needs (publishing & selection). The main functionality of service selector is to select the best Web service that satisfies requester's QoS and business offer constraints. The service publisher component facilitates the provider in registering the business, service, QoS and offer related information; and publishes them into the respective registries. The main objective of the service manager is to obtain feedback from the requesters to estimate the response specific QoS properties like Reputation, Compliance and successability [13]. This component also allows updating and deletion of the business, service, QoS and business offer related information. Fig. 4 shows a sequence of activities for the composite service publishing.

5   The Service Selection Model for Compositions

The main objective of service selection in composition is to identify the best assignment of the services to the tasks to maximize the profit in terms of QoS and business offers.

5.1   Evaluation of Business Offers

The Web service provider can publish multiple business offers for a service. We define a parameter to compare different types of business offers called *Profit Index (PI)* which is computed as the ratio of profit amount to the payable amount. The value of PI is computed based on the type of offer as follows. The PI value for value and commodity based business offers is computed as, PI=Offer value/Execution cost. The value of PI for conditional business offer is, PI=Offer value/(Execution cost x C) where C is the number of predefined executions of a service. The PI value for lucky coupon offer is, PI=Coupon value/(Execution cost x Offer period). The necessary information for the computation of PI is recorded in the BO registry by the service publisher component for all business offers.

5.2   Business Offer (BO) Aggregation

We aggregate PI of the business offer for each composition pattern during pattern wise selection of the services. For the parallel tasks with XOR pattern, the largest PI value is found and for AND/OR patterns, sum of the PI values of all tasks (sub-set of tasks in case of OR) is obtained. For looping patterns, PI value is multiplied with looping frequency. In case of sequence patterns, sum of the PI values of all sequentially arranged tasks are obtained as the aggregated value. Note that, PI is a *beneficial* business criteria i.e. higher the value of PI, higher the profit.

5.3   Composition Structure

We arrange the individual tasks of the composition in the form of a graph called *Composition Graph* as follows. A Composition graph (CG) is a directed hetero-geneous graph G=(V, E) where V is a set of nodes and E is an edge set. The set V consists of two types of nodes called *Pattern nodes* and *Task nodes*. The set E is a set of directed edges between various nodes which indicate the flow. The pattern node with zero out-degree is referred as *end* node and the pattern node with zero in-degree is called as *start* node. Fig. 1 represents composition graph for the conference arrangement service.

Let G be the composition graph with N tasks. Let t_1, $t_2 \ldots t_N$ be the task nodes of G. Let w_1, $w_2 \ldots w_M$ be the candidates found through the discovery for

N tasks. Let R be the CRT of height H having Q simple quality requirements and B simple offer requirements. Now for the each discovered candidate service, we associate two vectors called *QoS-Vector* and *BO-Vector* with values u_i and v_j, $i \in \{1,\ldots,Q\}$ and $j \in \{1,\ldots,B\}$ that represent the values of QoS and PI.

5.4   The Selection Algorithm

The selection algorithm finds suitable candidate services for all the tasks present in the composition graph. The algorithm is presented below.

1. For all M candidate services (w_1, $w_2 \ldots w_M$), scale the associated QoS-Vector and BO-Vector using min-max normalization technique [15] as follows. For each entry of the vectors, find the maximum (max) and minimum (min) across M vectors. Replace the value y with the new value (ny) as follows. If y corresponds to lossy criteria then, replace y by ny=(max-y)/(max-min). For beneficial criteria, replace y by ny=(y-min)/(max-min). Also ny=1, if (max-min)=0.

2. For each task T of the composition graph G, perform the selection as follows. (a) For each leaf node (level 0) of R, Let X be the QoS property or business offer defined at the leaf node. Attach the candidate services and the normalized values of X to the leaf node. Now multiply the node preference to the attached values to get the score of the Web service at that node.
   (b) For i = 1 to H of R perform the following actions
   For each internal node at tree level i, if the node is AND node then select the services present in all the child nodes of AND node. Compute the new score for the selected service by adding the service scores at the child nodes and multiply the sum with the node preference. Attach the service and its computed score to the AND node. If the node is OR node then select the distinct services from child nodes in the descending order of thier scores (eliminating the duplicates). Compute the score for the selected Web service by multiplying its score with the node preference. Attach the services and their scores to the OR node.
   (c) At the root node, sort the selected services in the descending order of their score.
   (d) The first service and its QoS-Vector and BO-Vector is attached to the task node T as the best (locally optimal) candidate for that task.

5.5   QoS and BO Computation

Once the best candidates are assigned with QoS values and PI values to all tasks of the composition graph, we compute the QoS and PI value of the resulting composition. The estimation of QoS and PI is presented below.

1. The algorithm walks into G and labels the pattern nodes with positive number using multiple stacks (labeling method is not described due to page restrictions).

2. For each pattern in the descending order of the label, the QoS and PI is aggregated using pattern based aggregation as explained in subsection 1.3 and subsection 5.2. The aggregated QoS and PI is assigned for the pattern.

After the computation of QoS and PI value for the composition, the CSP considers the computation effort and other factors related to the performance and business, to estimate (experimentally & analytically) the new QoS values for the composite service. The CSP can use any strategy to register the QoS and business offers for the composite service.

5.6   Illustration for Selection

As an illustration for the QoS and BO-aware composition, we consider the composition graph presented in Fig. 5(a) and the CSP's requirements as in Fig. 2. The service discovery for each task finds the candidate Web services. The task candidate services and their QoS, BO values are shown in Fig. 5(b). Fig. 6 shows the values of the candidate services at each node of CRT. From the figure, we observe that, the service W_1 is selected for task A, W_3 is selected for task B and so on. Now the QoS and BO computation algorithm computes the QoS and PI value for the composition as explained in section 5.5. The computed QoS and BO (PI & Discount) of composition is as follows. The maximum execution duration is 79 and the maximum execution cost is $178. The minimum discount amount is $12.96 and aggregated PI value (minimum) is 1.11. The best execution path

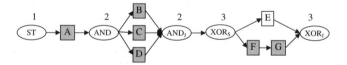

(a) Composition Graph

Task	Services	ET	EC	RP	UD (%)	PI
A	W_1	21	$30	8	8	0.08
	W_2	16	$25	7	4	0.04
B	W_3	18	$16	6	22	0.22
	W_4	14	$22	5	16	0.16
C	W_5	10	$40	3	18	0.18
D	W_6	28	$30	5	21	0.21
	W_7	44	$28	4	26	0.26
	W_8	38	$42	8	14	0.14
E	W_9	20	$50	3	1	0.01
	W_{10}	25	$70	2	2	0.02
F	W_{11}	10	$20	6	10	0.1
	W_{12}	30	$30	7	15	0.15
	W_{13}	26	$40	5	7	0.07
G	W_{14}	20	$35	4	12	0.12
	W_{15}	5	$15	8	20	0.2

(b) Candidate Services and their QoS & BO Values (PI) for Various Tasks

Fig. 5. Composition Graph and Candidate Services

Task	Services	Node D	Node E	Node F	Node G	Node B	Node C	Node A
A	W_1	0.589744	0.727273	1	0.28	0.436364	0.6	**0.518182**
	W_2	0.717949	0.818182	0.833333	0.12	0.490909	0.5	0.495455
B	W_3	0.666667	0.981818	0.666667	0.84	0.589091	0.4	**0.494545**
	W_4	0.769231	0.872727	0.5	0.6	0.523636	0.3	0.411818
C	W_5	0.871795	0.545455	0.166667	0.68	0.348718	0.272	**0.310359**
D	W_6	0.410256	0.727273	0.5	0.8	0.436364	0.32	0.378182
	W_7	0	0.763636	0.333333	1	0.458182	0.4	0.429091
	W_8	0.153846	0.509091	1	0.52	0.305455	0.6	**0.452727**
E	W_9	0.615385	0.363636	0.166667	0	0.246154	0.1	**0.173077**
	W_{10}	0.487179	0	0	0.04	0.194872	0.016	0.105436
F	W_{11}	0.871795	0.909091	0.666667	0.36	0.545455	0.4	**0.472727**
	W_{12}	0.358974	0.727273	0.833333	0.56	0.436364	0.5	0.468182
	W_{13}	0.461538	0.545455	0.5	0.24	0.327273	0.3	0.313636
G	W_{14}	0.615385	0.636364	0.333333	0.44	0.381818	0.2	0.290909
	W_{15}	1	1	1	0.76	0.6	0.6	**0.6**

Fig. 6. The Trace of Selection Algorithm

for the CSP's requirements is shown in Fig. 5(a) by darkening the task nodes. The values of execution time, execution cost and discount amount for the best execution (optimal) path is 74, \$163 and \$14.76 respectively.

The existing selection methods for composition described in the literature [11], [12], [10], [5], selects the best candidates based on the local and/or global optimization of requirements defined on the multiple QoS properties with preferences. The proposed method finds the best candidates for the tasks based on the requirements defined on the multiple QoS properties and business offers involving AND/OR operators with varied preferences. If such a requirement arises, the existing (classical) methods have to be executed several times (exponential) which depends on the structure of the requirement.

Let Z be the number of nodes of CRT, R of height H. Let J be the degree (order) of R. Let K be the number of Web services selected as candidates for a task t. All (K) candidates are attached to each node of the tree (R). The complexity of the proposed selection algorithm is: $\Theta(T*Z*K)$ or $\Theta(T*J^{H+1}*K/(J-1))$.

6   Conclusion

The paper explores the selection mechanism to optimally assign the Web services to the individual tasks of the composition plan. The selection algorithm reads the composition requirement tree (CRT) that represents composite service provider's composition requirements defined on multiple QoS properties and business offers involving AND and OR operators with veried preferences. The paper also discusses the mechanism to compute (aggregate) the QoS and profit value (PI) of the composite service. The profit value (aggregated PI) information is useful to the CSP in offering a new business offer to the service consumers.

References

1. Tatte, S. (ed.): Business Process Execution Language for Web Services, Version 1.1. Technical Report, IBM Corp., Microsoft Corp. (February 2005), http://www-106.ibm.com/developerworks/webservices/library/ws-bpel/
2. Snell, J.: Introducing the Web Services Flow Language (June 2001), http://www.ibm.com/developerworks/library/ws-ref4/index.html
3. Grønmo, R., Jaeger, M.C.: Model-driven methodology for building qoS-optimised web service compositions. In: Kutvonen, L., Alonistioti, N. (eds.) DAIS 2005. LNCS, vol. 3543, pp. 68–82. Springer, Heidelberg (2005)
4. Liu, A., Huang, L., Li, Q.: QoS-aware web services composition using transactional composition operator. In: Yu, J.X., Kitsuregawa, M., Leong, H.-V. (eds.) WAIM 2006. LNCS, vol. 4016, pp. 217–228. Springer, Heidelberg (2006)
5. Jaeger, M.C., Rojec-Goldmann, G.: SENECA – simulation of algorithms for the selection of web services for compositions. In: Bussler, C.J., Shan, M.-C. (eds.) TES 2005. LNCS, vol. 3811, pp. 84–97. Springer, Heidelberg (2006)
6. Yang, L., Dai, Y., Zhang, B., Gao, Y.: A dynamic web service composite platform based on qoS of services. In: Shen, H.T., Li, J., Li, M., Ni, J., Wang, W. (eds.) AP-Web Workshops 2006. LNCS, vol. 3842, pp. 709–716. Springer, Heidelberg (2006)
7. Jaeger, M.C., Ladner, H.: Improving the QoS of WS Compositions based on Redundant Services. In: Proceedings of the International Conference on Next generation Web Services Practices (NWeSP 2005). IEEE, Los Alamitos (2005)
8. Menasce, D.A.: Composing Web Services: A QoS View. IEEE Internet Computing 8(6), 88–90 (2004)
9. Jaeger, M.C., Goldmann, G.R., Muhl, G.: QoS Aggregation for Web Service Composition using Workflow Patterns. In: Proceedings of the 8th IEEE Intl. Enterprise Distributed Object Computing Conf. (EDOC 2004). IEEE, Los Alamitos (2004)
10. Jaeger, M.C., Mühl, G., Golze, S.: QoS-aware composition of web services: An evaluation of selection algorithms. In: Meersman, R., Tari, Z. (eds.) OTM 2005. LNCS, vol. 3760, pp. 646–661. Springer, Heidelberg (2005)
11. Zeng, L., Benatallah, B., Ngu, A., Dumas, M., Kalagnanam, J., Chang, H.: QoS-Aware Middleware for Web Services Composition. IEEE Transactions on Software Engineering 30(5), 311–327
12. Yu, T., Zhang, Y., Lin, K.: Efficient Algorithms for Web Services Selection with End-to-End QoS Constraints. ACM Transactions on the Web, Article 6 1(1), 1–26 (2007)
13. D'Mello, D.A., Ananthanarayana, V.S.: A QoS Model and Selection Mechanism for QoS-aware Web Services. In: Proceedings of the International Conference on Data Management (ICDM 2008), Delhi, February 25-27, pp. 611–621 (2008)
14. Kreger: Web Services Conceptual Architecture (WSCA 1.0), IBM Corporation Specification (2001), www.ibm.com/software/solutions/webservices/pdf/wsca.pdf
15. D'Mello, D.A., Ananthanarayana, V.S., Thilagam, S.: A QoS Broker Based Architecture for Dynamic Web Service Selection. In: Proceedings of the Second Asia International Conference on Modelling and Simulation (AMS 2008), Kuala Lumpur, Malaysia, pp. 101–106. IEEE Computer Society, Los Alamitos (2008)

Managing Sustainability with the Support of Business Intelligence Methods and Tools

Maira Petrini and Marlei Pozzebon

FGV- EAESP, Sao Paulo, Brazil and HEC Montreal, Canada,
3000 chemin de la cote-sainte-catherine, Montreal, Quebec, Canada, H3T2A7
Maira.petrini@fgv.br, marlei.pozzebon@hec.ca

Abstract. In this paper we explore the role of business intelligence (BI) in help-ing to support the management of sustainability in contemporary firms. The concepts of sustainability and corporate social responsibility (CSR) are among the most important themes to have emerged in the last decade at the global level. We suggest that BI methods and tools have an important but not yet well studied role to play in helping organizations implement and monitor sustainable and socially responsible business practices. Using grounded theory, the main contribution of our study is to propose a conceptual model that seeks to support the process of definition and monitoring of socio-environmental indicators and the relationship between their management and business strategy.

Keywords: Business intelligence, corporate social responsibility, sustainability, business strategy, information planning, performance indicators, socio-environmental indicators.

1 Introduction

The main purpose of this research is to explore the role of business intelligence (BI) in helping to support the management of sustainability in contemporary firms. We suggest that BI methods and tools have an important but not yet well studied role to play in helping organizations implement and monitor sustainable and socially respon-sible business practices. The contribution of this paper is to shed light on *what role* and *how*. On the one hand, after years of significant investment to put in place a tech-nological platform that supports business processes and strengthens the efficiency of the operational structure, most organizations have reached a point where the use of tools to support the decision making process at the strategic level emerges as more important than ever. Herein lies the importance of the area known as business intelli-gence (BI), seen as a response to current needs in terms of access to relevant informa-tion for decision-making through intensive utilization of information technology (IT). BI systems have the potential to maximize the use of information by improving the company's capacity to structure large volumes of information and make accessible those that can create competitive advantage through improvement of business man-agement and decision making [1], what Davenport calls "competing on analytics" [2].

S.K. Prasad et al. (Eds.): ICISTM 2009, CCIS 31, pp. 88–99, 2009.

On the other hand, the concepts of sustainability and corporate social responsibility (CSR) have been among the most important themes to emerge in the last decade at the global level. Sustainability and CSR are considered equivalent concepts in this paper, since both take into consideration the environmental, social and economic dimensions. In addition, both refer to a long-term perspective based on the requirements necessary to provide for the present without compromising the needs of future generations. The increasing importance of sustainability and CSR should be evaluated within the complex context of globalization, deregulation, and privatization, where social, environmental, and economic inequalities continue to increase [3]. In the light of this contextual perspective, managers have to take into consideration not only increased sales and profits and/or decreased costs, but also sustainable development of the business itself and of the surrounding context. Therefore, a growing number of companies worldwide have engaged in serious efforts to integrate sustainability into their business practices [4]. Despite the explosion of interest in and concern with sustainable practices, their effective implementation faces serious obstacles. Up to now, most firms have kept the question of sustainability separated from considerations of business strategy and performance evaluation, areas that are often dominated by purely "economic" performance indicators [5].

Viewed separately, *sustainability/social responsibility* and *business intelligence* each represent relevant themes for investigation. Curiously, few studies have considered these two themes in conjunction. We argue that the purposive use of BI tools and methods can improve the definition, gathering, analysis and dissemination of socio-eco-financial information among employees, clients, suppliers, partners and community. The research question guiding this inquiry is: *How can BI tools and methods help support the integration of sustainability and social responsibility practices into business strategy and corporate management?* The term BI tools refers to technological applications and platforms (including software, hardware, network, etc.), while BI methods refers to methodologies concerned with different BI project phases. We put particular attention to one phase, the information planning, i.e., the systematic way of defining indicators, metrics and other relevant information in order to integrate them in monitoring and reporting activities. The paper is structured as follows: section 2 presents a review of literature around the two main themes: BI and sustainability/CSR; section 3 presents the research method; section 4 presents the results; section 5 shows the discussion and preliminary conclusions.

2   Sustainability and CSR Issues

From a historical point of view, the seminal work of Bowen, in 1953, was one of the starting points of the field known as corporate social responsibility (CSR) [6]. In our research, we define CSR as a comprehensive set of policies, practices and programs that are integrated into business operations, supply chains, and decision-making processes throughout a company, with the aim of inculcating responsibility for current and past actions as well as future impacts [7]. Likewise, although issues around sustainability have a long history, the predominant interpretation of sustainable development was introduced by the Brundtland Commission's report in 1987: meeting the needs of the present without compromising the ability of future generations to meet their own needs. A true sustainable enterprise contributes to sustainable development by delivering economic, social and environmental benefits simultaneously – the so called *triple bottom line* [8] [9] [12].

The two terms, sustainability and CSR, have progressively converged and nowadays they encompass similar dimensions and are often applied as synonymous or equivalent terms [10] [11]. First, both concepts involve multiple levels of analysis – individual, groups, firms, communities, etc. – and multiple stakeholders – employees, shareholders, clients, suppliers, partners, community members, etc. Second, CSR and sustainability deal with issues related to three distinct spheres that sometimes overlap: social, environmental and economic. It is important to note that the economic sphere is not limited to short-term performance indicators like return on investment (ROI), but also refers to elements that contribute to long-term financial success, like firm reputation and firm relationships. Consequently, managing sustainability and CSR implies seeking a balance between short- and long-term considerations, and among the interests of a larger group of stakeholders than those addressed by "traditional" management [3].

CSR and sustainability are becoming an important dimension of corporate strategy and an increasing number of firms are trying to determine and monitor the social and environmental impacts of their operations [13] [14]. Despite such an explosion of interest, the effective incorporation of sustainability into business strategy and management faces serious obstacles. Definition and monitoring of indicators that take into account the three dimensions – social, economic and environmental – and the various stakeholders – employees, clients, suppliers, shareholders and community members – seems to be one of the keys, but how can they be integrated with business strategy?

3   BI (Business Intelligence) Methods and Tools

From a historical standpoint, what we call business intelligence (BI) has been evolving during the last 35 years. In the 1970s, initial versions of analytical software packages appeared on the market. The 1980s saw the release of spreadsheet software, e.g., Excel, which is still widely used today. By mid-1980s and early 1990s, so-called executive information systems (EIS) were introduced and quickly grew in popularity by promising to provide top management with easy access to internal and external information relevant to decision-making needs, placing key information on the desktops of executives [15].

In the 1990s, three technological improvements brought about a revolution in analytical applications scenarios, accounting for the emergence of business intelligence (BI) systems: data warehouse technologies [16] [17], ETL tools (extraction, transformation and loading) and powerful end-user analytical software with OLAP capabilities (on-line analytical processing) [18]. Furthermore, the impact of Internet is far from negligible: current versions of analytical products are web-based and, through Internet or intranet connections, users can investigate and analyze data from home, while traveling or from any other location [19]. Today, terms like DSS and EIS have virtually disappeared, and BI is the accepted term for analytical and strategic information systems, including an array of applications classified under three broad headings:

analytics (data mining and OLAP), monitoring (dashboards, scorecards and alert systems) and reporting.

From a technical point of view, the stage is set for rapid growth in the adoption of BI applications in the coming decades. Nowadays, BI technologies integrate a large set of diversified resources (packages, tools and platforms), and various products are being released aimed at fulfilling different needs related to search for and use of information, ranging from report extractors (used on a more detailed informational level) to dashboards applications (used to consolidate within a single control panel the information linked to performance factors in a largely summarized level) and sophisticated mining applications (used to build predictive business models). The large variety of tools helps explain why a huge array of apparently dissimilar applications is commonly labeled "business intelligence". However, knowledge advancements in one stream, the technical, have been higher than in the other, the managerial side of BI [1]. Although the volume of information available in data warehouses is increasing and the functionalities of analytical tools are becoming more and more sophisticated, this does not automatically mean that firms and people are able to derive value from them [20].

3.1 Looking for Information Planning Methods That Integrate Sustainability

While general knowledge of the technological features and functionalities of BI tools is increasing unabated, the same is not true of knowledge of BI *methods* that guide the conception and operation of these tools. One example of BI methods involves those methodologies focusing on identification of strategic information needs to be integrated into data warehouses and BI applications [21]. There is a lack of clear frameworks in the BI literature that could serve as a guide in such an important phase: the definition of user requirements and information needs. The literature on BI is short on methodological approaches to information planning. A recent work showed that most of the methodologies applied in real-world BI projects are proposed by vendors and consulting firms, and those methodologies are not examined, validated or tested by academic research [22]. Faced with this absence of academic frameworks in BI literature, we reviewed literature on three related areas: data warehousing, IT strategic planning and accounting.

In the data warehousing literature, one finds a huge number of studies based on two authors, Inmon and Kimball, who are directly concerned with defining the structure and content of a data warehouse. However, both authors provide few guidelines regarding the definition of corporate performance indicators, which is the heart of a BI system from a corporate level perspective [16] [17]. The IT strategic planning literature identifies that the information definition phase should be linked to corporate strategic planning since BI systems are supposed to link operational and strategic dimensions of an organization through the flow of information [23]. Finally, accounting and general management literature provided an important contribution to our debate through the wide range of studies based on the balanced scorecard (BSC) approach. The core of BSC is translating the corporate strategic view into a set of measurable results that provide executives with a quick and understandable view of the business [24]. Its development was motivated by dissatisfaction with traditional performance measurements that were concerned only with financial metrics and that focused on the past rather than the future. As a result, BSC associates indicators and measures with the monitoring of the company's strategic objectives by using four different perspectives: financial, clients, internal processes and organizational learning and innovation [24]. In addition to integrating multiple perspectives, BSC also includes other mechanisms that appear important to our inquiry, particularly the linkage between operational and strategic levels [25]. There are several attempts to integrate a social and environmental dimension within a BSC model. We recognize four main strategies that gradually transform a BSC into a "sustainable" BSC or SBSC: partial SBSC, transversal SBSC, additive SBSC and total SBSC [26] [27].

In sum, although the integration of sustainability within management models is present in accounting and management literatures through different models of SBSC, these models are not yet integrated into the BI literature. The almost total absence of the theme sustainability within BI literature stimulated us to embark on a grounded approach, trying to learn from experiences of enterprises (that excel in sustainability practices) about what role of BI systems are playing or could play.

4 Research Approach

A summary of the steps guiding our empirical research is presented in Table 1. The main particularity of this adapted version, as proposed by Pandit [28] is the inclusion of step 4.

Table 1. An adapted version of grounded theory in 9 steps

Research Design Phase		
Step 1	Review of technical literature	Definition of research question.
Step 2	Selecting cases	Theoretical, not random, sampling.
Step 3	Develop rigorous data collection protocol	Defining themes that delineate the boundaries of the research question; Building an initial guide for interviews.
Step 4	Entering the field	Overlap data collection and analysis flexible and opportunistic data collection.
Data Collection Phase		
Step 5	Data ordering	Arraying events chronologically.
Data Analysis Phase		
Step 6	Analyzing data	Use coding and memo.
Step 7	Theoretical sampling	Theoretical replication across cases; Return to step 4 (until theoretical saturation).
Step 8	Reaching closure	Theoretical saturation when possible.
Literature Comparison Phase		
Step 9	Compare emergent theory with extant literature	Comparisons with conflicting and/or similar frameworks.

Our empirical work was guided by a methodological approach that is an adapted version of grounded theory that begins with a truly inductive approach, meaning one that starts from scratch without any theoretical model guiding data collection and analysis, but which integrates existing theoretical models in the last phase of data analysis [28]. Grounded theory is a well known research approach in social science and related fields [29]. Important characteristics of grounded theory are: (1) the construction of theory, and not just the codification and analysis of data, is a critical aspect; (2) as a general rule, the researcher should not define a conceptual framework before the beginning of the research (this is aimed at allowing concepts to emerge without being influenced by pre-defined "biased" concepts); and (3) the analysis and the conceptualization are arrived at primarily through data collection and constant comparison. In sum, it seeks to enhance an existing category and/or to form a new one, or to establish new points of relationship among them [29] [30] [31].

Instead of learning from an existing theoretical view (deductive approach), grounded theory emphasizes learning from the data (interactive and inductive approach). In addition, what mainly distinguishes grounded theory from other qualitative methods is its specific focus on theoretical development through continuous interdependence between data collection and analysis [31] [32].

4.1 Cases Selection

The field work was carried out in a Brazilian region, the state of Sao Paulo, which features an effervescent business movement towards sustainability. Grounded theory points out the importance of a theoretical non-random sample selection (step 2) [28]. With this goal in mind, we established the three following criteria for selecting Brazilian firms located in Sao Paulo (the selected companies should meet at least two of them): (1) the company should be an advocate or signatory of at least one of the various principles, norms, certifications or reports related to sustainability (GRI, ISO14001, SA 8000, etc.); (2) the company should be indexed by the Dow Jones Sustainability Index and/or the Bovespa Sustainable Business Index (Brazilian index); (3) the company should have received awards or public recognition for actions related to sustainability. Based on those criteria, five large companies recognized as leaders in sustainability practices were selected, denominated as FIN1, FIN2, FIN3, IND1 and COS1. FIN1, FIN2 and FIN3 are banks, IND1 and COS1 are manufacturers of plastic piper and cosmetics, respectively (a detailed description of the companies may be obtained upon request).

4.2 Data Collection

In all cases selected, data were collected from semi-structured interviews and supporting documentation (i.e., annual reports, social balances and websites). A research protocol was developed to guide the entire data collection process (step 3). The data collection was structured around three broad themes: (1) organizational structure, strategic planning; (2) sustainability and CSR vision and indicators' definitions; and (3) BI project maturity. It is important to clarify that these three themes do not correspond to a theoretical framework (grounded theory approach precludes a theoretical framework), but they were defined in order to set the boundaries of the research question and to guide the fieldwork of the researcher.

According to grounded theory, data should be collected and simultaneously analyzed. The questions included in the interview protocol (also available upon request) ranged from broad to more specific, and they changed as concepts, categories and relationships among them emerged from the data collected and analyzed. Because of the existing overlap between data collection and analysis, as the study evolved, we had to make a more purposive selection of new respondents. For example, due to new concepts (the seed of a theoretical model) that emerged from the collection/analysis process, we have identified categories that should be further developed. Consequently, we were forced to select new respondents (saturation was not yet attained at that point) or ask for a second round of interviews with respondents already interviewed (step 4).

In sum, in the first round of interviews, we conducted 16 interviews and in the second round, 5 interviews, totaling 21 interviews. All the interviews were conducted between March and September 2006. The interviewees were basically selected regarding their deep knowledge on the firms' sustainability processes and practices, including all phases: planning, operationalization and evaluation (the interviewees were business managers, sustainability coordinators, senior planning analysts and IT managers).

4.3 Data Analysis

Our first step was to prepare a detailed description of each company, based on interviews and supporting documentation (step 5). After that, several techniques suggested by grounded theory were employed (steps 6 and 7). First, using the coding technique, we identified concepts, possible categories and related properties from the collected data, i.e., categories were drawn from the data itself. Properties or sub-categories were also identified, establishing relationships between them (axial coding). Once all the cases were analyzed, data was reassessed and re-coded, using the scheme of categories and properties identified, according to the constant comparison method. The interaction between data and concepts ended when reassessments generated no new categories, sub-categories, or questioning of existing ones – in other words, when the theoretical saturation point was reached (step 8). Lastly, we compared the model grounded from the analysis to conceptual models previously reviewed in the literature, particularly the four SBSC models (step 9).

5 Research Results from a Grounded Theory Approach

Figure 1 illustrates the main contribution of this study, a conceptual model that seeks to support the process of definition and monitoring of socio-environmental indicators and their integration within the firms' business strategy and practices. Grounding from the systematic execution of the 9 steps of the adapted grounded approach, this model is original and helps in the integration of sustainability and social responsibility practices into business strategy and corporate management using the support of BI tools and methods.

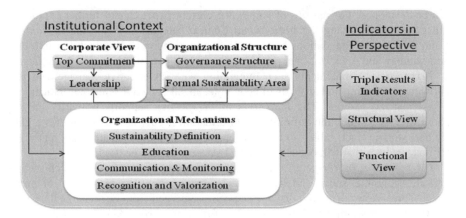

Fig. 1. A conceptual model that integrates sustainability

As any typical conceptual model, it is build of categories and properties or relationships between categories. The model encompasses two complementary building blocks: *Institutional Context* and *Indicators in Perspective*. The first block identifies a group of institutional categories that promote and allow the incorporation of sustainability into business strategy and management. The second block identifies a structure that integrates socio-environmental indicators with financial indicators and categorizes them in a way that provides a multidimensional perspective of the organizational performance.

The *Institutional Context* encompasses three views: corporate vision, organizational structure and organizational mechanisms. It confers huge importance on top management's commitment to the principles of sustainability as the starting point for the process of integrating sustainability into business strategy and practices. Top management's commitment is seen as a driver for building leadership and for changing the organizational structure in a way that allows effective incorporation of sustainability, particularly through creation of a governance structure and of a formal sustainability area with responsibility for promoting sustainability in all other areas.

Corporate vision and organizational structure are the drivers for a set of organizational mechanisms that will legitimate and consolidate the integration of sustainability: the clear definition of the role of sustainability within the firm, the implementation of an education program to promote sustainability internally and externally, the implementation of clear mechanisms for communication and monitoring, and finally, the implementation of a system of recognition and valorization of sustainable practices and initiatives. Although these four mechanisms form together the basis for integrating sustainability, one of them is directly related to BI projects: communication and monitoring. To be able to communicate and to monitor sustainable practices and advances in those practices, we need to put in place a set of sustainability-related indicators, which in turn cannot be in place without the support of a BI platform.

The identification of the *Institutional Context* categories and their interdependence represents an important part of our model, offering a frame within which a BI project can be developed. For reasons of length, the current paper cannot further explore the full richness underlying the *Institutional Context* categories. While the *Institutional Context* offers the foundations for the effective integration of sustainability into business strategy and organizational practices, the second part of the model, the *Indicators in Perspective,* complements such integration with an effective management tool, allowing monitoring and analysis of economic, social and environmental indicators in a systematized manner. The *Indicators in Perspective* supports primarily the first phase of a BI project: the definition of information needs. It also constitutes a guide for data storage (facts tables and dimensions of a data warehouse and data marts) and for data analysis that will occur once the BI system starts to be used.

The *Indicators in Perspective* rests on a multidimensional structure which seeks to allocate economic, social and environmental indicators at the same level of importance. Three views of *Indicators in Perspective* emerged from our data analysis: Triple Results, Structural and Functional. The first view, *Triple Results,* corresponds to the well-known concept of "triple bottom-line", and its purpose is to guarantee equal weight to economic, social and environmental indicators.

The second view is called *Structural View*, and is represented by four macro-level dimensions used to assess business performance: (1) business strategy; (2) stakeholders; (3) processes; and (4) training and education. Although these four dimensions have emerged inductively from systematic data collection/analysis as prescribed by grounded theory approach, they are comparable to the four dimensions set out in the BSC: finance, customers, process and learning & growing. We discuss the reasons and the implications of this parallel in the next section.

As described in the case selection, the five firms included in our investigation are recognized as leaders in sustainability. Despite their leadership, in only one organization we could identify a mature BI platform fully integrated with the corporate management system. In the other four firms, they succeed in performing well in terms of sustainability practices, but they do have neither a methodology nor an automated process for the systematization of indicators. This situation indicates that the processes of collecting, consolidating, analyzing and distributing such indicators are carried out manually or partially automated, and without guiding underlying assumptions. This absence of mature BI systems allowed us, initially, to question the role and importance of BI systems in supporting sustainability. However, as the empirical work evolved, we could perceive that even though the four firms have taken seriously the integration of sustainability practices in their management, the lack of a more sophisticated informational support is recognized as a significant weakness and a barrier to be overcome in the near future.

The lack of support to information management decreases the reliability of the information collected – socio-environmental indicators – and also impedes their integration with other indicators required for effective decision-making. In addition, factors deeply affecting management of the business and implementation of new strategies include the time spent to gather and validate the required indicators, the lack of real-time and updated information at given moments, the lack of thorough understanding of the indicators' meaning due to the absence of clear conceptualization (e.g., what is the definition of the indicator and what is its purpose?) and the lack of transparency.

BI systems are seen by these firms as an important alternative, a technological platform that will offer an open and permanent channel of communication and information diffusion for supporting sustainability practices.

6 Discussion and Preliminary Conclusions

The major contribution of this paper is to propose a conceptual model that helps to support the process of definition and monitoring of socio-environmental indicators that, combined to financial indicators, are integrated within the firms' business strategy and practices. One of strengths of the conceptual model is the synergy of two building blocks: *Institutional Context* and *Indicators in Perspective*. While the *Institutional Context* defines categories that, once articulated in a consistent way, help to promote the integration of sustainability into business strategy and management, the *Indicators in Perspective* identifies a structure that brings socio-environmental indicators and traditional financial indicators closer together and categorizes them in a way that provides a multidimensional perspective of the organizational performance. Proposing this model, we provide contributions to research and practice by filling a number of gaps revealed by literatures reviews on both themes: sustainability and BI.

First, one of the interesting capabilities of this model is the awareness that the implementation of a BI project aiming at supporting the management of sustainability cannot be conceived merely as a technological project, independent of important institutional initiatives. To work well, a BI project, particularly in the phase of definition of relevant information needs, must take place in a close relationship with other institutional categories, such as top commitment and leadership, governance structure and organizational mechanisms like recognition and valorization. A number of BI projects fail because they are taken just as "technical" projects. Our results help to shed some light on what role BI projects can play when the goal is to help integrate sustainability – essentially a role of support throughout BI methods and tools – and how – BI projects becomes instrumentally important but context-dependent. Our results show that BI systems are seen by firms that have not them yet that BI represents a technological platform that offers an open and permanent channel of communication and information diffusion for supporting sustainability practices. However, the development of a BI platform should be framed within a conceptual model that guides the definition and integration of socio-environmental indicators without neglecting the importance of the institutional context.

Second, by chosen an adapted approach of grounded theory, our conceptual model can be considered original – it was built from iterative data gathering and analysis using firms that excel in terms of sustainability practices – but that do not neglect existing theoretical models. The step 9 of the adapted version of grounded theory suggests a literature comparison phase, where conflicting or similar frameworks are systematically compared to the emergent conceptual model. In our case, we compared our conceptual models with existing SBSC models. Our conclusions were quite insightful. For instance, the *Indicators in Perspective* block rests on a multidimensional structure which seeks to allocate economic, social and environmental indicators at the same level of importance. This alignment is in line with the literature review, which has suggested the importance of multiple views and the central place occupied by the concept of triple bottom line.

In addition, the comparison of the four dimensions of the *structural view* with the four axes of the BSC was quite intriguing. The four dimensions of our conceptual model – business strategy, stakeholders, processes, training and education – have emerged inductively from systematic data collection/analysis as prescribed by grounded theory approach. Surprisingly, these four dimensions are quite comparable to the four axes of BSC – finance, customers, process, learning and growing. After carefully analysis, we considered our four dimensions quite appropriate in the context of sustainability. By replacing "finance" by "business strategy", our model enlarges the scope of the strategic goals of the firm, which can easily encompass sustainability goals. By replacing "customers" by "stakeholders", we are taken into consideration the various social actors intrinsically involved when sustainability matters: employees, suppliers, shareholders and community members in addition to customers.

A third element to be discussed is regarding literature on BI. Despite undeniable technological advances in BI implementation, we could recognize that the major difficulties are not technical but methodological and conceptual in nature. The major barriers are concerned with how to conceive and implement BI systems which effectively support strategic purposes and are integrated into corporate management systems. Our model contributes to integrate the definition and monitoring of eco-socio-financial indicators within the institutional context, showing that the implementation of a BI system requires a conceptual model guiding its development that take both the technical and managerial dimensions into account.

Acknowledgments. This research was supported by the FQRSC. A preliminary version of this paper was presented at the 2007 Academy of Management Conference.

References

1. Petrini, M., Pozzebon, M.: What role is "Business Intelligence" playing in developing countries? A picture of Brazilian companies. In: Rahman, H. (ed.) Data Mining Applications for Empowering Knowledge Societies, ch. XIII, pp. 237–257. IGI Global (2008)
2. Davenport, T.: Competing on Analytics. Harvard Business Review (2005)
3. Raynard, P., Forstarter, M.: Corporate Social Responsibility: Implications for Small and Medium Enterprises in Developing Countries. United Nations Industrial Development Organization, Viena, http://www.unido.org/doc/5162
4. Jones, T.M.: An integrating framework for research in business and society. Academy of Management Review 8(4), 55–564 (2003)
5. Clarkson, M.B.E.: A Stakeholder Framework for Analyzing and Evaluating Corporate Social Performance. The Academy of Management Review 20(1), 92–117 (1995)
6. Bowen, H.R.: The Social Responsibilities of the Businessman. Harper and Row, New York (1953)
7. Business for Social Responsibility, http://www.bsr.org
8. Hockerts, K.: Greening of Industry Network Conference, Bangkok (2001)
9. Elkington, J.: Cannibals with Forks: the Triple Bottom Line of 21st Century Business. New Society Publishers, Gabriola Island (1998)

10. Emerson, J.: The Blended Value Proposition: Integrating Social and Financial Returns. California Management Review 45(4), 35–51 (2003)
11. Mazon, R.: Uma Abordagem Conceitual aos Negócios Sustentáveis. Manual de Negócios Sustentáveis. FGV-EAESP (2004)
12. Dyllick, T., Hockerts, K.: Beyond the business case for corporate sustainability. Business Strategy and the Environment 11(2), 130–141 (2002)
13. Neto, F.P.M., Froes, C.: Gestão da Responsabilidade Social Corporativa: Um Caso Brasileiro. Qualitymark, Rio de Janeiro (2001)
14. Zadek, S.: Doing Good by Doing Well: Making the Business Case for Corporate Citizenship, New York, http://www.conference-board.org
15. Rasmussen, N.H., Goldy, P.S., Solli, P.O.: Financial Business Intelligence. John Wiley & Sons Inc., Chichester (2002)
16. Inmon, W.H.: The Data Warehouse and Data Mining. Communications of The ACM – Data Mining 39(11), 49–50 (1996)
17. Kimball, R.: The data warehouse toolkit: practical techniques for building dimensional data warehouses. Wiley, New York (2000)
18. Body, M., Miquel, M., Bédard, Y., Tchounikine, A.: A multidimensional and multiversion structure for OLAP applications. In: Proceedings 5th ACM DataWarehouse and OLAP (2002)
19. Carlsson, C., Turban, E.: DSS: Directions for the Next Decade. Decision Support Systems 33(2), 105–110 (2002)
20. Burn, J.M., Loch, K.D.: The societal impact of the World Wide Web – Key challenges for the 21st century. Information Resources Management Journal 14(4), 4–14 (2001)
21. Miranda, S.: Beyond BI: Benefiting from Corporate Performance Management Solutions. Financial Executive (2004)
22. Taydi, N.: Une approche intégrée et dynamique pour les projets d'intelligence d'affaires: une synthèse des méthodologies, techniques et outils. Master dissertation, HEC Montréal (2006)
23. Eisenhardt, K., Sull, D.: Strategy as simple rules, pp. 10–116. Harvard Business Review (2001)
24. Kaplan, R.S., Norton, D.P.: The balanced scorecard: translating strategy into action. Harvard Business School Press, Boston (1996)
25. Chenhall, R.H.: Integrative strategic performance measurement systems, strategic alignment of manufacturing, learning and strategic outcomes: an exploratory study. Accounting, Organizations & Society 30(5), 395–423 (2005)
26. Bieker, T., Dyllick, T., Gminder, C., Hockerts, K.: Towards a Sustainability Balanced Scorecard Linking Environmental and Social Sustainability to Business Strategy. In: Conference Proceedings of Business Strategy and the Environment, Leeds, UK (2001)
27. Figge, F., Hahn, T., Schaltegger, S., Wagner, M.: The Sustainability Balanced Scorecard - linking sustainability management to business strategy. Business Strategy and the Environment 11(5), 269–284 (2002)
28. Pandit, N.R.: The Creation of Theory: a Recent Application of the Grounded Theory Method. The Qualitative Report 2(4), 1–14 (1996)
29. Glaser, B., Strauss, A.: The Discovery of Grounded Theory. Aldine, Chicago (1967)
30. Glaser, B.: Doing grounded theory: issues and discussions. Sociology Press, Mill Valley (1998)
31. Strauss, A., Corbin, J.: Grounded Theory in Practice. Sage Publications, London (1997)
32. Calloway, L., Knaap, C.: Using Grounded Theory to Interpret Interviews, http://csis.pace.edu/~knapp/AIS95.htm

A Novel Particle Swarm Optimization Approach for Grid Job Scheduling

Hesam Izakian[1], Behrouz Tork Ladani[2], Kamran Zamanifar[2], and Ajith Abraham[3]

[1] Islamic Azad University, Ramsar branch, Ramsar, Iran
hesam.izakian@gmail.com
[2] Department of Computer Engineering, University of Isfahan, Isfahan, Iran
{ladani, zamanifar}@eng.ui.ac.ir
[3] Norwegian Center of Excellence, Center of Excellence for Quantifiable Quality of Service,
Norwegian University of Science and Technology, Trondheim, Norway
ajith.abraham@ieee.org

Abstract. This paper represents a Particle Swarm Optimization (PSO) algorithm, for grid job scheduling. PSO is a population-based search algorithm based on the simulation of the social behavior of bird flocking and fish schooling. Particles fly in problem search space to find optimal or near-optimal solutions. In this paper we used a PSO approach for grid job scheduling. The scheduler aims at minimizing makespan and flowtime simultaneously. Experimental studies show that the proposed novel approach is more efficient than the PSO approach reported in the literature.

1 Introduction

Computational Grid [1] is composed of a set of virtual organizations (VOs). Any VO has its various resources and services and on the basis of its policies provides access to them and hence grid resources and services are much different and heterogeneous and are distributed in different geographically areas. At any moment, different resources and services are added to or removed from grid and as a result, grid environment is highly dynamic.

Service is an important concept in many distributed computations and communications. Service is used to depict the details of a resource within the grid [2]. Grid services and resources are registered within one or more Grid Information Servers (GISs). The end users submit their requests to the Grid Resource Broker (GRB). Different requests demand different requirements and available resources have different capabilities. GRB discovers proper resources for executing these requests by querying in GIS and then schedules them on the discovered resources. Until now a lot of works has been done in order to schedule jobs in a computational grid. Yet according to the new nature of the subject further research is required. Cao [3] used agents to schedule grid. In this method different resources and services are regarded as different agents and grid resource discovery and advertisement are performed by these agents. Buyya [4] used economic based concepts including commodity market, posted price modeling, contract net models, bargaining modeling etc for grid scheduling.

S.K. Prasad et al. (Eds.): ICISTM 2009, CCIS 31, pp. 100–109, 2009.

As mentioned in [8] scheduling is NP-complete. Meta-heuristic methods have been used to solve well known NP-complete problems. In [10] Yarkhanan and Dongarra used simulated annealing for grid job scheduling. GAs for grid job scheduling is addressed in several works [12], [13], [14] and [16]. Abraham et al. [15] used fuzzy PSO for grid job scheduling.

Different criteria can be used for evaluating the efficacy of scheduling algorithms and the most important of which are makespan and flowtime. Makespan is the time when grid finishes the latest job and flowtime is the sum of finalization times of all the jobs. An optimal schedule will be the one that optimizes the flowtime and makespan [15]. The method proposed in [15] aims at simultaneously minimizing makespan and flowtime. In this paper, a version of discrete particle swarm optimization (DPSO) is proposed for grid job scheduling and the goal of scheduler is to minimize the two parameters mentioned above simultaneously. This method is compared to the method presented in [15] in order to evaluate its efficacy. The experimental results show the presented method is more efficient and this method can be effectively used for grid scheduling. The remainder of this paper is organized in the following manner. In Section 2, we formulate the problem, in Section 3 the PSO paradigm is briefly discussed and Section 4 describes the proposed method for grid job scheduling, and Section 5 reports the experimental results. Finally Section 6 concludes this work.

2 Problem Formulation

GRB is responsible for scheduling by receiving the jobs from the users and querying their required services in GIS and then allocating these jobs to the discovered services. Suppose in a specific time interval, n jobs $\{J_1, J_2, ..., J_n\}$ are submitted to GRB. Also assume the jobs are independent of each other (with no inter-task data dependencies) and preemption is not allowed (they cannot change the resource they has been assigned to). At the time of receiving the jobs by GRB, m nodes $\{N_1, N_2, ..., N_m\}$ and k services, $\{S_1, S_2, ..., S_k\}$ are within the grid. Each node has one or more services and each job requires one service. If a job requires more than one independent service, then we can consider it as a set of sub-jobs each requiring a service.

In this paper, scheduling is done at node level and it is assumed that each node uses First-Come, First-Served (FCFS) method for performing the received jobs. We assume that each node in the grid can estimate how much time is needed to perform each service it includes. In addition each node includes a time as previous workload which is the time required for performing the jobs given to it in the previous steps. We used the ETC model to estimate the required time for executing a job in a node. In ETC model we take the usual assumption that we know the computing capacity of each resource, an estimation or prediction of the computational needs of each job, and the load of prior work of each resource.

Table 1 shows a simple example of a set of received jobs by GRB in a specific time interval and status of available nodes and services in the grid. In this Table GRB received 5 jobs in a time interval and the status of available nodes and resources in the grid is as follows:

Table 1. A simple example of a set of jobs and grid status

			Nodes status:				
$Jobs: \{J_1, J_2, J_3, J_4, J_5\}$							
$Services: \{S_1, S_2, S_3, S_4\}$		Previous workload		S_1	S_2	S_3	S_4
$Nodes: \{N_1, N_2, N_3\}$							
Jobs Requirements:		N_1	15	18	5	75	∞
$J_1 \rightarrow S_1$	$J_4 \rightarrow S_3$						
		N_2	60	∞	17	∞	15
$J_2 \rightarrow S_2$	$J_5 \rightarrow S_4$						
$J_3 \rightarrow S_1$		N_3	36	12	∞	98	∞

There are 4 services and 3 nodes; N_1 includes 15 time units of previous workload which means that it requires 15 time units to complete the tasks already submitted to it. This node requires 18 time units to perform S_1, 5 time units to perform S_2 and 75 time units to perform S_3. Since this node does not include S_4, it is not able to perform it. Therefore the required time to perform S_4 by this node is considered as ∞. In this Table job J_1 requires service S_1, J_2 requires S_2, J_3 requires S_1, J_4 requires S_3, and J_5 requires S_4. Scheduling algorithm should be designed in a way that each job is allocated to a node which includes the services required by that job.

Assume that $C_{i,j}$ ($i \in \{1,2,...,m\}, j \in \{1,2,...,n\}$) is the completion time for performing jth job in ith node and W_i ($i \in \{1,2,...,m\}$) is the previous workload of N_i, then Eq. (1) shows the time required for N_i to complete the jobs included in it. According to the aforementioned definition, makespan and flowtime can be estimated using Equations. (2) and (3) respectively.

$$\sum C_i + W_i \tag{1}$$

$$makespan = \max\{\sum C_i + W_i\}, \\ i \in \{1,2,...,m\} \tag{2}$$

$$flowtime = \sum_{i=1}^{m} C_i \tag{3}$$

As mentioned in the previous section the goal of the scheduler is to minimize makespan and flowtime simultaneously.

3 Particle Swarm Optimization

Particle Swarm Optimization (PSO) is a population based search algorithm inspired by bird flocking and fish schooling originally designed and introduced by Kennedy and Eberhart [9] in 1995. In contrast to evolutionary computation paradigms such as genetic algorithm, a *swarm* is similar to a population, while a *particle* is similar to an *individual*. The particles fly through a multidimensional search space in which the position of each particle is adjusted according to its own experience and the experience of its neighbors. PSO system combines local search methods (through self experience) with global search methods (through neighboring experience), attempting to balance exploration and exploitation [5].

In 1997 the binary version of this algorithm was presented by Kennedy and Eberhart [6] for discrete optimization problems. In this method, each particle is composed of D elements, which indicate a potential solution. In order to evaluate the appropriateness of solutions a fitness function is always used. Each particle is considered as a position in a D-dimensional space and each element of a particle position can take the binary value of 0 or 1 in which 1 means "included" and 0 means "not included". Each element can change from 0 to 1 and vise versa. Also each particle has a D-dimensional velocity vector the elements of which are in range $[-V_{max}, V_{max}]$. Velocities are defined in terms of probabilities that a bit will be in one state or the other. At the beginning of the algorithm, a number of particles and their velocity vectors are generated randomly. Then in some iteration the algorithm aims at obtaining the optimal or near-optimal solutions based on its predefined fitness function. The velocity vector is updated in each time step using two best positions, *pbest* and *nbest*, and then the position of the particles is updated using velocity vectors.

Pbest and *nbest* are D-dimensional, the elements of which are composed of 0 and 1 the same as particles position and operate as the memory of the algorithm. The personal best position, *pbest*, is the best position the particle has visited and *nbest* is the best position the particle and its neighbors have visited since the first time step. When all of the population size of the swarm is considered as the neighbor of a particle, *nbest* is called global best (star neighborhood topology) and if the smaller neighborhoods are defined for each particle (e.g. ring neighborhood topology), then *nbest* is called local. Equations 4 and 5 are used to update the velocity and position vectors of the particles respectively.

$$V_i^{(t+1)}(j) = w.V_i^t(j) + c_1 r_1(pbest_i^t(j) - X_i^t(j)) + c_2 r_2(nbest_i^t(j) - X_i^t(j)) \qquad (4)$$

$$X_i^{(t+1)}(j) = \begin{cases} 1 & if \ sig(V_i^{(t+1)}(j)) > r_{ij} \\ 0 & otherwise \end{cases} \qquad (5)$$

where,

$$sig(V_i^{(t+1)}(j)) = \frac{1}{1 + \exp(-V_i^{(t+1)}(j))} \qquad (6)$$

In Eq. (4) $X_i^t(j)$ is jth element of ith particle in tth step of the algorithm and $V_i^t(j)$ is the jth element of the velocity vector of the ith particle in tth step. c_1 and c_2 are positive acceleration constants which control the influence of *pbest* and *nbest* on the search process. Also r_1 and r_2 are random values in range $[0, 1]$ sampled from a uniform distribution. w which is called inertia weight was introduced by Shi and Eberhart [7] as a mechanism to control the exploration and exploitation abilities of the swarm. Usually w starts with large values (e.g. 0.9) which decreases over time to smaller values so that in the last iteration it ends to a small value (e.g. 0.1). r_{ij} in Eq. (5) is a random number in range $[0, 1]$ and Eq. (6) shows sigmoid function.

4 Proposed PSO Algorithm for Grid Job Scheduling

In this section we propose a version of discrete particle swarm optimization for grid job scheduling. Particle needs to be designed to present a sequence of jobs in available grid nodes. Also the velocity has to be redefined. Details are given what follows.

4.1 Position of Particles

One of the key issues in designing a successful PSO algorithm is the representation step which aims at finding an appropriate mapping between problem solution and PSO particle. In our method solutions are encoded in a $m \times n$ matrix, called position matrix, in which m is the number of available nodes at the time of scheduling and n is the number of jobs. The position matrix of each particle has the two following properties:

> 1) All the elements of the matrices have either the value of 0 or 1. In other words, if X_k is the position matrix of kth particles, then:

$$X_k(i,j) \in \{0,1\} \quad (\forall i,j), i \in \{1,2,...m\}, j \in \{1,2,...,n\} \quad (7)$$

> 2) In each column of these matrices only one element is 1 and others are 0.

In position matrix each column represents a job allocation and each row represents allocated jobs in a node. In each column it is determined that a job should be performed by which node. Assume that X_k shows the position matrix of kth particle. If $X_k(i,j) = 1$ then the jth job will be performed by ith node. Figure 1 shows a position matrix in the example mentioned in Table 1. This position matrix shows that J_2 and J_4 will be performed in N_1; J_3 and J_5 will be performed in N_2 and J_1 will be performed in N_3.

4.2 Particles Velocity, *pbest* and *nbest*

Velocity of each particle is considered as an $m \times n$ matrix whose elements are in range $[-V_{\max}, V_{\max}]$. In other words if V_k is the velocity matrix of kth particle, then:

	J_1	J_2	J_3	J_4	J_5
N_1	0	1	0	1	0
N_2	0	0	1	0	1
N_3	1	0	0	0	0

Fig. 1. Position matrix in the example mentioned in Table 1

$$V_k(i, j) \in [-V_{max}, V_{max}] \quad (\forall i, j), i \in \{1,2,...m\}, j \in \{1,2,...,n\} \tag{8}$$

Also *pbest* and *nbest* are m.n matrices and their elements are 0 or 1 the same as position matrices. $pbest_k$ represents the best position that kth particle has visited since the first time step and $nbest_k$ represents the best position that kth particle and its neighbors have visited from the beginning of the algorithm. In this paper we used star neighborhood topology for *nbest* . In each time step *pbest* and *nbest* should be updated; first fitness value of each particle (for example X_k) is estimated and in case its value is greater than the fitness value of $pbest_k$ (*pbest* associated with X_k), $pbest_k$ is replaced with X_k . For updating *nbest* in each neighborhood, *pbests* are used so that if in a neighborhood, the fitness value of the best *pbest* (*pbest* with max fitness value in neighborhood) is greater than *nbest*, then *nbest* is replaced with it.

4.3 Particle Updating

Eq. (9) is used for updating the velocity matrix and then Eq. (10) is used for position matrix of each particle.

$$V_k^{(t+1)}(i, j) = w.V_k^t(i, j) + c_1 r_1 (pbest_k^t(i, j) - X_k^t(i, j)) + c_2 r_2 (nbest_k^t(i, j) - X_k^t(i, j)) \tag{9}$$

$$X_k^{(t+1)}(i, j) = \begin{cases} 1 & if (V_k^{(t+1)}(i, j) = \max\{V_k^{(t+1)}(i, j)\}), \forall\ i \in \{1,2,...m\} \\ 0 & otherwise \end{cases} \tag{10}$$

In Eq. (9) $V_k^t(i, j)$ is the element in ith row and jth column of the kth velocity matrix in tth time step of the algorithm and $X_k^t(i, j)$ denotes the element in ith row and jth column of the kth position matrix in tth time step. Eq. (10) means that in each column of position matrix value 1 is assigned to the element whose corresponding element in velocity matrix has the max value in its corresponding column. If in a column of velocity matrix there is more than one element with max value, then one of these elements is selected randomly and 1 assigned to its corresponding element in the position matrix.

4.4 Fitness Evaluation

In this paper, makespan and flowtime are used to evaluate the performance of scheduler simultaneously. Because makespan and flowtime values are in incomparable ranges and the flowtime has a higher magnitude order over the makespan, the value of mean flowtime, *flowtime / m*, is used to evaluate flowtime where *m* is the number of available nodes. The Fitness value of each solution can be estimated using Eq. (11).

$$fitness = (\lambda . makespan + (1-\lambda).mean\_flowtime)^{-1}, \quad \lambda \in [0, 1] \tag{11}$$

λ in Eq. (11) is used to regulate the effectiveness of parameters used in this equation. The greater λ, more attention is paid by the scheduler in minimizing makespan and vise versa.

4.5 Proposed PSO Algorithm

The pseudo code of the proposed PSO algorithm is stated as follows:

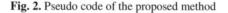

Create and initialize an $m \times n$ dimensional swarm with P particles
repeat
    for *each particle i=1,...,P* **do**
        if $f(X_i) > f(pbest_i)$ **then** // $f(\ )$ is the fitness function
            $pbest_i = X_i$;
        end
        if $f(pbest_i) > f(nbest_i)$ **then**
            $nbest_i = pbest_i$;
        end
    end
    for *each particle i=1,...,P* **do**
        update the velocity matrix using Eq. (9)
        update the position matrix using Eq. (10)
    end
until *stopping condition is true*;

Fig. 2. Pseudo code of the proposed method

5 Implementation and Experimental Results

In this Section, the proposed algorithm is compared to the method presented in [15]. Both approaches were implemented using VC++ and run on a Pentium IV 3.2 GHz PC. In the preliminary experiment the following ranges of parameter values were tested: $\lambda = [0, 1]$, c_1 and $c_2 = [1, 3]$, $w = [1.4 \rightarrow 0.01]$, $P = [10, 30]$, $V_{max} = [5, 50]$, and maximum iterations = $[20 \times m, 200 \times m]$ in which m is the number of nodes. Based on experimental results the proposed PSO algorithm and the method presented in [15] perform best under the following settings: $\lambda = 0.35$, $c_1 = c_2 = 1.5$, $w = 0.9 \rightarrow 0.1$, $P = 28$, $V_{max} = 30$, and maximum iteration $= 100 \times m$.

5.1   Comparison of Results with the Method Proposed in [15]

Abraham et al. [15] used Fuzzy discrete particle swarm optimization [11] for grid job scheduling. In their method, the position of each particle is presented as $m \times n$ matrices in which m is the number of available nodes and n is the number of received jobs. Each matrix represents a potential solution whose elements are in [0, 1] intervals in which the total sum of the elements of each column is equal to 1. The value of s_{ij}, the element in ith row and jth column of the position matrix, means the degree of membership that the grid node N_j would process the job J_i in the feasible schedule solution [15]. In the first time step of the algorithm one position matrix is generated using LJFR-SJFR heuristic [16] that minimizes the makespan and the flowtime simultaneously and others are generated randomly and then in each time step these matrices are updated using velocity matrix whose elements are real numbers in range $[-V_{max}, V_{max}]$. After updating each position matrix, it is normalized in a way that each element is in range [0, 1] and the sum of values of each column equals 1 and then using these obtained matrices schedules are generated.

In this paper, for comparison and evaluation of the scheduler, makespan and mean flowtime are used simultaneously. A random number in the range $[0, 500]$, sampled from uniform distribution, is assigned to the previous workload of each node in our tests. One or more services (at most k services) of $\{S_1, S_2, ..., S_k\}$ are randomly selected for each node. The time for executing services is randomly selected in range $[1, 100]$ if the node has these services; otherwise it is selected as ∞.

For each job one service among k services is selected randomly as the required service of that job. To improve the efficiency of our proposed method and the method presented in [15] we generate only feasible solutions in initial step as well as each iteration/generation.   In other words each job is allocated to the node which has the service required by that job. If in grid there is a job that its corresponding service does not exist in any of the nodes, then its allocated node is considered as -1 and this means that this job is not performable in the grid at that

Table 2. Comparison of statistical results between our method and FPSO proposed in [15]

Case Study	Number of (Jobs, Nodes, Services)	Number of iterations: $(100 \times m)$	LJFR-SJFR heuristic		FPSO [15]		Proposed DPSO	
			make-span	flow-time	make-span	flow-time	make-span	flow-time
I	(50,10,40)	1000	607.9	1337.4	530.5	1252.2	**500.6**	**1186.8**
II	(100,10,80)	1000	750.3	2440.1	658.2	2309.1	**581.4**	**2139.5**
III	(300,10,160)	1000	1989.0	6956.8	1359.7	6769.3	**1226.7**	**6483.9**
IV	(50,20,40)	2000	482.7	1230.4	470.9	1057.0	**462.8**	**891.3**
V	(100,20,80)	2000	550.2	1881.6	511.3	1443.2	**497.4**	**1278.5**
VI	(300,20,160)	2000	886.9	4863.5	667.1	4215.7	**535.8**	**3830.9**
VII	(50,30,40)	3000	467.3	1177.1	468.6	821.5	**459.0**	**691.3**
VIII	(100,30,80)	3000	487.7	1603.4	468.9	1124.8	**443.5**	**983.1**
IX	(300,30,160)	3000	554.6	3691.2	533.5	3324.3	**490.2**	**2912.4**

specific time. In this case ∞, the required time for performing the job, is not taken into account in fitness estimation so that the efficiency of the method does not fade. Nine grid status of different sizes with number of jobs, n=50, 100, 300 number of nodes, m=10, 20, 30 and number of services, k=40, 80, 160 are generated. The statistical results of over 50 independent runs are illustrated in Table 2.

As evident, the proposed method performs better than the Fuzzy PSO proposed in [15]. Figures 3 and 4 show a comparison of CPU time required to achieve results and the fitness values of each method for different case studies as shown in Table 2.

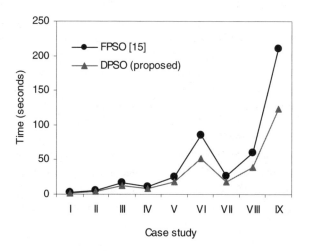

Fig. 3. Comparison of convergence time between our proposed method and FPSO [15]

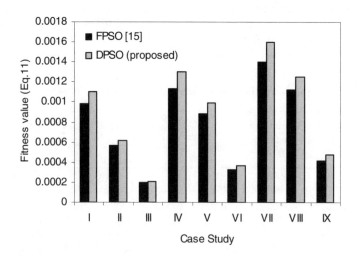

Fig. 4. Comparison of fitness values between our proposed method and FPSO [15]

6 Conclusions

This paper presented a version of Discrete Particle Swarm Optimization (DPSO) algorithm for grid job scheduling. Scheduler aims at generating feasible solutions while minimizing makespan and flowtime simultaneously. The performance of the proposed method was compared with the fuzzy PSO through carrying out exhaustive simulation tests and different settings. Experimental results show that the proposed method outperforms fuzzy PSO. In the future, we plan to use the proposed method for grid job scheduling with more quality of service constraints.

References

1. Foster, I., Kesselman, C., Tuecke, S.: The Anatomy of the Grid: Enabling Scalable Virtual Organizations. International Journal of High Performance Computing Applications 15, 200–222 (2001)
2. Cao, J., Kerbyson, D.J., Nudd, G.R.: Performance Evaluation of an Agent-Based Resource Management Infrastructure for Grid Computing. In: Proceedings of 1st IEEE/ACM International Symposium on Cluster Computing and the Grid, pp. 311–318 (2001)
3. Cao, J.: Agent-based Resource Management for Grid Computing. Ph.D. Thesis, Department of Computer Science University of Warwick, London (2001)
4. Buyya, R.: Economic-based Distributed Resource Management and Scheduling for Grid Computing. Ph.D. Thesis, School of Computer Science and Software Engineering Monash University, Melbourne (2002)
5. Salman, A., Ahmad, I., Al-Madani, S.: Particle Swarm Optimization for Task Assignment Problem. Microprocessors and Microsystems 26, 363–371 (2002)
6. Kennedy, J., Eberhart, R.C.: A Discrete Binary Version of the Particle Swarm Algorithm. In: IEEE International Conference on Systems, Man, and Cybernetics, pp. 4104–4108 (1997)
7. Shi, Y., Eberhart, R.C.: A Modified Particle Swarm Optimizer. In: Proceedings of the IEEE Congress on Evolutionary Computation, pp. 69–73 (1998)
8. Coffman Jr., E.G. (ed.): Computer and Job-Shop Scheduling Theory. Wiley, New York (1976)
9. Kennedy, J., Eberhart, R.C.: Particle Swarm Optimization. In: Proceedings of the IEEE International Conference on Neural Networks, pp. 1942–1948 (1995)
10. YarKhan, A., Dongarra, J.: Experiments with scheduling using simulated annealing in a grid environment. In: Parashar, M. (ed.) GRID 2002. LNCS, vol. 2536, pp. 232–242. Springer, Heidelberg (2002)
11. Pang, W., Wang, K., Zhou, C., Dong, L.: Fuzzy Discrete Particle Swarm Optimization for Solving Traveling Salesman Problem. In: Proceedings of the Fourth International Conference on Computer and Information Technology, pp. 796–800. IEEE CS Press, Los Alamitos (2004)
12. Di Martino, V., Mililotti, M.: Sub Optimal Scheduling in a Grid Using Genetic Algorithms. Parallel Computing 30, 553–565 (2004)
13. Liu, D., Ca, Y.: CGA: Chaotic Genetic Algorithm for Fuzzy Job Scheduling in Grid Environment, pp. 133–143. Springer, Heidelberg (2007)
14. Gao, Y., Ron, H., Huangc, J.Z.: Adaptive Grid Job Scheduling with Genetic Algorithms. Future Generation Computer Systems 21, 151–161 (2005)
15. Abraham, A., Liu, H., Zhang, W., Chang, T.G.: Scheduling Jobs on Computational Grids Using Fuzzy Particle Swarm Algorithm, pp. 500–507. Springer, Heidelberg (2006)
16. Abraham, A., Buyya, R., Nath, B.: Nature's Heuristics for Scheduling Jobs on Computational Grids. In: 8th IEEE International Conference on Advanced Computing and Communications, pp. 45–52 (2000)

Priority-Based Job Scheduling in Distributed Systems

Sunita Bansal[1] and Chittaranjan Hota[2]

[1] Computer Science & Information Systems Group
Research & Consultancy Division
Birla Institute of Technology & Science
Pilani, Rajasthan, 333031, India
sunita_bansal@bits-pilani.ac.in
[2] Computer Science and Information Systems Group
Birla Institute of Technology & Science, Pilani
Hyderabad Campus, Hyderabd, AP, India
hota@bits-hyderabad.ac.in

Abstract. Global computing systems like SETI@home tie together the unused CPU cycles, buffer space and secondary storage resources over the Internet for solving large scale computing problems like weather forecasting, and image processing that require high volume of computing power. In this paper we address issues that are critical to distributed scheduling environments such as job priorities, length of jobs, and resource heterogeneity. However, researchers have used metrics like resource availability at the new location, and response time of jobs in deciding upon the job transfer. Our load sharing algorithms use dynamic sender initiated approach to transfer a job. We implemented distributed algorithms using a centralized approach that improves average response time of jobs while considering their priorities. The job arrival process and the CPU service times are modeled using M/M/1 queuing model. We compared the performance of our algorithms with similar algorithms in the literature. We evaluated our algorithms using simulation and presented the results that show the effectiveness of our approach.

Keywords: Job Scheduling, Distributed Systems, Priorities.

1 Introduction

Job scheduling is more complicated in distributed systems because of non availability of shared states amongst nodes, high communication costs, heterogeneous hosts and networks, and low node resilience. Also, it is practically difficult to arrive at an optimal solution to the distributed scheduling problem. In a multi-computer environment, allocating resources like CPU, memory, and peripheral devices to jobs is a complex task. This is because of the physical distribution of resources, and insufficient global state information available at individual nodes. Current research addresses these issues in distributed scheduling.

In [1], authors have analyzed the use of a group of workstations over five months and found that only 30% of the workstation capacity was utilized. On the contrary,

S.K. Prasad et al. (Eds.): ICISTM 2009, CCIS 31, pp. 110–118, 2009.
© Springer-Verlag Berlin Heidelberg 2009

some users possess workstations or computing resources that are two small to meet their processing demands. To alleviate these problems, resource allocation component of distributed multi-computer systems judiciously decides to move a process from heavily loaded machine to lightly loaded machine which in turn is called as load sharing. The performances of load sharing algorithms are measured in terms of average system throughput, delay or response time of the overall system or that of a single user.

Load distributing algorithms are classified as either static, dynamic or adaptive [2]. In a static algorithm, the scheduling is carried out according to a predetermined policy. The state of the system at the time of the scheduling is not taken into consideration. On the other hand, a dynamic algorithm adapts its decision to the state of the system. Adaptive algorithms are special type of dynamic algorithms where the parameters of the algorithm are changed based on the current state of the system.

The policies adopted by any load-sharing algorithm are as follows: Transfer policy decides when to initiate the job transfer across the system and hence, whether the node is a sender or a receiver. This is decided by a threshold called as load index [3]. Selection policy that determines which job is to be transferred i.e. either the newly arrived job or a job that has been executed partially. Information policy collects the load information at other sites. Location policy determines the node to which a process is to be transferred.

Job scheduling policies are broadly classified as sender-initiated, receiver-initiated [4, 5] and symmetrically initiated [6]. An algorithm is sender initiated if the node at which a job has arrived determines as to where the arrived job is to be transferred. In a receiver-initiated algorithm, a receiver or the target node determines from which sender it will receive a job for execution at its end. In a symmetric policy, both receiver and the sender try to locate each other [7].

To date there has been very little work done on priority-based dynamic scheduling in multi-computer environments with resource availability at receivers varying randomly over time. This work addresses the challenges in managing and utilizing computing resources in distributed systems efficiently with job priorities as a critical metric for making efficient scheduling decisions in a sender initiated approach. We propose novel priority-based dynamic scheduling algorithms and compare the performance of our algorithms with the sender initiated algorithms from the literature.

Section 2 presents the related work. Section 3, we present the system model. Section 4 describes the design of our algorithms. Section 5, we describe the implementation of proposed our algorithms and comparison of simulation results with existing sender initiated algorithm. Finally, Section 6 concludes our paper.

2   Related Work

Lu and Zomaya [8] introduced a hybrid policy to reduce the average job response time of a system. In their work, clusters are organized into regional grids around a set of well-known broker sites in terms of the transfer delay. When a job arrives at a site it calculates the expected completion time of the job within the region, based on which it transfers the job either to any other site or to a global queue. Each broker gets periodically update information from the remote sites and then it sends information about the lightly loaded nodes to its local site in order to minimize average response time of the system.

Shah et al. [9] proposed dynamic, adaptive, and decentralized load balancing algorithms for computational grids. Several constraints such as communication delay due to underlying network, processing delays at the processors, job migration cost and arbitrary topology for grid systems are considered. They proposed two algorithms, MELISA and LBA wherein MELISA works with large-scale grid environments and LBA works with small-scale grids. These algorithms used metrics like migration cost, resource and network heterogeneity, and expected completion time of jobs for the purpose of balancing the load.

Viswanathan et al. [10] developed incremental balancing and buffer estimation strategies to derive optimal load distribution for non-critical load. The job scheduler will first obtain the information about the available memory capacity and computing speed of the sinks and the size of the jobs from the sources. The scheduler will then calculate the optimum load fractions and notify each sink. Then sinks ask the source to transfer the load.

Han et al. [11] introduced a new type of client scheduling proxy that mediates job scheduling between server and clients. Scheduling proxy is elected amongst the clients clustered using network proximity. This proxy serves as a leader of the clients. Scheduling proxy downloads coarse-grained work unit from the server and distributes fine-grained work units to the clients. In order to prevent the failure of scheduling proxy, workers must preserve result until they receive confirmation from the server.

Zhang et al. [12] described a distributed scheduling service framework for real time CORBA. The global scheduling decisions considered the end system's requirements and the entire system. Their results showed the benefits of their framework when the resource management is necessary.

3 Model

3.1 System Model

Our scenario is based on independent computing sites with existing local workload and management systems. We examine the impact on job scheduling results when the computing sites participate in a distributed environment.. For the sake of simplicity, in this paper we consider a central distributed scheduler that efficiently uses unused resources. Note, that for a real world application a single central scheduler would be a critical limitation. We also assume central scheduler is dedicated and reliable. It is responsible to allocate the job efficiently.

We assume that each participating site in the distributed environment has a node (Single processor, memory and secondary storage). The nodes are linked with a fast interconnection network that does not favor any specific communication pattern inside the machine [6]. The machines use space sharing and run the jobs in an exclusive fashion. Moreover, the jobs are not preempted nor time-sharing is used. Therefore, once started a job runs until completion. We assume that there is common threshold and priority policy. Range of queue length and priority is same. Processing speed, Memory space and Job size is normalized.

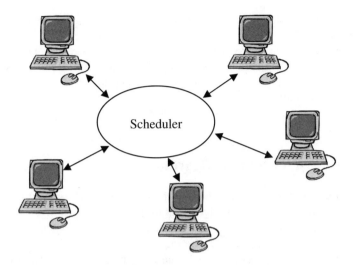

Fig. 1. Overview of Grid System

Now, we will formally define the system. As described above, we consider a Grid system with N sources denoted as $S1, S2, ..., SN$ and M sinks denoted as $K1, K2, ..., KM$. Source and sink has direct link to scheduler.

3.2 Communication Model

In our examination our assumption, input data is only considered to be transferred to the machine before the job execution while output data of the job is transferred back afterwards. Network communication is neglected in our evaluation. As shown later in this paper, the average wait time of a job before its execution is at least several

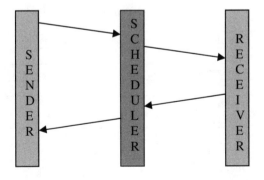

Fig. 2. Overview of Communication System

minutes. Assuming that resources are usually well connected, the data overhead will
have no impact on most jobs.

Each node communicates to scheduler when status is change. Sender nodes send
actual job with following parameters priority (P_i), length (L_i). Receivers sends message with processing speed (C_j), buffer space (B_j) and queue length (Q_j). There is at
most four communications held.

We describe terminology and notation that is use through out the paper below
(refer to Table 1).

4   Enhance Sender Initiate Algorithm

Jobs are submitted by independent users at the local sites. The scheduling problem is
an on-line scenario without knowledge of future job submissions. Our evaluations are
restricted to batch jobs, as this job type is dominant on most systems. A job request
consists of several parameters as e.g. the requested load size and priority of job. It is
the task of the scheduling system to allocate the jobs to receiver. The jobs are executed without further user interaction.

Sender initiated (SI) algorithm does not consider the status of job and receiver.
Sender simply finds receiver and transfer the newly arrived job. What if there is more
than one receiver. As soon as it gets response transfer a job. Enhance Sender initiated
(ESI) considers status of receiver and sender. It finds the receivers which complete
the task with α period. Assign the highest priority job to smallest queue length
receiver.

Nodes are periodically informs to scheduler. Nodes queue length can be less than
the threshold.

The working of ESI algorithm shown in algorithm 2. It makes group of job
which has highest P_i and lowest Q_j. Calls the fitness function algorithm 1. That
tries to assign a job which complete with α period based on close allocation
policy.

Table 1. List of notation and terminology

Source	List of source
Sink	List of sink
M	Number of sinks
N	Number of Sources
L_i	Length of job
P_i	Priority of job
B_j	Virtual Memory size of sink
C_j	Processing speed of sink
Q_j	Queue length
T_h	Threshold
α	Time period

Algorithm 1 (Fitness Function(X, Y, α))
{

       ⊬ $x_i \in X \{L_i, P_{i,}\}$
       ⊬ $y_j \in Y \{Q_j, C_j, B_j)$
       For each x_i
              For each y_j
                     1. $T = x_i. L_i / y_j. C_j$
                     2. $T \mathrel{+}= x_i. L_i / y_j. B_j$
                     If $(T < = \alpha + 1)$
                            Assign job to y_j
                 Else
                            Find new receiver

}

Algorithm 2 (Scheduling Procedure)

Input:
⊬ $S_i \in$ Source $\{L_i, P_{i,}\}$
⊬ $K_j \in$ Sink $\{Q_j, C_j, B_j)$
α is complete time.
While (Source $!= \{\Phi\}$ or Sink $!= \{\Phi\}$)
                1.   n=Find the senders which has same P_i
                2.   m=Find the receiver which has same Q_j
                3.   if (n==m)
                4.   j=FitessFuction (n, m, α)
                5.   else if (n < m)
                6.   j=FitessFunction (n, n, α)
                7.   else
                8.   j=Fitessfunction (m,m, α)
                9.   if(j==true)
                10.   Transfer the job
                11. else
                12. Process locally
End while
End procedure

5 Simulation Results

We present the result of our simulation study and compare the performance of our proposed algorithms with SI algorithm. We have considered heterogeneous nodes linked with a fast interconnection network. This is dynamic load sharing algorithm. It tries to share the load, assigning each job to the computing site that yields the shorter completion time. It is assumed that each grid site retrieves the current load value of scheduler without cost, but the transfer cost of the job is considered. During the procedure of job transfer, the load in destination site may be changed.

Simulation Model

In this section, we study the performance of the algorithms under different system parameters via simulations. We assume that the simulated system includes a fixed 100
number of nodes. The nodes are connected to scheduler with fast network. Nodes are
single computer. Sender node will assign priority of job on common policy. It sends
priority and job size to scheduler. Receiver also sends queue length, processor speed
and buffer size to scheduler. We assume all values are normalized. We have shown
different graphs. Fig. 3: shows, when senders are more than receiver. Scheduler allo-
cates job which has higher priority job rest job is locally execute. Fig. 4: shows, when
senders are less than receivers. Scheduler will allocate the jobs to best receivers. Fig. 5:
shows, different periods load sharing. Total impact of system is increase by 27%.

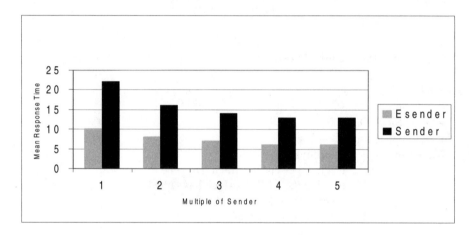

Fig. 3. Comparison of SI algorithm with ESI algorithm when senders are multiple of receivers

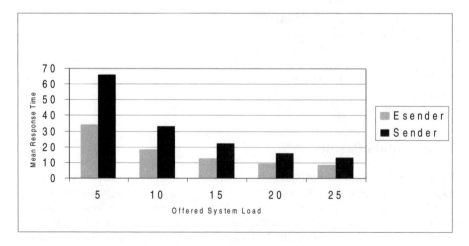

Fig. 4. Comparison of SI algorithm with ESI algorithm when receivers are more than senders

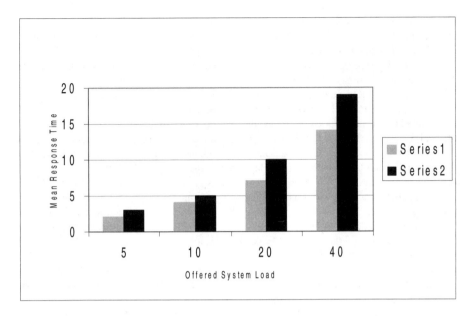

Fig. 5. Comparison of SI algorithm with ESI for different period

6 Conclusion

We presented centralized scalable algorithm for distributed network. Our objective is to minimize total execution time and shorter the completion time for jobs that arrive to a grid system for processing. We considered priority, job size, queue length, processing speed and buffer space of receiver.

We plan to investigate this algorithm in distributed fashion with network parameters and node failure.

References

[1] Mukat, M.W., Livny, M.: Profiling Workstations Available Capacity for Remote Execution, Performance. In: Proceedings of the 12 th IFIP WG 7.3 Symposium on Computer Performance, Brussels, Belgium
[2] Shivaratri, N.G., Kruger, P., Singhal, M.: Load Distributing in Locally Distributed System. IEEE Computer 25(12), 33–44 (1992)
[3] Kureger, P., Livny, M.: The Diverse Objectives of Distributed Scheduling Policies. In: Proceedings of the 7th IEEE International Conference on Distributed Computing Systems, Berlin, Germany (September 1987)
[4] Edward, R.Z.: Attacking the Process Migration Bottleneck. In: Proceedings of the 11th ACM Symposium on Operating Systems Principle, Austin, Texas, United States, November 8-11 (1987)

[5] Carsten, E., Volker, H., Ramin, Y.: Benefits of Global Grid Computing for Job Scheduling. In: Proceedings of the Fifth IEEE/ACM International Workshop on Grid Computing, Pittsburgh, USA, November 8 (2004)
[6] Krueger, P., Finkel, V.: An Adaptive Load Balancing Algorithm for a Multicomputer, Technical Report No. 539, University of Wisconsin-Madison (April 1984)
[7] Srinivasan, S., Kettimuthu, R., Subramani, Sdayappan, P.: Characterization of Backfilling Strategies for Parallel Job Scheduling. In: Proceedings of International Conference on Parallel Processing Workshop, Vancouver, Canada, August 18-21 (2002)
[8] Kai, L., Albert, Y.Z.: A Hybrid Policy for Job Scheduling and Load Balancing in Heterogeneous Computational Grids. In: Sixth International Symposium on IEEE Transactions on Parallel and Distributed System, Hagen berg, Austria, July 5-8 (2007)
[9] Ruchir, S., Bhardwaj, V., Manoj, M.: On the Design of Adaptive and De-centralized Load Balancing Algorithms with Load Estimation for Computational Grid Environments
[10] Sivakumar, V., Thomas, G.R., Dantong, Y., Bharadwaj, V.: Design and Analysis of a Dynamic Scheduling Strategy with Resource Estimation for Large-Scale Grid Systems. In: Proceedings of the Fifth IEEE/ACM International Workshop on Grid Computing, Pittsburgh, USA, November 8 (2004)
[11] Jaesun, H., Daeyeon, P.: Scheduling Proxy: Enabling Adaptive Grained Scheduling for Global Computing System. In: Proceedings of the Fifth IEEE/ACM International Workshop on Grid Computing, Pittsburgh, USA, November 8 (2004)
[12] Zhang, J., DiPippo, L., Fay-Wolfe, V., Bryan, K., Murphy, M.: A Real-Time Distributed Scheduling Service For Middleware Systems. In: IEEE WORDS 2005 (2005)

An Efficient Search to Improve Neighbour Selection Mechanism in P2P Network

C.R. Totekar and P. Santhi Thilagam

Department of Computer Engineering,
National Institute of Technology Karnataka - Surathkal,
India - 575025
{chinmaytoekar@gmail.com, santhi_soci@yahoo.co.in}

Abstract. One of the key challenging aspects of peer-to-peer systems has been efficient search for objects. For this, we need to minimize the number of nodes that have to be searched, by using minimum number of messages during the search process. This can be done by selectively sending requests to nodes having higher probability of a hit for queried object. In this paper, we present an enhanced selective walk searching algorithm along with low cost replication schemes. Our algorithm is based on the fact that most users in peer-to-peer network share various types of data in different proportions. This knowledge of amount of different kinds of data shared by each node is used to selectively forward the query to a node having higher hit-ratio for the data of requested type, based on history of recently succeeded queries. Replication scheme replicates frequently accessed data objects on the nodes which get high number of similar queries or closer to the peers from where most of the queries are being issued. Two simple replication schemes have been discussed and their performances are compared. Experimental results prove that our searching algorithm performs better than the selective walk searching algorithm.

Keywords: Peer-Peer systems, selective walk, query table, replication schemes.

1 Introduction

Peer-to-Peer systems are increasingly being used nowadays with increasing speed of communication links and high computational power of nodes. Decentralized P2P systems have gained enormous popularity particularly Gnutella [11], Freenet [12] etc. Compared to Centralized P2P systems, which have high probability of single or multiple point of failure. On the other hand purely decentralized system achieves higher fault tolerance by dynamically re-configuring the network as the nodes join and leave.

Structured P2P systems such as CAN [14], Pastry [13], and Chord [15] guarantee to find existing data and provide bounded data lookup efficiency. However, it suffers from high overhead to handle node churn, which is a frequent occurrence of node joining/leaving. Unstructured P2P systems such as gnutella[11] are more flexible in that there is no need to maintain special network structure, and they can easily support complex queries like keyword/full text search. The drawback is that their routing

S.K. Prasad et al. (Eds.): ICISTM 2009, CCIS 31, pp. 119–127, 2009.
© Springer-Verlag Berlin Heidelberg 2009

efficiency is low because a large number of peer nodes have to be visited during the search process. The flooding mechanism used by gnutella[11] like systems, results in increased amount of traffic as exponential number of redundant messages are generated during the search process, as the value of TTL is increased.

To address the problems of the original flooding several alternative schemes have been proposed. These algorithms include iterative deepening[4], directed BFS[4], local indices based search[4], random walk[7], Probabilistic search[9], popularity-biased random walk [6], adaptive probabilistic search[5], and dynamic index allocation scheme[2], Selective walk searching[8]. All these techniques try to reduce the number of redundant messages generated because of flooding during search and make search more efficient. These methods either try to maintain the indices of neighboring nodes by each node, or by selecting only those nodes having high probability of having hits, based on recent past.

Selective search scheme uses hints from neighboring peers such as number of shared files, most recent succeeded queries etc to select the most promising candidate for forwarding the queries. However it does not consider the type of data objects that are shared by each node in the algorithm. Most users share files which they themselves are interested in, some might exclusively share movies and some may share only music, and may not share even a single file of picture or document, or archive type. Thus forwarding the query for music file type, to a neighbor having higher share of data as done in Selective Walk searching, might not give proper results if the selected neighbor is sharing no files of type audio, but is having higher share of overall data. Similarly each peer in Selective Walk technique gets rating based on information about most recent succeeded queries maintained in its query table. We can extend this by having a query counter at each node which stores the most recent succeeded queries of each type, thus node having higher value of counter for requested file type gets higher chance of getting selected.

There are very few replication schemes in literature, such as owner replication as used by gnutella and path replication as used by freenet. Gnutella creates copy on the requester node whereas path replication replicates on the path between requester and provider. The third replication algorithm is random replication which is harder to implement. There is need for a simpler approach to find replication site and how many copies to be created.

In this paper, we propose an enhanced selective search which rates each peer based on information about the type of data files it provided recently and number of hits it had. It also uses other details of most recent successful queries at each peer such as list of frequently accessed files and details of total number of different types of files shared by each of the neighboring peers, while selecting the neighbor.

We would like to draw your attention to the fact that it is not possible to modify a paper in any way, once it has been published. This applies to both the printed book and the online version of the publication. Every detail, including the order of the names of the authors, should be checked before the paper is sent to the Volume Editors.

2 Related Work

Several attempts have been made to reduce the number of redundant messages produced during the search which uses flooding. Broadly they can be classified in to two categories: breadth first search and depth first search. Directed BFS [4], iterative deepening [4] improve upon Breadth first search but still have the disadvantage that the number of messages will increase exponentially with increasing TTL. Whereas DFS Based methods such as Adaptive Probabilistic Search [9], Random Walks [7], Selective Walk Searching[8] algorithms perform better as they produce linearly increasing messages with respect to increasing TTL.

Selective walk searching algorithm emphasizes on peer selection criteria, which is based on several hints such as number of files shared by each peer, information about most recent succeeded queries and specific peer that provided the answer. The querying node selects a set of neighbors to forward the query to according to the selection criteria based on these hints. Each of those intermediate peers then selects one peer each to propagate the query until the TTL becomes zero. Each node maintains the details of files shared by its neighbors; neighbor with higher number of files is considered having higher probability of having hit to a query. But this does not consider the type of data file that is queries for, hence there might be situation where one neighbor might have higher number of audio files and another neighbor having higher number of video files, then the neighbor to be selected should be decided based on the type of file requested. If it is request for audio file then the neighbor having highest number of files of type audio should be selected.

Secondly Selective Walk searching algorithm uses a query table at each node to store the most recent succeeded queries, so that if similar query comes at a later stage it will be forwarded according to the past experience. But only if similar query exists then the node will be selected else node with higher shared data will be selected. We use counters at each peer which indicate the number successful queries of each type in the recent past, based on that the neighbor having higher count of recent hits for particular type can be selected.

3 Problem Description

There are a lot of existing techniques for improving search by minimizing the number of redundant messages by selecting most promising neighbor at each step. But no algorithm takes into consideration the type of data shared by each node so that peers can be rated based on this information. Our goal is to design a search scheme for Unstructured P2P system which will decrease the search time, minimize the number of redundant messages, use hints from past to improve selection of neighbor, minimize the number of duplicate messages at each node, and bring the frequently accessed objects closer to the peers from where most queries originate.

4 Proposed Schemes

Here is brief description of the proposed searching technique and the two replication schemes:

4.1 Improved Selective Walk Searching

In selective walk search method type of data file requested is not considered hence no details about the file type is maintained at the query tables, and details about the different types of data shared by the neighbors. Improved selective walk searching scheme considers these details to provide better hints and an improved selection criteria to use these hints.

Information Maintained by nodes: Each node maintains three types of information: most recent succeeded queries in Query Table, a counter for each data type which stores the count of recent successful queries for each type of data, and total number of files shared of each type. Each of these information units will be exchanged with neighbors periodically, so that each node will have these details of each of their neighbors. The Query Table maintains the details of most recent successful queries such as query id, number of hits, Search criteria and the file type. Each node maintains a counter for each type of data which keeps the count of recent hits for each type of data and total number of files of each type that are shared. The structure of the query table is given in Fig. 1.

Fig. 1. Format of Query Table

Selection of neighbor: Selection of neighbor is done based on hints such as most recent successful queries listed in the query tables of neighbors, Counter for successful queries for each type, and total number of files shared of each type. The searching node will select a fraction of the neighbors to forward the query to. Suppose the total number of neighbors is N then n neighbors will be selected based on following criteria:

(i) First all the nodes that have similar object in their Query tables will be selected in order of the number of hits, until number of selected neighbors does not exceed n.

(ii) If the number of selected neighbors is still less than 'n' than select the neighbors having higher number of shared files of requested file type until their number reaches n.

Once the query is passed to neighboring nodes they will intern forward the query to exactly one node by using following criteria: Look up the query tables of neighbors, if there is an entry already existing in neighbors query table then select the neighbor having highest number of hits for that object, else select the neighbor having highest number of recent hits for the requested file-type. The queries will be forwarded until the TTL becomes zero. This is depicted in Fig. 2.

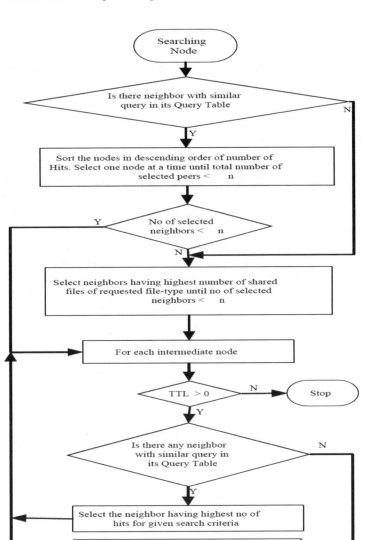

Fig. 2. Criteria for Neighbor Selection

Updating of Query Table: When the query is formed, each query has unique Query-Id, which gets recoded at each of the intermediate nodes. When there is a hit at a particular node that node sends Query-Hit message in response in the reverse path back to the query source, thus each node updates its query table on this path. This is done by having Unique Id for Query-Hit same as that in the Query message. Similarly the Hit counter for that particular data type is to be incremented at each of the nodes.

5 Experimental Study

Assumptions: Some assumptions have been taken during the implementation of the proposed algorithm. Some of them are:

(1) All nodes are having equal bandwidth and processing-power.
(2) The communication delay between each pair of nodes is same and equal to 1.
(3) All the nodes will be up and running throughout the search process.

Experimental Setup: Simulation environment consists of 64 nodes, each containing between 1000-4000 files, each node connected to 1 - 8 nodes, with an average 3 connection per node. The cost of links between peers is fixed and equal to 1. Distance between the farthest nodes is 9. The file sizes are between 1 and 5MB for audio, 5 to 700MB for Video, 10KB to 1MB for text and 5 to 200MB for archives. The size of Query Table is taken as 500, which means 500 recent succeeding Queries are maintained in Query Table. Each of the nodes are having a separate available disk space, of varying capacities, where the frequently accessed data can be replicated. A randomly chosen peer will generate search query for a file using TTL equal to 9. The total cost of searching a file is calculated. A total of 8000 such searches are initiated and total cost of 10000 searches is computed. For determining the performance of the replication schemes initially there exists exactly one copy of file in the entire network. Replication will be happening in background during the search process, whenever any node gets overloaded. The performances are measured at the end of search process.

Performance Metrics: Scalability, efficiency and responsiveness are the main properties of good searching algorithm. An efficient search algorithm should generate minimum number of messages and at the same time should have high hit rate too. Below are some of the performance metrics used in evaluation and comparison of our algorithms.

Query Efficiency: It can be defined as ratio of query hits to messages per node.

Query Efficiency = Query Hits / Msgs Per Node.

Search responsiveness: A second criterion for search performance is search responsiveness, which reflects how quickly the system responds to user requests.

Responsiveness = SuccessRate / No.of.hops.

Therefore *Search Efficiency* can be defined as Query Efficiency x Search Responsiveness.

Average Time for first hit: Average Time for first hit is the time taken by searching node to get the first response.

Number of duplicate messages: Number of duplicate messages reflect the number of query messages dropped by nodes when same query is received more than once.

6 Results and Analysis

The searching algorithm is implemented and compared with Selective Walk search. Simulation was performed with a total of 8000 searches, and performance recorded as shown below in the graphs. Query efficiency and search responsiveness of our search scheme are both better compared to the Selective Walk (SW) which is shown in Fig. 3 and Fig.4. Search efficiency of our algorithm is 0.65 as against 0.4 of Selective walk Searching scheme which is shown in Fig. 5. Our algorithm again results in better success rate, approximately 15% more than the SW scheme. Average time taken by queries is also improved using our scheme which is shown in Fig. 6 and Fig.7.

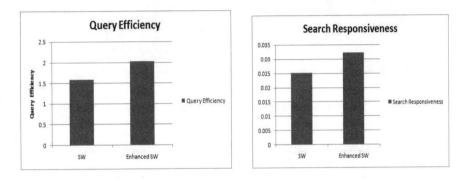

Fig. 3. Query Efficiency

Fig. 4. Search Responsiveness

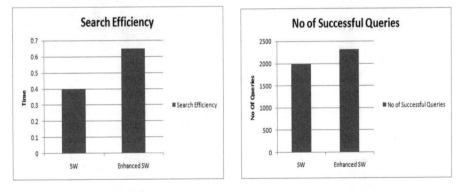

Fig. 5. Search Efficiency

Fig. 6. Number of successful queries

Average Time for first hit is better in our case as shown in Fig. 8. Our algorithm performs better in reducing number of such duplicate messages as shown in Fig. 9 and Fig.10, which in turn will help to reduce loss of messages at node due to overloading, when a node gets too many queries. Search efficiency of our algorithm is 0.65 as against 0.4 of Selective walk Searching scheme which is shown in Fig. 5. Our algorithm again results in better success rate, approximately 15% more than the SW scheme. Average time taken by queries is also improved using our scheme which is shown in Fig. 6 and Fig.7. Average Time for first hit is better in our case as shown in Fig. 8. Our algorithm performs better in reducing number of such duplicate messages as shown in Fig. 9 and Fig. 10, which in turn will help to reduce loss of messages at node due to overloading, when a node gets too many queries.

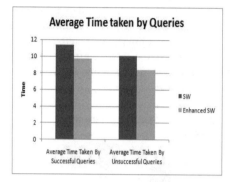

Fig. 7. Average Time for successful queries

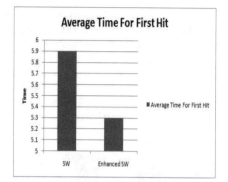

Fig. 8. Average Time for first Hit

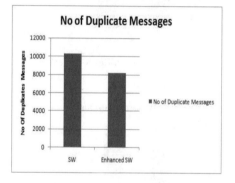

Fig. 9. Number of duplicate messages

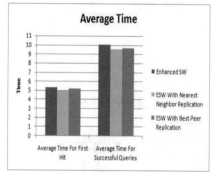

Fig. 10. Comparison of search with the two-replication schemes and without replication

Finally the comparison of the two replication schemes Nearest Neighbor replication and Best Neighbor replication show that Nearest Neighbor replication has slight better performance than Best Neighbor in terms of average time for successful queries and average time for first query. Both the algorithms result in decrease in Average time for first hit and Average time for successful queries.

7 Conclusions

The paper presents the design and evaluation of an efficient search scheme which tries to improve neighbor selection mechanism by rating each peer based on past history and contents shared by node. Our searching scheme performs better, at least as good as Selective Walk scheme in most of the criteria's used for evaluation. It has more success rate, better search response and decreased number of duplicate messages. We have also presented and evaluated two replication schemes which in turn help to reduce the search time by replicating the frequently accessed objects closer to query issuers. Because of the simplicity of these replication schemes and their performance benefits they can be easily incorporated in real world P2P networks.

References

1. Lv, Q., Cao, P., Cohen, E., Li, K., Shenker, S.: Search and Replication in Unstructured Peer-to-Peer Networks. In: Proceedings of the 16th International Conference on Supercomputing, pp. 84–95 (2002)
2. Ohta, T., Masuda, Y., Mitsukawa, K., Kakuda, Y., Ito, A.: A Dynamic Index Allocation Scheme for Peer-to-Peer Networks. In: ISADS 2005, pp. 667–672 (2005)
3. Zheng, Q., Lu, X., Zhu, P., Peng, W.: An Efficient Random Walks Based Approach to Reducing File Locating Delay in Unstructured P2P Network. In: Global Telecommunications Conference, vol. 2(5). IEEE, Los Alamitos (2005)
4. Yang, B., Garcia-Molina, H.: Improving Search in Peer-to-Peer Networks. In: Proceedings of the 22nd International Conference on Distributed Computing Systems (2002)
5. Tsoumakos, D., Roussopoulos, N.: Adaptive Probabilistic Search for Peer-to-Peer Networks. In: Proceedings of the Third International Conference on Peer-to-Peer Computing (2003)
6. Zhong, M., Shen, K.: Popularity-biased random walks for peer to peer search under the square-root principle. In: Proceedings of the 5th International Workshop on Peer-to-Peer Systems, CA (2006)
7. Gkantsidis, C., Mihail, M., Saberi, A.: Random Walks in Peer-to-Peer Networks. In: Twenty-third Annual Joint Conference of the IEEE Computer and Communications Societies, vol. 1(12) (2004)
8. Xu, Y., XiaoJun, Wang, M.C.: Selective Walk Searching Algorithm for Gnutella Network. In: Consumer Communications and Networking Conference, CCNC, pp. 746–750 (2007)
9. Cheng, A.-H., Joung, Y.-J.: Probabilistic File Indexing and Searching in Unstructured Peer-to-Peer Network. In: CCGrid 2004, pp. 9–18 (2004)
10. Zhao, J., Lu, J.: Overlay Mismatching of Unstructured P2P Networks using Physical Locality Information. In: Proceedings of the Sixth IEEE International Conference on Peer-to-Peer Computing, pp. 75–76 (2006)
11. Official Gnutella website, http://www.gnutella.com
12. Official FreeNet website, http://www.freenetproject.org
13. Rowstron, A., Druschel, P.: Pastry: Scalable, decentralized object location, and routing for large-scale peer-to-peer systems. In: Guerraoui, R. (ed.) Middleware 2001. LNCS, vol. 2218, pp. 329–350. Springer, Heidelberg (2001)
14. Ratnasamy, S., Francis, P., Handley, M., Karp, R., Shenker, S.: A Scalable Content-Addressable Network. In: Proceedings of ACM SIGCOMM, pp. 161–172 (2001)
15. Stoica, I., Morris, R., Karger, D., Kaashoek, M.F., Balakrishnan, H.: Chord: A Scalable Peertopeer Lookup Service for Internet Applications. In: Proceedings of ACM SIGCOMM, CA (2001)

Application of Neural Networks in Software Engineering: A Review

Yogesh Singh[1], Pradeep Kumar Bhatia[2], Arvinder Kaur[1], and Omprakash Sangwan[3]

[1] University School of Information and Technology, G.G.S. IP University, Delhi, India
[2] Dept. of CSE, GJU of Science & Technology, Hisar, Haryana
[3] Amity Institute of Information Technology, Amity University, Uttar Pradesh

Abstract. The software engineering is comparatively new and ever changing field. The challenge of meeting tight project schedules with quality software requires that the field of software engineering be automated to large extent and human intervention be minimized to optimum level. To achieve this goal the researchers have explored the potential of machine learning approaches as they are adaptable, have learning capabilities and non-parametric. In this paper, we take a look at how Neural Network (NN) can be used to build tools for software development and maintenance tasks.

Keywords: Neural Network, Software Testing, Software Metrics.

1 Introduction

The challenge of modeling software systems in a fast moving scenario gives rise to a number of demanding situations. First situation is where software systems must dynamically adapt to changing conditions. The second one is where the domains involved may be poorly understood and the last one is where there may be no knowledge (though there may be raw data available) to develop effective algorithmic solutions. To answer the challenge, a number of approaches can be utilized [13]. One such approach is the transformational programming. Under the transformational programming, software is developed, modified, and maintained at specification level, and then automatically transformed into production-quality software through automatic program synthesis [21]. This software development paradigm will enable software engineering to become the discipline capturing and automating currently undocumented domain and design knowledge [31].

In order to realize its full potential, there are tools and methodologies needed for the various tasks inherent to the transformational programming. In this paper, we take a look at how Neural Network (NN) can be used to build tools for software development and maintenance tasks as NNs have learning abilities and are particularly useful for (a) poorly understood problem domains where little knowledge exists for the humans to develop effective algorithms; (b) domains where there are large databases containing valuable implicit regularities to be discovered; or (c) domains where

S.K. Prasad et al. (Eds.): ICISTM 2009, CCIS 31, pp. 128–137, 2009.

programs must adapt to changing conditions. In this paper we survey the existing work on applications of NNs in software engineering and provide research directions for future work in this area.

2 Neural Network (NN) Methodology

NN methods were developed to model the neural architecture and computation of the human brain [33]. A NN consists of simple *neuron-like* processing elements. Processing elements are interconnected by a network of weighted connections that encode network knowledge. NNs are highly parallel and exercise distributed control. They emphasize automatic learning. NNs have been used as memories, pattern recall devices, pattern classifiers, and general function mapping engines [27,42]. A classifier maps input vectors to output vectors in two phases. The network learns the input-output classification from a set of training vectors. After training, the network acts as a classifier for new vectors.

Figure 1 shows the anatomy of a processing element (also called a node, unit, processing unit, or neuron). Node output or activation, $o(\mathbf{x}, \mathbf{w})$, is a function of the weighted sum (dot product) of the input vector x and the interconnection weights w . A common activation (output) function is the unipolar sigmoid with a range from 0.0 to 1.0. The activation equations are

$$\Phi(a) = \frac{1}{1 + e^{-a}}$$

$$a = \sum_{i=1}^{N} \mathbf{w}\,\mathbf{x}$$

The number of input units and the number of output units are problem dependent. The number of hidden units is usually not known. Hidden units allow the NN to learn by forcing the network to develop its own internal representation of the input space. The network that produces the best classification with the fewest units is selected as the best topology. A net with too few hidden units cannot learn the mapping to the required accuracy. Too many hidden units allow the net to "memorize" the training data and do not generalize well to new data. *Backpropagation* is the most popular training algorithm for multilayer NNs [29,42]. NNs have been established to be an effective tool for pattern classification and clustering [32].

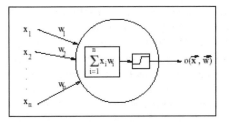

Fig. 1. Architecture of Neural Network

A set of data (Input-Output information) is used for training the network. Once the network has been trained it can be given any input (from the input space of the map to be approximated) and it will produce an output, which would correspond to the expected output from the approximated mapping. The quality of this output has been established to correspond arbitrarily close to the actual output desired owing to the generalization capabilities of these networks

3   Software Development Life Cycle (SDLC) and Application of NNs in Software Engineering

A variety of life cycle models have been proposed and are based on task involved in developing and maintaining software [4]. Figure 2 shows SDLC phases and the phase wise application of NNs in software engineering.

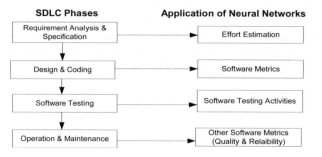

Fig. 2. SDLC Phrases and Application of Neural Networks

4   Application of NNs in Software Engineering

Several areas in software development have already witnessed the use of NNs. In this section, we take a look at some reported results of applications of NN in the field of software engineering. The list is definitely not a complete one. It only serves as an indication that people realize the potential of NNs and begin to reap the benefits from applying them in software development and maintenance.

4.1   Software Project Effort Estimation

Accurate estimation of software development effort is critical in software engineering. The research literature contains various models for estimating programming effort. Traditional, well-understood, mature approaches include Putnam's SLIM [36], Albrecht's function point method of estimation [7], as well as COCOMO [12] and COCOMO II [2,14]. These traditional methods use well-defined equations for estimating effort and need to be calibrated in local environment. So besides these traditional approaches, various machine learning techniques, including NNs, [23,24,37] have evolved as these techniques have learning capability to learn from past experience and are also dynamic in nature.

In [19], Gavin R. Finnie and Gerhard E. Witting proposed models for predicting effort using neural networks and Case Based Reasoning (CBR). They compared analogy based method using CBR with different versions of FP-based Regression models and NN. The data used consisted of 299 projects from 17 different organizations and concluded that NN performed better than analogy followed by regression models. Their performance to large degree dependent on the data set on which they are trained, and extent to which suitable project data is available. According to Gray and McDonell [20], NNs are the most common software estimation model-building technique used as an alternative to mean least squares regression. These are estimation models that can be "trained" using historical data to produce ever better results by automatically adjusting their algorithmic parameter values to reduce the delta between known actual and model predictions.

Anita Lee et. al integrated NN with cluster analysis for software development cost estimation and concluded that it proved to be a promising approach to provide more accurate results on the forecasting of software development costs. It furthermore showed that the integration of NNs with cluster analysis increases the training efficacy of the network, which in the long run results in more accurate cost estimation than using just NNs [9,30].

K.K. Shukla presented a genetically trained NN predictor trained on historical data [38]. It demonstrated substantial improvement in prediction accuracy as compared to both a regression-tree-based approach, as well as back propagation trained NN approach. Data set used is from 78 real-life software projects. A multi-layered feed-forward NN with 39 input neurons is used for effort prediction. The neurons correspond to Boehm's features [4]. A genetic algorithm (GA) is utilized. A large number of simulation experiments comparing Backpropagation(BP), Quickdrop (a fast version of BP) and the genetically trained NNs were performed. The GNN outperformed BPNN and QPNN in all folds. Also, there was not much variation in productivity accuracy with GNN. Their experimental result showed that GNN was a significant improvement over the other methods.

Abbas Heiat [1] compared artificial NN and regression models for estimating software development effort. The aim of this paper was to identify whether or not it is possible to increase the prediction performance of software developments either utilizing multilayer perceptron and radial basis function NNs compared to usual regression analysis. Many different models were developed to deal with the fundamental question of how to most accurately deal with the estimation of software developments. With three different software development projects that were analyzed by either a classical regression analysis approach or modified artificial NNs approaches, the authors found clear evidence for an improvement in accuracy using especially RBF NNs which were using LOC input data over RBF NNs using FPA input data.

Krishnamoorthy Srinivasan and Douglas Fisher compared backpropagation learning methods to traditional approaches for software effort estimation [39]. Their results suggested that the approach was competitive with SLIM, COCOMO and function points. Their experiments also showed the sensitivity of learning to various aspects of data selection and representation. Mohanty and Kemerer indicated that traditional models are quite sensitive as well [26,34]. Apart from estimating effort, NN has also been used to determine the most suitable metrics for effort estimation. The work by G. Boetticher et.al. demonstrated a process for extracting a set of software metrics from a program, associating the metrics with a program's effort, then constructing a NN model. An automated process created over 33,000 different NN models. Product metrics were grouped into four categories: size, vocabulary, objects, and complexity, then analyzed for training and test cases. By using a powerset combination of the input metric categories 15 different input configurations were categorized. These categories were "Size", "Vocabulary", "Object", "Complexity", "Size and Vocabulary", "Size and Objects", "Size and Complexity", "Vocabulary and Objects", "Vocabulary and Complexity", "Objects and Complexity", "Size, Vocabulary, and Objects", "Size, Vocabulary, and Complexity", "Size, Objects, and Complexity", "Vocabulary, Objects, and Complexity", and "Size, Vocabulary, Objects and Complexity". Compound metrics, in the form of combinations of individual measures, generally outperformed individual measures. Overall, the input configuration, which included all the metrics, produced the most accurate model against the test data than any subset combination. The cross-validation experiments produced estimates within 30% of the actual 73.26% of the time [15].

4.2 Software Metrics

Software metrics are numerical data related to software development. It provides effective methods for characterizing software. Metrics have traditionally been composed through the definition of an equation, but this approach is limited by the fact that all the interrelationships among all the parameters be fully understood. The goal of our work is to find alternative methods for generating software metrics. Deriving a metric using a NN has several advantages. The developer needs only to determine the endpoints (inputs and output) and can disregard the path taken. Unlike the traditional approach, where the developer is saddled with the burden of relating terms, a NN automatically creates relationships among metric terms.

The paper by G. Boetticher, K. Srinivas and D. Eichmann explored an alternative, NN approach to generating metrics [17]. Experiments performed on two widely known metrics, McCabe and Halstead, indicated that the approach is sound, thus serving as the groundwork for further exploration into the analysis and design of software metrics. The experiment ranged over seven different NN architectures broken into three groups: "broad, shallow architectures (4-5-3, 4-7-3, and 4-10-3)", "narrow, deep architectures (4-7-7-3 and 4-7-7-7-3)", and "narrow, deep architectures with hidden layers that connected to all previous layers (4+7-7-3 and 4+7+7-7-3)". These three groups were formed in order to discover whether there was any connection between the complexity of architecture and its ability to model a metric. Both the broad and deep architectures did moderately well at matching the actual Halstead volume metric, but the connected architecture performed significantly better. Furthermore, there is no significant advantage for a five versus four layer connected architecture, indicating that connecting multiple layers may be a sufficient condition for adequately modeling the metric.

The researchers described further an alternative approach to the assessment of component reusability based upon the training of NNs to mimic a set of human evaluators from an Ada package specification with a corresponding set of reusability ratings for that package [16]. Three types of experiments viz. Black Box, White box and Grey Box were performed. In the black box experimental context they attempted to train a NN to associate direct measures as number of procedures or functions, number of uncommented lines etc to reusability measure. The white box experiments attempted to correlate measures extracted from Ada bodies as cyclomatic complexity, coupling, volume etc. The Grey Box experiments combined the measures of White Box and Black Box. The intent of this experiment was to determine whether combining black and white input parameters yielded better results. The grey box experiments produced the best statistical results to corresponding assessments. The result showed that a neural approach is not only feasible, but can achieve good results without requiring inputs other than those readily available with metrics evaluation packages. This approach supports intuitive assessment of artifacts without excessive concern for the derivation of polynomials to model the desired function.

4.3 Software Testing Activities

Software testing is the process of executing a program with the intention of finding errors. Software testing consumes major resources in software product's lifecycle [4, 5]. As test case automation increases, the volume of tests can become a problem. Further, it may not be immediately obvious whether the test generation tool generates effective test cases. A NN is neither system nor test case metric specific. Therefore, it can be used across a variety of testing criteria, test case and coverage metrics, and fault severity levels. The neural net formalizes and objectively evaluates some of the testing folklore and rules-of-thumb that is system specific and often requires many years of testing experience. Indeed, it might be useful to have a mechanism that is able to learn, based on past history, which test cases are likely to yield more failures versus those that are not likely to uncover any. They presented experimental results on using a NN for pruning a test case set while preserving its effectiveness. Charles Anderson, Nneliese Von and Rick Mraz have shown NN to be a useful approach to test case effectiveness prediction. The test data was generated from "SLEUTH". Neural Net was trained to predict from different severity levels viz S1, S2, S3 and S4. NN was able to predict test case with Severity S1 with 91.7% accuracy, S2 with 89.4% accuracy, S3 with 82.85 accuracy and S4 with 94.4% accuracy [8].

Another area of software Testing in which NN has found attention is its use as Test Oracle. The essence of software testing is to determine a set of test cases for the item being tested. One of the most difficult and expensive parts of testing is the generation of test data, which has traditionally been done manually [18]. Test automation is software that automates any aspect of testing of an application system. It includes capabilities to generate test inputs and expected results, to run test suites without manual intervention, and to evaluate pass/no pass [11]. A test case has a set of inputs, a list of expected outputs for the Implementation Under Test (IUT) [4,11]. Generating test inputs automatically is relatively easy, but generating expected results is a tough job. We cannot hope to perform extensive automated testing without expected results. In testing, an Oracle is a trusted source of expected results. The oracle can be a program specification, a table of examples, or simply the programmer's knowledge of how a program should operate. A perfect oracle would be behaviorally equivalent to the implementation under test (IUT) and completely trusted. It would accept every input specified for the IUT and would always produce a correct result. Developing a perfect Oracle is therefore as difficult as solving the original design problem. If a perfect Oracle was available and portable to the target environment, we could dispense with the IUT and fulfill the application environment with the ideal oracle. The perfect oracle however is something like a philosopher's stone for software. We cannot guarantee the ability of any algorithm to decide that an algorithm is correct in all possible cases. The random sampling of input space along with an Oracle that provides the corresponding output of the software forms an automated test generation mechanism [8,32]. Our work described a preliminary set of experiments to determine whether NNs can be used as test Oracle. The appeal of a neural approach lied in a NN's ability to model a function without the need to have knowledge of that function. With the help of triangle classification problem (TRIYP) our experimental results show that NN can be used to give automated outputs with a reasonable degree of accuracy[3,4,6,18].

4.4   Other Software Metrics (Software Quality and Software Reliability)

W. Pedrycz et.al. has given a fuzzy neural relational model of software quality derived from the McCall hierarchical software quality measurement framework (hsqf) that has learning capability. This software quality model expresses the intuition of developers in selecting particular factors and criteria leading to overall software quality measures. The introduction of fuzzy sets into the hsqf makes it possible to capture information granules, to give meaningful "shape" to otherwise daunting collection of factors, criteria and metrics their relationship and relative importance. The model has a measure of "intelligence" as it can learn the biases of developers for particular software development environments [35].

Thwin and Quah have used Ward and Generalized Regression Neural Networks for predicting software development faults and software readiness [41]. There models are based on object-oriented metrics as predictor variables and number of faults as response variables. More recently, Gyimothy et al. [22] have used a NN developed at their university to predict the fault proneness of classes in open source software.

Importance of software quality is increasing leading to development of new sophisticated techniques, which can be used in constructing models for predicting quality attributes. Artificial Neural Networks (ANN) is one such technique, and successfully trained for predicting maintainability using object-oriented metrics. The ANN model demonstrate that they were able to estimate maintenance effort within 30 percent of the actual maintenance effort in more than 70 percent of the classes in the validate set, and with MARE of 0.265 [43].

It is shown by Nachimuthu Karunanithi et al. that NN reliability growth models have a significant advantage over analytic models in that they require only failure history as input and not assumptions about either the development environment or external parameters. Using the failure history, the neural-network model automatically develops its own internal model of the failure process and predicts future failures. Because it adjusts model complexity to match the complexity of the failure history, it can be more accurate than some commonly used analytic models. Results with actual testing and debugging data suggested that neural-network models are better at endpoint predictions than analytic models [25].

Another study presented an empirical investigation of the modeling techniques for identifying fault-prone software components early in the software life cycle. Using software complexity measures, the techniques built models, which classified components as likely to contain faults, or not. The modeling techniques applied in the study cover the main classification paradigms, including principal component analysis, discriminant analysis, logistic regression, logical classification models, layered NNs and holographic networks. Experimental results were obtained from 27 academic software projects. The models were evaluated with respect to four criteria: predictive validity, misclassification rate, achieved quality, and verification cost. Their result suggested that no model was able to discriminate between components with faults and components without faults [28]. NNs have also been used for assessment of software reliability in distributed development environment. It is difficult to assess the software reliability in distributed environment as the complexity of software increases in distributed environment. The reliability assessment method using NN has been compared with several conventional software reliability assessment methods and verified that this method fitted best to the actual data among the compared methods in terms of all goodness to fit evaluation criteria [40].

5 Conclusions and Research Directions

In this paper, we show how NNs have been used in tackling software engineering problems. The NNs have been used in various phases of software development right from planning phase for effort estimation to software testing, software quality assurance as well as reliability prediction. It has also been used for developing new metrics. These developments will definitely help maturing software engineering discipline. One very important advantage of using NN is that it helps in systematically modeling expert knowledge. There is a need to develop NN based tools that become a part of software development and help in automating the software development process to optimum level. So there is an urge for NN community to come forward in help of software engineering discipline, so that full potential of NN can be utilized in solving the problems faced by software professionals. More similar type of studies must be carried out with large data sets to validate object-oriented design metrics across different fault severities. Also focus should be given on cost benefit analysis of models using NN method. This will help to determine whether a given fault proneness model would be economically viable.

References

1. Heiat, A.: Comparison Of Artificial Neural Network And Regression Models For Estimating Software Development Effort. Information and software Technology 44(15), 911–922 (2002)
2. Abts, C., Clark, B., Devnani Chulani, S., Horowitz, E., Madachy, R., Reifer, D., Selby, R., Steece, B.: COCOMO II Model Definition Manual," Center for Software Engineering, University of Southern California (1998)
3. Acree, A.T., Budd, T.A., DeMillo, R.A., Lipton, R.J., Sayward, F.G.: "Mutation Analysis" School of Information and Computer Science, Georgia Institute Technology, Atlanta, Tech. Rep. GIT-ICS-79/08 (September 1979)
4. Aggarwal, K.K., Singh, Y.: Software Engineering, 3rd edn. New Age International Publishers (2008)

5. Aggarwal, K.K., Singh, Y., Kaur, A.: Code Coverage Based Technique For Prioritizing Test Cases For Regression Testing. In: ACM SIGSOFT, vol. 29(5) (September 2004)
6. Aggarwal, K.K., Singh, Y., Kaur, A., Sangwan, O.P.: A Neural Net Base Approach to Test Oracle. In: ACM SIGSOFT (May 2004)
7. Albrecht, A.J., Gaffney Jr., J.E.: Software Function, Source Lines of Code and Development Effort Prediction: Software Science Validation. IEEE Transactions on Software Engineering 24, 345–361 (1978)
8. Anderson, C., Mayrhauser, A., Mraz, R.: On The Use of Neural Networks to Guide Software Testing Activities. In: Proceeding International Test Conference, Washington, DC (October 1995)
9. Lee, A., Cheng, C.H., Balakrishan, J.: Software Development Cost Estimation: Integrating Neural Network With Cluster Analysis. Information and Management 34(1), 1–9 (1998)
10. Beizer, B.: Software Testing Techniques. Van Nostrand Reinhold, New York (1990)
11. Binder Robert, V.: Testing Object-Oriented Systems-Models, Patterns and Tools. Addison Wesley, Reading (1999)
12. Boehm, B.: Software Engineering Economics. Prentice-Hall, Englewood Cliffs (1981)
13. Boehm, B.: Requirements that handle IKIWISI, COTS, and rapid change. IEEE Computer 33(7), 99–102 (2000)
14. Boehm, B., et al.: Cost Models for Future Software Life Cycle Process: COCOMO-II. Annals of Software Engineering (1995)
15. Boetticher, G.: An Assessment Of Metric Contribution In The Construction Of Neural Network-Based Effort Estimator. In: Second International Workshop on Soft Computing Applied to Software Engineering (2001)
16. Boetticher, G., Eichmann, D.: A Neural Net Paradigm for Characterizing Reusable Software. In: Proceeding of the First Australian Conference on Software Metrics, pp. 41–49 (1993)
17. Boetticher, G., Srinivas, K., Eichmann, D.: A neural Net -Based approach to Software Metrics. In: Proceeding of the Fifth International Conference of Software Engineering and Knowledge Engineering, San Francisco, CA, June 16-18, 1993, pp. 271–274 (1993)
18. DeMillo, R.A., And Offutt, A.J.: Constraint-Based Automatic Test Data Generation. IEEE Transactions on Software Engineering SE-17(9), 900–910 (1991)
19. Finnie, G.R., Wittig, G.E.: AI tools for software development effort estimation. In: Software Engineering and Education and Practice Conference, pp. 346–353. IEEE Computer society Press, Los Alamitos (1996)
20. Gray, A.R., MacDonnell, S.G.: A Comparison of Techniques for Developing Predictive Models of Software Metrics. Information and software technology, 425–437 (1997)
21. Green, C., et al.: Report On A Knowledge-Based Software Assistant. In: Rich, C., Waters, R.C. (eds.) Readings in Artificial Intelligence and Software Engineering, pp. 377–428. Morgan Kaufmann, San Francisco (1986)
22. Gyimothy, T., Ferenc, R., Siket, I.: Empirical validation of object-oriented metrics on open source software for fault prediction. IEEE Trans. Software Engineering 31, 897–910 (2005)
23. Haykin, S.: Neural Networks, A Comprehensive Foundation. Prentice Hall India, Englewood Cliffs (2003)
24. Hodgkinson, A.C., Garratt, P.W.: A Neuro fuzzy Cost Estimator. In: Proc. 3rd International Conf. Software Engineering and Applications (SAE), pp. 401–406 (1999)
25. Karunanithi, N., Whiyley, D., Malaiya, Y.K.: Using neural networks in reliability prediction. IEEE Software 9(4), 53–59 (1992)
26. Kemerer, C.F.: An empirical validation of Software Cost Estimation Models. Comm. ACM 30, 416–429 (1987)
27. Kohonen, T.: Self Organizing Maps, 2nd edn. Springer, Berlin (1997)

28. Lanubile, F., Lonigro, A., Visaggio, G.: Comparing models for identifying fault-prone software components. In: Proc. of the 7th Int'l. Conf. Software Eng. and Knowledge Eng., pp. 312–319 (June 1995)
29. Fausett, L.: Fundamentals of Neural Networks. Prentice Hall, Englewood Cliffs (1994)
30. Anita, L., cheng, C.H., Jaydeep, B.: Software development cost estimation: Integrating neural network with cluster analysis. Information and Management 34(1), 1–9 (1998)
31. Lowry, M.: Software Engineering in the Twenty First Century. AI Magazine 14(3), 71–87 (Fall, 1992)
32. Von Mayrhauser, A., Anderson, C., Mraz, R.: Using A Neural Network To Predict Test Case Effectiveness. In: Proceedings IEEE Aerospace Applications Conference, Snowmass, CO (February 1995)
33. McClelland, J.L., Rumelhart, D.E., The PDP Research Group: Parallel Distributed Processing: Exploration in the Microstructure of Cognition, vol. 1. MIT Press, Cambridge (1986)
34. Mohanti, S.: Software cost Estimation: Present and future. Software practice and Experience 11, 103–121 (1981)
35. Pedrycz, W., Peters, J.F., Ramanna, S.: Software quality Measurement: Concept and fuzzy neural relational model. IEEE (1998)
36. Putnam, L.H.: A General Empirical Solution to the Macro Software Sizing and Estimating Problem. IEEE Transactions on Software Engineering 2(4), 345–361 (1978)
37. Samson, B., Ellison, D., Dugard, P.: Software Cost Estimation Using an Albus Perceptron. Information and Software Technology, 55–60 (1997)
38. Shukla, K.K.: Neuro-Genetic prediction of software development effort. International Journal of Information and Software Technology 42(10), 701–703 (2000)
39. Srinivasan, K., Fisher, D.: Machine learning approaches to estimating software development effort. IEEE Trans. Soft. Eng. 21(2), 126–137 (1995)
40. Tamura, Y., Yamada, S., Kimura, M.: Comparison of Software Reliability assessment methods based on neural network for distributed development environment (August 2003)
41. Thwin, M.M.T., Quah, T.-S.: Application of Neural Networks for predicting Software Development faults using Object Oriented Design Metrics. In: Proceedings of the 9th International Conference on Neural Information Processing, pp. 2312–2316 (November 2002)
42. Wittig, G., Finnie, G.: Estimating software development effort with connectionist models. Information and Software Technology, 469–476 (1997)
43. Aggarwal, K.K., Singh, Y., Malhotra, A.K.R.: Application of Neural Network for Predicting Maintainability using Object-Oriented Metrics. Transaction on Engineering, Computing and Technology 15 (October 2006)

Model-Based Software Regression Testing for Software Components

Gagandeep Batra[1], Yogesh Kumar Arora[2], and Jyotsna Sengupta[1]

[1] Department of Computer Sci. Punjabi University, Patiala
[2] Gian Jyoti Inst. of Mgmt. & Technology., Mohali
B_gagan@mailcity.com, yogesh_84kumar@yahoo.com,
jyotsna@pbi.ac.in

Abstract. This paper presents a novel approach of generating regression test cases from UML design diagrams. Regression testing can be systematically applied at the software components architecture level so as to reduce the effort and cost of retesting modified systems. Our approach consists of transforming a UML sequence diagram of a component into a graphical structure called the control flow graph (CFG) and its revised version into an Extended control flow graph (ECFG) The nodes of the two graphs are augmented with information necessary to compose test suites in terms of test case scenarios. This information is collected from use case templates and class diagrams. The graphs are traversed in depth-first-order to generate test scenarios. Further, the two are compared for change identification. Based on change information, test cases are identified as reusable, obsolete or newly added. The regression test suite thus generated is suitable to detect any interaction and scenario faults.

Keywords: Regression testing, UML, Components.

1 Introduction

Regression testing, as quoted from [5], "attempts to validate modified software and ensure that no new errors are introduced into previously tested code". Regression testing attempts to revalidate the old functionality inherited from the old version. The new version should behave as the older one except where new behavior is intended. In component-based software systems, regression test case selection is a challenge since vendors of commercial-off-the-shelf (COTS) components do not release source code. One characteristic distinguishing regression testing from developmental testing is the availability of existing test suites at regression test time. If we reuse such test suites to retest a modified program, we can reduce the effort required to perform testing. Unfortunately, test suites can be large, and we may not have time to rerun all tests in such suites. Thus, we must often restrict our efforts to a subset of the previously existing tests. The problem of choosing an appropriate subset of an existing test suite is known as selective retest problem [10].

Unified Modeling Language (UML) is used by number of software developers as a basis for design and implementation of their component-based applications. The

S.K. Prasad et al. (Eds.): ICISTM 2009, CCIS 31, pp. 138–149, 2009.

Unified Modeling Language (UML) is a standard   language for specifying, visualizing, constructing, and documenting the artifacts of software systems, as well as for business modeling and other non-software systems. Many researchers are using UML diagrams such as state-chart diagrams, use-case diagrams, sequence diagrams, etc to generate test cases and it has led to Model based test case generation. The UML Testing Profile bridges the gap between designers and testers by providing a means to use UML for test specification and modeling. This allows the reuse of UML design documents for testing and enables testing earlier in the development   phase. Our focus in this paper is on regression testing of components at design level from UML class and sequence diagrams. The UML and its diagrams are widely used to visually depict the static structure and more importantly, the dynamic behavior of applications. This trend provides us with an excellent opportunity to meld our approach with UML and give developers and testers the capability of automatically generating black-box conformance tests.

The proposed approach is based on the UML 2.0 standard [2, 7] and is supported by Net Beans UML tool [11]. We introduce a representation called Control Flow Graph (CFG) and Extended Control Flow Graph (ECFG) that integrates information from class and sequence diagrams. The CFG is used to derive test input constraints. The novelty of our approach is that it considers change identification form CFG and ECFG, to derive regression test suite. The rest of the paper is organized as follows: section 2 presents some related work in the field of UML based regression testing; section 3 describes the proposed approach; section 4 gives application of approach to show its efficacy. Finally in Section 5, we conclude our paper and the future research directions.

2   UML and Regression Testing

This section describes the existing work on regression testing based on UML notations. Regression testing [4] attempts to revalidate the old functionality inherited from the old version. The new version should behave exactly as the old except where new behavior is intended. Therefore, regression tests for a system may be viewed as partial operational requirements for new versions of the system. The problem of regression testing [3] may be broken down into two sub problems: the test selection problem and the test plan update problem.

Lihua et al. [6] use model checkers to compare the functionality of two different versions of GUI components. It covers specification based criteria to generate regression test suite . Their approach is somewhat different as their major focus in on the specification based test cases, so mapping is between change identification at analysis level, which implies that any wrong specification can have a negative impact on the test data.

Scheetz et al. [9] describe an approach to generate system test inputs from UML class diagrams. The class diagrams are restricted to contain only classes, associations and specifications and specification structures. Test objectives are derived from defining desired states of the class instances after the test is executed. They used the UML and its associated object constraint language (OCL) to describe the information about the changes from one version to next which was provided by component developer and constitutes modified methods and directly/indirectly affected methods. It is utilized to select test cases. Though the method is quite simple and can be easily implemented in practice, but generates some ir-relevant test cases as well.

Pilskalns et al. [8] adopted Graph-based approach to combine the information from structural and behavioral diagrams (Class Diagram and Sequence Diagrams). In this approach, each sequence diagram is transformed into an Object-Method Directed Acyclic Graph (OMDAG). The OMDAG is created by mapping object and sequence method calls from a Sequence Diagram to vertices and arcs in a directed acyclic graph. This approach does not take into account the pre and post conditions of the operations which affect behavior of a class. Also, it is inefficient to handle concurrency. Briand et al. [5] describe the TOTEM (Testing Object Oriented Systems with Unified Modeling Language) methodology. Test requirements are derived from UML use case diagrams and sequence diagrams. These Diagrams store the sequential dependencies between use cases in the form of an activity diagram. This approach, however is a semi-automatic way of scenario coverage with pre-specified initial conditions and test oracles.

Atifah Ali et al. [1] propose a methodology for identifying changes and test case selection based on the UML designs of the system. Their work is based on the existing work on control flow analysis with some extensions to regression testing. They included class diagram to get more information while testing and capturing the changes. Though the approach is simple, it needs to be automated. Our technique is based on the UML 2.0 models. It considers the effects of dynamic binding, which is a problematic area for code as well as design based techniques and has not yet been exploited. It also incorporates method constraints and their changes. It works at the level of class attributes and methods, and method constraints i.e., pre and post conditions are also mapped from the class diagram to ECFG.

3   The Proposed Approach

For a given software component, our approach is to generate its Regression Test suite based on the UML sequence diagram storage graph drawn with the help of Net Beans Tool. We have used two flow graphs; control flow graph CFG and extended current flow graph ECFG to store the UML sequence Diagram of the software components and its enhanced version. The two flow graphs are compared to identify the changes and hence to generate test cases. Figure 1 shows the proposed approach.

The different steps of the approach are as follows:

- To construct Sequence diagrams for software components.
- To construct the CFG and ECFG to store sequence diagram and its newer version.
- Traversal of CFG to generate test suites and its comparison with ECFG for change identification hence, regression test suite generation.

3.1   Construction of Sequence Diagram for Software Components

Sequence Diagrams display the time sequence of the objects participating in the interaction. It consists of the vertical dimension (time) and horizontal dimension (different objects). It models system behavior by specifying how objects interact to

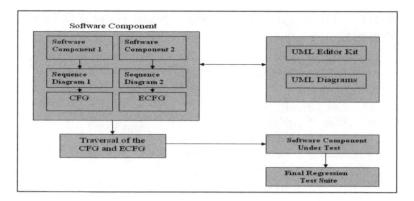

Fig. 1. Control flow graph based criteria for generation of regression test suite

complete the task. An interaction is expressed by messages between lifelines. A lifeline is participant in an interaction. A lifeline represents a class instance. A message can represent a message invocation, a reply message or creation and deletion of a class instance.

The UML sequence diagrams are designed with the help of Net Beans Tool working as a reverse engineering tool. Figure 2 shows the Basic Sequence Diagram of Enrolment procedure of students in a course.

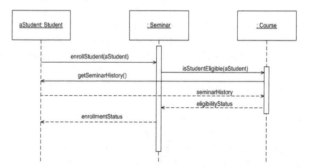

Fig. 2. Sequence Diagrams of Enrolment of students in a course

3.2 Construction of Control Flow Graph (CFG) to Store Sequence Diagram

Sequence Diagrams are represented by Control Flow Graph. A Control Flow Graph of Sequence diagram consists of states; each state is assumed to be vertex of graph. Transition from one state to another depending on the input/applicable method with signatures is given by an edge of graph. Sequence diagram states are represented by the vertices of graph and methods by the edges through which the object changes from one state to another. Method constraints i.e. precondition and post conditions are also mapped from class diagram to CFG. It makes it possible to maintain information about operational constraints. Any changes in these constraints can make a significant difference to the implementation of operation.

3.2.1 Structure of Edge and Vertices: Each Edge and Vertex of CFG Is Represented as a State

Struct VertexState
{
Char StateId [10];
int IsInitial, IsFinal;
VertexState *NextVertex;
EdgeState *EdgeRoot; }
 VertexState ()
 { IsInitial=IsFinal=0;
  EdgeRoot=NULL;}

The composition of vertex node is as follows

State Name	Next State	Method Pointer	Pre condition	Post condition

The composition of Edge node is as follows

Metho d Id	State From	State To	Sender	Receiver	Attributes	Next metho d

State name:It represents different states of sequence diagram
Next State: It represents the target state of transition
Method pointer: It contains information about method transitions between different states
Pre condition and post condition represents the conditions to be satisfied for transition between states
Method id: It represents the name of the method
State from: It represents the initial state of method transition
State To   : It represents the final state of transition
Sender   : It represents the object of component that transmits the message
Receiver   : It represents the destination object of message
Attributes : It represents the attributes of method Each attribute is abstract type and consists of data type and attribute name
Next Method : It contains id of another method transitioned from the same sender.

Algorithm: Generation of Storage Graph from Sequence Diagram.
Sequence Diagram states are represented by vertex node of Control flow graph.

1. Let INIT be the initial state. If it is null, it indicates absence of sequence diagram.
2. Initialize INIT = S; where S represents the starting state of sequence diagram.
3. Sequence Diagram states are added iteratively in the CFG by traversing it until last state is extracted.

Method: Adding methods/ transitions into States of the control flow graph.

A method represents the transition from state s1 to state s2 in graph. It has attributes like method id, parameters, sender, receiver, and return type of the method etc.

Let   s1 = source state of the transition

s2 = destination state of the transition

M = method of each transition

G1 = CFG of initial version of sequence diagram.

1. Let X = G1.findsourceState (s1)
2. If   X = NULL then

Display "invalid source state"

Else

t1 = X.lasttransition ( )   // last transition in the list of transitions from state X will be extracted

t1-> NextTransition = M   // new method will be added in the transition list.
3. Stop

3.3   Construction of ECFG for the Newer Version of Software Component

An extended control flow graph is constructed to store sequence diagram for the newer version of software component. The construction procedure is the same as for CFG. ECFG may have some new states or methods with changed operations in terms of parameter type, number of parameters passed, return type visibility, or scope of a method.

3.4   Traversal of Storage Graph to Generate Regression Test Suite

Both the CFG and ECFG are traversed to generate operational test scenarios. The changes between the two graphs can be:

a) Operational Changes in terms of pre/post condition, scope of method, return type visibility, or name of the method.
b) Attributes associated Changes in terms of   type or scope of a variable.
c) Conditional/guards Changes in terms of   change in the control flow structure.

Finally, test cases are generated by change identification from the test scenarios. Based on the changes identified, test cases from the CFG test suite are categorized as *obsolete, re-testable* and *reusable*. Obsolete test cases are discarded from the new test suite while re-testable and reusable test cases are incorporated in the new test suite. Depth first traversal technique is used for   traversing   CFG and ECFG.

3.4.1   Generation of Regression Test Suite

Regression test suite is generated by traversing the storage graphs for both versions of sequence diagrams for components. Initially Extended Control Flow Graph (ECFG) is traversed to search for all methods applicable to different states and comparing them with that in CFG. Finally based on the comparison, test cases are identified as reusable, re-testable and obsolete.

G1= Graph for first Sequence Diagram
G2= Graph for Modified Version of G1
T1= Traversed Stats of G1
T2= Traversed States of G2
E1= Edges of the Graph G1
E2= Edges of the Graph G2

Method: Traversal of ECFG
1. T2= G2[INIT]   // initial states of Graph G2
2. while T2 != NULL   // repeat for all states of graph while
their exist a transition
3.   E2= T2 [METHOD POINETR]
4.   While E2 ! = NULL   // repeat for all transition of
current states
5.   CALL Check Match(G1, E2)   // search the method E2 in G2
6.   E2=E2[NEXTMETHOD]   // jump to next transition method
        End while
7.   T2=T2 [NEXTSTATE]   // jump to next state
    End while
8. End

Method: Check Match(G1,E2)
1. T1 = G1[INIT]
2. While (T1!=NULL)   // starts searching for match for each state
and method within state
3.   E1 = T1[Method Pointer]
4.   While (E1!= NULL)
5   If(E1 == E2) then Display " Reusable Test case" // Attributes of E1 and E2
matche
6.   Else if (E1[Method Id] == E2 [Method Id]
7.   Match = True
                Endif
8.   E1= E1 [Next Method]
        End while
9. If Match = True then Display "Obsolete Test cases"
10. Else Display "Re-Testable Test cases"
11. End

4 Case Study

We have used Net Beans UML Tool to draw the UML design artifacts. We implement our approach using C++ programming language in Windows OS and run the
program in Intel machine with P-IV processor at 2.6 GHz. We make use of Library
system as the running example for this paper. Figure 3 and Figure 4 represent sequence diagrams of two different version of Library component. Tables 1 and 2
represent test case scenarios for these two components and Table 3 show regression
test suite.

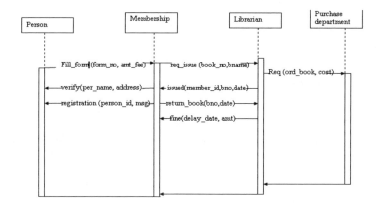

Fig. 3. Sequence Diagram for 1st version of Library System

Table 1. Test case scenarios of the first version of the Library Record System

Source	Destination	Method(parameter)	Sender	Receiver
S1	S2	Fill_form(int form_no,int AmtFee)	Person	Membership
S2	S3	Req_issue(int BookNo,int BName)	Membership	Librarian
S2	S1	verify(string PerName,string Address)	Person	Membership
S2	S1	Registration(int PerId,string msg)	Membership	Person
S2	S3	Return_Book(int BookNo,int Date)	Membership	Librarian
S3	S2	Issued(int MembId,int BookNo)	Librarian	Membership
S3	S2	Fine(int delay, float amount)	Librarian	Membership
S3	S4	Request(list order, float cost)	Librarian	PurDept

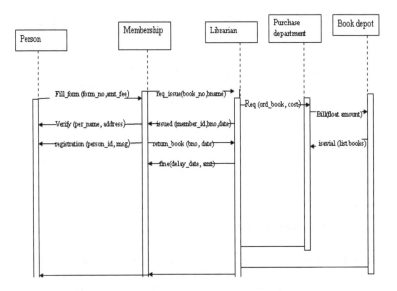

Fig. 4. Sequence diagram of the modified version for the Library Record system

Table 2. Test case scenarios of the modified version of the Library Record System

Source	Destination	Method(parameter)	Sender	Receiver
S1	S2	Fill_Form(int form_no,int AmtFee)	Person	Membership
S2	S3	Req_Issue(int BookNo,int BName)	Membership	Librarian
S2	S1	Verify(string PerName,string Address)	Person	Membership
S2	S1	Registration(int PerId,string msg)	Membership	Person
S2	S3	Return_Book(int BookNo,int Date)	Membership	Librarian
S3	S2	Issued(int MembId,int BookNo)	Librarian	Membership

Table 2. (*continued*)

S3	S2	Fine(int delay, float amount)	Librarian	Membership
S3	S4	Request(list order,float cost)	Librarian	PurDept
S4	S5	Order(list olderls)	PurDept	BookDepo
S4	S5	IsAvail(list books)	PurDept	BookDepo
S5	S4	Status()	BookDepo	PurDept
S5	S4	Bill(float amount)	BookDepo	PurDept

Table 3. Final regression test suite for Library record system

Test_ID	State	Method(parameter)	Sender	Receiver	Comments
01	S2	FillForm(int formno ,int AmtFee)	Person	Membership	Reusable
02	S3	Req_Issue(int BookNo, int BName)	Membership	Librarian	Reusable
03	S1	Verify(string PerName,string Address)	Membership	Person	Obsolete
04	S1	Registration(int PerId, string msg)	Membership	Person	Reusable
05	S3	Ret_Book(int BookNo,string Date)	Membership	Librarian	Obsolete
06	S2	Issued(int MembId, int BookNo)	Librarian	Membership	Reusable
07	S2	Fine(int delay, float	Librarian	Membership	Reusable

Table 3. *(continued)*

		amount)			
08	S4	Request(list order, float cost)	Librarian	Purdept	Reusable
09	S5	Order(list olderls)	Purdept	Bookdepo	New Added
10	S5	IsAvail(list books)	Purdept	Bookdepo	New Added
11	S4	status ()	Bookdepo	Purdept	New Added
12	S4	Bill(float amount)	Bookdepo	Purdept	New Added

5 Conclusions

We focused on automatic generation of regression test suite from sequence diagrams. This work on model-based regression testing seems to be very promising. As it can be systematically applied at the software design level in order to anticipate the testing phase as soon as possible, while reducing the cost of retesting modified systems.

However, many issues still need to be resolved. This approach considers only the impact of design changes on the test case re-selection process, while not considering other changes. Moreover, we have assumed to trace the components from their architecture. In future work, we do plan to generate the regression test suite by incorporating information from other UML diagrams like activity and use case models also. Also, we would like to improve tool support to store and traverse this graph to handle complex applications.

References

[1] Ali, A., Nadeem, A., Iqbal, M.Z.Z., Usman, M.: Regression Testing based on UML Design Models. In: Proceeding of the 10th International Conference on Information Technology. IEEE press, Los Alamitos (2007)
[2] Pilone, D., Pitman, N.: UML 2.0 in a Nutshell. O'Reilly, NY (2005)
[3] Baradhi, G., Mansour, N.: A Comparative Study of Five Regression Testing Algorithms. In: Australian Software Engineering Conference (ASWEC 1997). IEEE, Los Alamitos (1997)
[4] Leung, H.K.N., White, L.: Insights into Regression Testing. Journal of Software Maintenance 2(4), 209–222 (1990)
[5] Briand, L., Labiche, Y.: A UML Based Approach to System Testing. Journal of Software and Systems Modeling 1, 10–42 (2002)

[6] Xu, L., Dias, M., Richardson, D.: Generating Regression Tests via Model Checking. In: Proceedings of 28th Annual Interbnational Computer Software and Applications Conference (COMPSAC 2004), pp. 336–341 (2004)

[7] Harrold, M.J.: Testing: A Roadmap. In: Finkelstein, A. (ed.) ACM ICSE 2000, The future of Software Engineering, pp. 61–72 (2000)

[8] Object Management Group (OMG): UML 2.0 Superstructure Final Adopted specification. ver. 2.1.1 (2003)

[9] Pilskalns, O., Williams, D., Aracic, D.: Security Consistency in UML Designs. In: Proceeding of 30th International Conference on Computer Software and Applications (COMPSAC 2006), pp. 351–358. IEEE press, Los Alamitos (2006)

[10] Scheetz, Cui, J.: Towards Automated Support For Deriving Test Data From UML Sequence Diagrams. In: Proceedings of 6th International Conference on Unified Modeling Language, San Francisco, USA, October 20-24 (2003)

[11] Chakrabarti, S.K., Srikant, Y.N.: Specification Based Regression Testing Using Explicit State Space Enumeration. In: Proceedings of the International Conference on Software Engineering Advances (ICSEA 2006). IEEE Press, Los Alamitos (2006)

[12] http://www.netbeans.org/features/uml/

An Approach towards Software Quality Assessment

Praveen Ranjan Srivastava and Krishan Kumar

[1]PhD student, [2]ME student
Computer Science & Information System Group, Bits Pilani – 333031 India
{praveenrsrivastava, appkakk}@gmail.com

Abstract. Software engineer needs to determine the real purpose of the software, which is a prime point to keep in mind: The customer's needs come first, and they include particular levels of quality, not just functionality. Thus, the software engineer has a responsibility to elicit quality requirements that may not even be explicit at the outset and to discuss their importance and the difficulty of attaining them. All processes associated with software quality (e.g. building, checking, improving quality) will be designed with these in mind and carry costs based on the design. Therefore, it is important to have in mind some of the possible attributes of quality. We start by identifying the metrics and measurement approaches that can be used to assess the quality of software product. Most of them can be measured subjectively because there is no solid statistics regarding them. Here, in this paper we propose an approach to measure the software quality statistically.

Keywords: quality assurance, quality metrics, quality factors.

1 Introduction

Software quality is defined as "the degree to which a system, system component, or process meets specified requirements", or "the degree to which a system, system component, or process meets customer or user needs or expectations." [1]

We all are aware of the critical problems encountered in the development of software systems: the estimated costs for development and operation are overrun; the deliveries are delayed; and the systems, once delivered, do not perform adequately. Software, as such, continues to be a critical element in most large-scale systems because of its cost and the critical functions it performs. Many of the excessive costs and performance inadequacies can be attributed to the fact that "software systems possess many qualities or attributes that are just as critical to the user as the function they perform" [2]. For this reason, considerable emphasis in the research community has been directed at the software quality area.

Quality, in general, is the totality of features and characteristics of a product or a service that bears on its ability to satisfy the given needs [3]. According to Ljerka Beus Dukic and Jorgen Boegh [4], software quality evaluation is defined as "the systematic examination of the software capability to fulfill specified quality requirements." A software quality model is defined as, "a set of characteristics and sub

S.K. Prasad et al. (Eds.): ICISTM 2009, CCIS 31, pp. 150–160, 2009.

characteristics, as well as the relationships between them that provide the basis for specifying quality requirements and evaluating quality." Therefore, a software quality model is a good tool to evaluate the quality of a software product.

Producing high quality software is an ultimate goal of a software development organization. But, the challenge is that how can the quality be measured of software product more accurately? Most of the quality metrics and attributes can be measured subjectively and the measurement process almost depends on the experience and positive intention of the evaluator.

The quality of software product can be seen through three different dimensions [5]:

1. The Economic one, represented by managers' viewpoint
2. The Social one, represented by user's viewpoint
3. The Technical one, represented by developer's viewpoint

Regarding to software quality we must consider all three dimensions. The management interested in the overall quality and specifically with time and cost. The user is interested in satisfaction of his/her needs which he/she bought the product for. The developer is interested in the performance and functional requirements.

In this paper we present an approach to evaluate the software quality by viewing all these dimensions. This paper is organised as follows. In Section 2 we first present the background work. Then in section 3 discuss our approach. Then, in Section 4, we illustrate our work using a case study. Finally, Section 5 presents conclusions and future work.

2 Background

There is some confusion about the business value of quality even outside the software development context. On the one hand, there are those who believe that it is economical to maximize quality. This is the "quality is free" perspective espoused by Crosby, Juran and Gryna, and others [6].

Software quality can either refer to the software product quality or the software process quality [1]. Examples used for product quality are the CMM (Capability Maturity Model) and CMM-I (Capability Maturity Model-Integrated) models. The ISO standards have been set for both product as well as process quality. For example, ISO/IEC JTC1 ISO9126 (1-5), talks of software product quality while ISO/IEC JTC1, ISO-15504-5 talks about software process quality [1].

Measuring the quality of a software product is really a challenging task. One problem in making this determination is the absence of a widely accepted definition of software quality. This leads to confusion when trying to specify quality goals for software. A limited understanding of the relationships among the factors that comprise software quality is a further drawback to making quality specifications for software [2].

A second current problem in producing high quality software is that only at delivery and into operations and maintenance is one able to determine how good the software system is. At this time, modifications or enhancements are very expensive. The user is usually forced to accept systems that cannot perform the mission adequately because of funding, contractual, or schedule constraints [2].

Various research letters and articles have been focused on measuring software quality but they all have some assumptions or measure the quality subjectively. Justin M Beaver and Guy A. Schiavone [7] talk only about the skill and experience of development team. According to [7] the assumption is that a more educated and experienced team will produce higher quality in their software product and artifacts. He does not take customer's viewpoint into account. Dr. Mark C. Paulk [8] again focuses on the product, technology, and programmer ability factors that may affect software quality in addition to the process factors.

SQA provides visibility to management that the software products and processes in the project life cycle conform to the specified requirements and established plans [9]. Various quality metrics and factors are given by J. A. McCall [10] but there is no clear procedure to evaluate them.

In this paper we propose an approach to measure software quality by keeping all quality factors and attributes in mind.

3 Our Approach

For the purposes of this paper, we distribute all quality attributes in two sub categories that are; external and internal quality attributes [11] and then we add a new sub category regarding to manger (as shown in Figure 1). Then we let evaluate all these attributes to the participants of a software project (these are user, developer, and manager). According to our methodology, we design five different tables (one for user, two for developer, one for manager, and one for total quality value, discussed in later sections) which have various attributes for evaluation. We look at these attributes from three different points of views, these are:

User's view (Also called external Quality)

We designate Correctness, Usability, efficiency, Integrity, Adaptability, Accuracy, Robustness and Reliability to be quality attributes (as listed in Table 1) that can be best evaluated by the software user and hence also call them the external quality attributes. The procedure to quantify external quality is as follows

For each of the attributes included Table 1, we designate a scale ranging from 0 to 1 on which the value of every attribute needs to be marked. The scale is further divided into categories Very Low, Low, Normal, High, and Very High, having corresponding values 0.2, 0.4, 0.6, 0.8, 1.0.

Now the user according to his/her evaluation of every quality attribute (in table 1) gives them one of the 5 defined values. Sum of values for all the marked attributes is taken and the average of this sum is computed to get the absolute quality value. This average is taken by Equation 1 (arithmetic mean)

$$\text{Average Quality} = \text{Sum of Quality Factors Values/ Total Number of Quality Factors} \qquad (1)$$

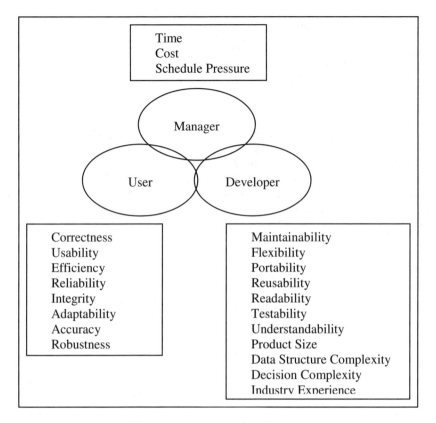

Fig. 1. Categorization of quality attributes

Note that the values are given here only for illustration purpose. We will propose a real life case study in next section.

Table 1. External quality attributes

Quality Factors	Very Low (0.2)	Low (0.4)	Normal (0.6)	High (0.8)	Very High (1.0)	Value
Correctness			X			0.6
Usability		X				0.4
Efficiency				X		0.8
Integrity		X				0.6
Adaptability				X		0.8
Accuracy					X	1.0
Robustness	X					0.2
Reliability		X				0.4
Total						4.8

So from the table we obtain:

Total External Quality Value (EQ) = 4.8
Total Number of External Quality Factors (NEQ) = 8
Hence Average External Quality Value = EQ / NEQ
$$= 0.6$$
Developer's View (Internal View)

Continuing with the approach explained earlier, we designate quality factors maintainability, flexibility, portability, reusability, readability, testability, understandability, and industry experience to be evaluated by the developer. And hence these factors are collectively known as internal software characteristics also. Since developer is not a single person, organizations have a team of programmers for a project though the values in developer's table would be filled by average view of developers.

The same methodology, used for evaluating external quality characteristics, is used here.

Table 2. Internal quality attributes

Quality Factors	Very Low (0.2)	Low (0.4)	Normal (0.6)	High (0.8)	Very High (1.0)	Value
Maintainability		X				0.4
Flexibility	X					0.2
Portability			X			0.6
Reusability				X		0.8
Readability					X	1.0
Testability				X		0.8
Understandability	X					0.2
Industry Experience	X					0.2
Total						4.2

The value of last attribute (Industry Experience) in Table 2 is decided on the basis of developer's experience in industry on different projects that is, as follows

0 - 1 years experience = Very Low (0.2)
1 - 3 years experience = Low (0.4)
3 - 5 years experience = Normal (0.6)
5 - 7 years experience = High (0.8)
7 years and above = Very High (1.0)

In addition to the above attributes listed in Table 2, we have picked some other attributes (product size, data structure complexity, and decision complexity) from Manish Agrawal, and Kaushal Chari [12] of developer's interest. They are listed in Table 3. They also fall under internal quality attributes. In the following attributes (listed in Table 3) low value indicates the high quality.

Table 3. Internal quality attributes

Quality Factors	Very Low (1)	Low (0.8)	Normal (0.6)	High (0.4)	Very High (0.2)	Value
Product Size	X					1.0
Data Structure Complexity				X		0.4
Decision Complexity		X				0.8
Total						2.2

So, from the table 2 and 3 we obtain:
Total Internal Quality Value from Table 2 and 3 (IQ) = 4.2 + 2.2 = 6.4
Total Number of Internal Quality Factors from Table 2 and 3 (IEQ) = 8 + 3 = 11
Hence Average Internal Quality Value = IQ / IEQ
$$= 0.58$$

Manager's View

Apart from the distinction of quality attributes into external and internal software characteristics we identify three more attributes – the software cost, time, and schedule pressure
of development. The last attribute of Table 4 we take from from Manish Agrawal, and
Kaushal Chari [12]. Clearly these factors are essential for the software development
organization to judge the performance of their effort. We designate the project manager
of the development team to evaluate these factors in the same manner defined earlier.

We assume that the manager would have an expected cost and time of completion
for the project and these parameters are then evaluated by comparison of there final
value with the estimate of the manager. Hence if final cost and time exceeds the estimated value they should be given high values. Very low and low values show that the
project has been completed before deadline and under expected cost, while high and
very high values show that project exceeds the deadline and expected cost. Normal
value shows that project has been completed in expected time and cost. This is
explained below as follows:

Table 4. Managerial quality attributes

Quality Factors	Very Low (1)	Low (0.8)	Normal (0.6)	High (0.4)	Very High (0.2)	Value
Time	X					1.0
Cost				X		0.4
Schedule Pressure			X			0.6
Total						2.0

Schedule pressure is defined as the relative compression of the development
schedule mandated by management compared to the initial estimate provided by the
development team based on project parameters [12]. We compute the value of schedule pressure by Equation 2.

Schedule Pressure = (Team estimated cycle-time -Management mandated
cycle-time) / Team estimated cycle-time   (2)

Cycle time is measured as the number of calendar days that elapse from the date the
baseline requirements are delivered until the date the software passes customer-
acceptance testing [12]. The maximum value of Schedule pressure can be 1 and the
minimum is 0 (but the zero value of schedule pressure is not possible in real life situations). Value of schedule pressure will get an entry in Table 4 as follows:

0.01– 0.2 = Very high (0.2)
0.21 – 0.4 = High (0.4)
0.41 – 0.6 = Normal (0.6)
0.61 – 0.8 = Low (0.8)
0.81 – 1.0 = Very Low (1.0)
Total Managerial Quality Value (MQ) = 2.0
Total Number of Managerial Quality Factors (MEQ) = 3
Hence Average Managerial Quality Value = MQ / MEQ
= 0.66

To obtain the value for total quality we assign weights to each of the internal, external
and managerial quality characteristics obtained previously, such that the sum of these
weights equals 1 and the weights are assigned on the basis of the relative importance
of these quality characteristics as explained below:(These weights should be decided
by the quality assurance team of the development organization) and the final value of
quality for software under consideration can be computed by following equation.

$$\text{Total Quality (in \%)} = \left(\sum_{i=1}^{N} (\text{Value}_i * \text{Weight}_i) \right) *100 \quad (3)$$

Table 5. Total quality value

Quality Characteristics	Value	Weight	Weighted Value (Value*Weight)
Internal Quality	0.58	0.3	0.171
External Quality	0.6	0.4	0.24
Managerial Quality	0.66	0.3	0.21
Total			0.621

Hence Total Quality = Weighted Sum of All Quality Characteristics * 100
= 61.2 %

4 Case Study

Small software is to be developed to count the frequency of given word. The program
takes a word *w* and a file name *f* as input and returns the number of occurrences of *w*
in the text contained in the file named *f*. An exception is raised if there is no file with
name *f*.

Now, from the above narration we find the following requirements or features.

Req-1: program should return the frequency of specified word.

Req-2: program should return an appropriate message if word string is not null but file does not exist.

Req-3: program should return an appropriate message if word string is not null but file is empty.

Req-4: program should return an appropriate message if file exists and containing data but word string is null.

Req-5: program should return an appropriate message when file does not exist and word string is null.

Req-6: program should return an appropriate message if word string is empty and file exists but does not contain data.

Now suppose the developer developed the following piece of pseudo-code for counting the number of words.

```
begin
    string w, f;
    input (w, f);
    if (!exists (f)) {
        raise exception;
        return -1; }
    if (length (w) == 0)
        return -2;
    if ( empty (f))
    return -3;
    return(getCount(w,f));
end
```

The code above contains eight distinct paths created by the presence of the three if statements. However, as each if statement could terminate the program, there are only six feasible paths. We can partition the input domain into six equivalence classes depending on which of the six paths is covered by a test case. These six equivalence classes are defined in the Table 6.

Table 6. Equivalence classes

Equivalence class	Word-string	File
E1	not-null	exists, nonempty
E2	not-null	does not exist
E3	not-null	exists, empty
E4	null	exists, nonempty
E5	null	does not exist
E6	null	exists, empty

We try to use the methodology developed above in evaluating the quality of one of our projects. The project developed is complete software that counts the frequency of given word. The software has been developed in JAVA programming language using object oriented approach with special importance given to its usability and interactive features. Following the above approach this software is evaluated as follows:

User's View

The software was evaluated from the users perspective by some of our colleagues not involved in the development process and the result were as follows

Table 7. External quality attributes

Quality Factors	Very Low (0.2)	Low (0.4)	Normal (0.6)	High (0.8)	Very High (1.0)	Value
Correctness				X		0.8
Usability				X		0.8
Efficiency			X			0.6
Integrity			X			0.6
Adaptability				X		0.8
Accuracy					X	1.0
Robustness		X				0.4
Reliability			X			0.6
Total						5.6

Hence External Quality Value = 5.6 / 8 = 0.7

As sufficient importance was given to usability and adaptability, this was seen in the software output as well and these were appropriately rated high by the users. Also the results calculated conformed highly with the manually calculated ones resulting in high values being awarded to correctness and reliability. However the software didn't provide erroneous messages for certain invalid inputs and thus was graded low on Robustness.

The overall external quality value came to an acceptable 0.7.

Developer's View

The software was evaluated on the internal factors by the developers as follows

Table 8. Internal quality attributes

Quality Factors	Very Low (0.2)	Low (0.4)	Normal (0.6)	High (0.8)	Very High (1.0)	Value
Maintainability			X			0.6
Flexibility		X				0.4
Portability	X					0.2
Reusability				X		0.8
Readability			X			0.6
Testability			X			0.6
Understandability				X		0.8
Industry Experience	X					0.2
Total						4.2

As the code was written in JAVA, using an object oriented approach its reusability and understand-ability was regarded high. Being of relatively small size it was not difficult to maintain or test and sufficient comments were provided to keep readability normal. However the software was developed only for windows operating system and hence its portability came out to be very low.

Table 9. Internal quality attributes

Quality Factors	Very Low (1)	Low (0.8)	Normal (0.6)	High (0.4)	Very High (0.2)	Value
Product Size	X					1.0
Data Structure Complexity				X		0.4
Decision Complexity		X				0.8
Total						2.2

Internal Quality Value (from Table 8 and 9) = 6.4 / 11
= .58
The overall Internal quality was evaluated to be .58

Managerial View
The software was evaluated on the cost and time front by the project manager as follows

Table 10. Managerial quality attributes

Quality factors	Very Low (1.0)	Low (0.8)	Normal (0.6)	High (0.4)	Very High (0.2)	Value
Time	X					1.0
Cost				X		0.4
Schedule Pressure			X			0.6
Total						2.0

Managerial Quality = 2.0 / 3
= .66
As the project was a small project, its cost was low and it was completed in the scheduled time resulting in managerial quality of .66

Overall Quality

Table 11. Total quality value

Quality Characteristics	Value	Weight	Weighted Value
Internal Quality	0.58	0.4	0.232
External Quality	0.7	0.4	0.28
Managerial Quality	0.66	0.2	0.132
Total			0.644

The project being not very large, lesser importance was placed on the managerial quality factors and more on internal and external quality giving the total quality of the project to be **64.4 %**.

5   Conclusion and Future Work

The proposed model starts from the need to obtain an integrated software quality measurement, which includes the three different views – user's view, developer's view, and manager's view. We start from a questionnaire table filled by users, developers and managerial staff; gathering the opinions of users, developers, and managers about their quality perceived about a certain product, it is possible to obtain a single numerical value useful for the software quality assurance team to take the decisions for future.

Our goal is to develop an approach through which we could be able to measure the quality of software statistically. Since there are very few attributes which can be measured or computed mathematically so we have to measure them on a scale and only then can we get to a numerical value.

Soft computing techniques can evaluate software quality assessment which may ultimately help software organization, end users in a greater extent. A number of extensions and applications of the model may be possible by using techniques like artificial neural networks, evolutionary computation and combination of neuro-fuzzy approach. In fact there is considerable need for applied research and strategy evaluation in this area using these techniques which is further scope of this research paper and will be dealt in the subsequent work.

References

1. Kalaimagal, S., Srinivasan, R.: A Retrospective on Software Component Quality Models. SIGSOFT Software Engineering Notes 33(5) (November 2008)
2. Cavano, J.P., Mccall, J.A.: A Framework for the Measurement of Software Quality
3. Sharma, A., Kumar, R., Grover, P.S.: Estimation of Quality for Software Components – an Empirical Approach. SIGSOFT Software Engineering 33(5) (November 2008)
4. Dukic, L.B., Boegh, J.: COTS Software Quality Evaluation. In: Erdogmus, H., Weng, T. (eds.) ICCBSS 2003. LNCS, vol. 2580, pp. 72–80. Springer, Heidelberg (2003)
5. Buglione, L., Abran, A.: A Quality Factor for Software
6. Slaughter, S.A., Harter, D.E., Krishnan, M.S.: Evaluating the Cost of Software Quality
7. Beaver, J.M., Schiavone, G.A.: The Effects of Development Team Skill on Software Product Quality. ACM SIGSOFT Software Engineering Notes 31(3) (2006)
8. Paulk, M.C.: Factors affecting Personal Software Quality
9. SPAWAR Systems center San Deigo.: Software Management for Executives Guide Book, PR-SPTO-03-v1.8a (December 1, 2003)
10. Gaffney, J.E.: Metrics in Software Quality Assurance
11. Fritz Henglien, DIKU.: An Introduction to Software Quality
12. Agrawal, M., Chari, K.: Software Effort, Quality, and Cycle Time: A Study of CMM Level 5 Projects. IEEE Transactions on Software Engineering 33(3) (2007)

A Study of the Antecedents and Consequences of Members' Helping Behaviors in Online Community

Kuo-Ming Chu

No. 840, Chengcing Rd., Niaosong Township,
Kaohsiung County, 833, Taiwan, R.O.C.
chu@csu.edu.tw

Abstract. Despite the growing popularity of online communities, there are a major gap between practitioners and academicians as to how to share information and knowledge among members of these groups. However, none of the previous studies have integrated these variables into a more comprehensive framework. Thus more validations are required the aim of this paper is to develop a theoretical model that enables us to examine the antecedents and consequences effects of members' helping behavior in online communities. The moderating effects of the sense of community on the relationships between members' helping behaviors on information sharing and knowledge contribution are also evaluated. A complete model is developed for empirical testing. Using Yahoo's members as the samples of this study, the empirical results suggested that online communities members' helping behavior represents a large pool of product know-how. They seem to be a promising source of innovation capabilities for new product development.

Keywords: online community, helping behavior, social capital, sense of community, information sharing, knowledge contribution.

1 Introduction

There are currently at least 10 million individual users of over 50,000 online communities on the internet. Porter (2004) and Wind & Mahajan (2002) all indicated that more than 84 percent of internet users have contacted or participated in an online community, and this growth in membership is expected to continue. Despite the growing popularity of online communities, there is a clear divergence of opinions on the concept, role and implications of online communities. Horrigan (2001) and Smith (2002) indicated that such forums have a high level of message traffic: over 150 million messages were posted in Usenet groups in 2000. The forums, though, with the highest level of message traffic were online communities. These messages are often a key resource for consumers facing even the smallest of buying decisions.

Raban & Rafaeli(2007) identified online communities as self-organizing systems of informal learning, that allow members to interact and learn together by engaging in joint activities and discussions, helping each other, and sharing information. Through these interactions, they build relationships and form a community around the domain. The members devote themselves to contributing and sharing information. A part of

S.K. Prasad et al. (Eds.): ICISTM 2009, CCIS 31, pp. 161–172, 2009.

the sharing involves offering and receiving emotional support in the online community. Within the world of online communities, there are numerous discussion themes posted on bulletin boards and developed with the interactive assistance amongst the members. In this way they make it easier for knowledge and information to be shared well. Online communities are the subjects of this research, and the researchers intended to investigate how members of these communities help each other in a way that gains momentum to enhance the process of knowledge sharing and information exchange.

Recently many researchers (Nahapiet & Ghoshal,1998; Lesse & Everest,2001; Fetterman,2002) have investigated the roles of social capital in diverse communities around the world. Social capital, in this context, refers to the invisible bonds that connect people into smaller and larger social groups and allow people to work together cooperatively, for the good of the group rather than the benefit of the individual. In this short study, we present the concept of social capital as a tool to pinpoint the major resources of online communities. The occurrence of helping behavior observed in these online communities raises several questions. This paper attempts to explore social capital, the nature of online community, helping behavior, interaction and browsing behavior differences among different members.

2 Helping Behavior in Online Communities

By online community we refer to communities in which information and communication technologies, particularly the Internet, facilitate the interaction of members, who share a specific objective (Blanchard, 2004). Taylor et al.(1997) indicated that helping behavior refers to an act performed voluntarily to help someone else when there is no expectation of receiving a reward of any form. However the literature examining helping behavior in online groups is relatively sparse. The studies of online helping in internet groups by Blanchard & Markus (2004), Subramani & Peddibholta (2004), Bulter et al. (2002) and Wasko & Faraj (2000) suggested that helping behavior, generalized reciprocity and community interest created by the ongoing interaction of the members of these online groups are important motivations for participation. Responses to posts reflect the willingness of members to help or to share their opinions with the original poster.

According to Wasko & Faraj (2000), Zhang & Storck (2001), Bulter et al. (2002) and Blanchard & Markus (2004) interaction in online groups comprises posts by individuals and subsequent responses to these posts by others. Rodon et al.(2005) also suggested that there are two dimensions that reflect the nature of helping behavior in online communities: information sharing and knowledge contribution.

2.1 Information Sharing

Helping behaviors in online communities are conventionally viewed as comprising of information sharing in response to requests for assistance. Raban & Rafaeli(2007), Burnett & Buerkle (2004), Subramani & Peddibholta (2004) and Andrews (2002) considered that helping behavior is only provided in response to posts requesting assistance in online communities. That is, information sharing occurs when questions pertain to issues that have already been considered and problems for which solutions exist.

Wang & Fesenmaier (2004) indicated that information interaction on the internet exists in cooperative forms and shows the strong willingness of members to share their own experience and information even if the parties do not know each other. Information sharing, as defined by Davenport & Prusak (1988), refers to members providing the information required for problem solving by other organization members. Blanchard & Markus (2004) and Hendriks (1999) indicated that one of the distinguishing characteristics of a successful online community are members' helping behavior and emotional attachment to the community and other individuals. Wang & Fesenmaier (2004) emphasized further that information sharing and its integrating mechanisms are an important foundation of community management. Raymond (2003), Kollock (1999) and Rheingold (1993) described generous online behavior where people are willing to give out free help and information, which encourages reciprocity between members.

Thus, improving the active involvement of community members is quite critical. Only through continuous on-line interaction by organization members can new information be created, and thereby make the online space more attractive through the sharing of increasingly abundant resources.

2.2 Knowledge Contribution

Rodon *et al.*(2005) defined that knowledge contribution is the capability to interpret data and information through a process of giving meaning to the data and an attitude aimed at wanting to do so. New information and knowledge are thus created, and tasks can be executed. The capability and the attitude of knowledge contribution are of course the result of available sources of information, experience, skills, culture, character, personality, feelings, etc. Others including, Clemmensen (2005), Wasko & Faraj (2000), Cross (2000) and Subramani & Peddibholta (2004) indicate that knowledge creation by providing a response that is *more than an answer* is relevant in the context of interpersonal helping behavior. Subramani & Peddibholta (2004) further explained that responses to posts involving issues that have no 'answers' are reflected in posts that have occurred in response to questions (i.e., acts of knowledge creation and exchange). That is, members may choose to comment on one of the responses rather than respond to the original message. According Nonaka (1990), the behaviors reflected in such responses reflect knowledge sharing and lead to the dissemination of knowledge.

2.3 The Influences of Social Capital on Helping Behaviors

Mathwick & Klebba (2003) and Iwan *et al.* (2004) indicated that social capital facilitates information exchange, knowledge sharing, and knowledge construction through continuous interaction, built on trust and maintained through shared understanding. Among studies of social capital in online communities are those by Subramani &

Peddibholta(2004), Wang & Fesenmaier(2004) and Blanchard & Horan(2004). They all indicated that social capital fosters the togetherness of community members and may encourage collaboration. According to Cook (2005), Hopkins *et al.* (2004), Ruuskanen (2004) and Lesser (2000), social capital is about trust, networks, belief, norms of reciprocity and exchange of information. In this short paper we draw on the concept of social capital as an agent that can facilitate the complex interaction of elements which contribute to the functioning of these communities in cyberspace, and explore some implications for them.

Thus, knowledge sharing is a type of outward behavior performed by knowledge owners through online communities and an inward behavior performed by people in search of knowledge. Also, through information media, knowledge can be transmitted through space or time to explain or mutually interact. The results can be easily seen by others, so that other members can commonly own the knowledge to form a context of helping others in the online community. We summarize these arguments in the following hypothesis:

H_1 : The levels of (a) network, (b) norms, (c) belief, and (d) trust relationship between members has a positive impact on the helping behaviors in the online community.

2.4 The Influences of the Nature of Online Community on Helping Behaviors

Prior research on biology, prosocial behavior and group dynamics suggests that a variety of features of online community are likely to influence members' helping behavior. Subramani & Peddibholta (2004) integrated the results of previous studies and pointed out that four factors of online community can influence helping behaviors: size of community, diversity of members, ancillary resources and role of members.

According to Hall *et al.*(2004), Galston (1999) and Batson (1998), group size is an important factor influencing the nature of helping behaviors in online communities. Butler *et al.*(2002) and Dholakia *et al.*(2004) suggested that the relationship of group size to outcomes is complex. Among others Bandura(1977), Hutchins (1990) and Pickering & King (1992) indicated that according to theories of social learning, the hitch-riding phenomenon and resource enhancement, a larger community will have a greater intensity of helping behaviors. Thus, to sum up these research results, we know that the size of online communities have both positive and negative influences on their members' behaviors to help others. Based on the above relationship, H_2 is posed as follows:

H_{2a} : The size of community has both positive and negative influences on helping behavior in the online community.

Subramani & Peddibhotla (2004) further suggested that the diversity of members contributes significantly to group outcomes. Katzenback & Smith (1993), Mohram (1995), Pelled et al. (1999) and Franke & Shah (2001) further pointed out that through interaction individual's encounter a variety of viewpoints other than their own and based on such a variety of viewpoints members in the online community learn, develop and refine their own perspectives in this way. However, according to Daft &

Lengel(1986), a highly diversified community members may cause negative influences on the interaction among members. The diversified professional knowledge among members may cause a positive effect for community behaviors tendency to help others, but the diversified range of opinions will negate this benefit. Based on the above discussion, the following hypothesis is proposed.

H_{2b} : *The diversity of members have both positive and negative impact on helping behavior in the online community.*

Kearsley (2000) and Hann et al. (2000) pointed out that communities can promote better knowledge sharing through the use of computer databases and network technology. In general, most online communities maintain repositories of ancillary information to supplement the information and knowledge sharing online. The most common tools are forums, message boards, chat rooms, mailing lists, instant messagings, guest books and group calendars. The forum is the most common tool. Forums are created by anticipating answers to questions frequently raised by members in the online community and are repositories of basic specific information stored in the proprietary format. But past studies pointed people invest substantial resources in creating or obtaining information, they were often willing to share it without response. In fact, some attribute humanity itself to the act of sharing information(Raban and Rafaeli, 2007; Dunbar, 1996). Based on the above discussion, the following hypothesis is proposed.

H_{2c} : *The degree of ancillary resources has a significant effect on helping behavior in the online community.*

Baym (1999) and Finholt & Sproull (1990) supposed that the internet world actually features numerous onlookers who always silently browse the information posted by other members. In the current jargon these silent members are known as *lurkers*. They typically have little or no involvement in discussion. Such members are more formally called *peripheral members* by Zhang & Storck (2001) and Wenger. In online communities, peripheral members are less visible in that if they do not participate; only they themselves know of their existence. Thus, when we analyze the community behaviors related to helping others, we have taken the role of members into consideration. According to the research findings of Zhang & Storck (2001), these marginal members are frequently scout for resources among communities in two groups: those who are intermittent users and those who are continuously interacting. In addition to their acquiring information and knowledge from online communities, these members will also perform mutually-beneficial behaviors within communities. However, the behaviors of helping others by these people are typically the alternation for marginal members influenced by core members or central members. Based on the above discussion, the following hypothesis is proposed.

H_{2d} : *The role of members (peripheral or central) has a significant effect on the helping behavior in the online community.*

2.5 The Sense of Community

According to the research of Uzzi *et al.*(2003) and Rodon *et al.*(2005), helping behaviors involve the notion of willingness to contribute. Hall *et al.* (2004) pointed out that a stronger social infrastructure might enhance knowledge creation and knowledge sharing capabilities. The sense of online community, as it described above, is different because it develops only in virtual settlements (Blanchard *et al.* 2004), which have particular characteristics, one of which is anonymity (Koh *et al.* 2003). Rheingold (1993) further asserts that members within the same community tend to have a sense of community that can enable members to easily exchange information. Therefore it is argue that in situations where members do not know each other, a sense of community will help in ones decision to give information. Therefore, it is suggested that a sense of community has a moderatoring effect on the willingness to exhibit helping behaviors. Therefore the following hypothesis is proposed.

> H_3 : *The influences of social capital and online community characteristics on members' helping behaviors are moderated by the sense of community.*

3 Research Conceptual Framework

The purpose of this study is to develop a theoretical model (Figure 1) that enables the researcher to examine the antecedents and moderating effects of the sense of community on members' helping behavior in online communities. Specifically, the four factors of social capital (i.e. network, norm, belief, and trust) and other factors of the nature of online community (i.e. the size of online community, diversity of online community members, ancillary resources, and the role of online community members) will have impacts on members' helping behaviors (i.e. information sharing and knowledge contribution). In addition the sense of community will have moderating effects on the influences of social capital and online community characteristics on members' helping behaviors.

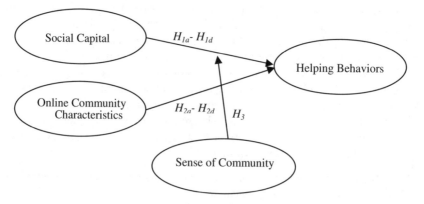

Fig. 1. The Conceptual Model of This Study

4 Empirical Results

4.1 Data Collection and Sample

The sample is 425 participants including nine online communities in Taiwan, including Kimo, CPB, Sony music etc. who were contacted and asked to participate in the study. Data were collected between August and December 2007 via the web for Internet users using a standardized questionnaire. Excluding those surveys that were undeliverable and those who believed that it was inappropriate to respond, the overall effective response rate was 84% (355 of 425). Our sample consists mainly of all 355 survey respondents, most are females (59%). 35% of respondents age from 20 to 24. The group in the age range from 25 to 29 accounts for 34%. 46% hold a bachelor degree, and second largest group is the high school graduates. With regard to the primary occupation of respondents, most are students (32%), and the second largest group is from the service industry.

4.2 Operationalization and Measurement

All indicators of constructs were measured using five-point multi-item scales. Multi-item measures were developed based on Cronbach's a and item-to-total correlations exceeding appropriate levels (Cronbach's α >0.60; item-to-total correlation > 0.30). Convergent validity was checked through exploratory factor analyses. The results are shown in Table 1.

Table 1. Results of the Confirmatory Factor Analysis

Construct	Item-to-total correlation	Cronbach's α	Construct reliability
Social Capital		0.8757	0.8761
Network(11)	0.8548		
Norm(10)	0.7936		
Belief(13)	0.8927		
Trust(10)	0.8557		
Online Community Characteristics		0.9013	0.8996
Size of online community(7)	0.8866		
Diversity of Members(10)	0.9060		
Ancillary Resource(13)	0.8584		
The Role of Members(9)	0.8333		
Helping Behaviors		0.8795	0.8747
Information Sharing(7)	0.8572		
Knowledge Contribution(12)	0.9050		
Sense of Community		0.8435	0.8769
Sense of Community(8)	0.8343		

The measurement model was tested for validity and reliability following the procedure suggested by Anderson and Gerbing (1988) and Homburg (1998). Regarding detail fit criteria (Table 1), very few measures fall short of desired thresholds, which is regarded as acceptable.

The study uses Gerbing and Anderson's (1988) two-stage approach, starting with the measurement phase, followed by the structural model estimation phase. Before analyzing the entire model, the descriptive statistics as well as correlation analysis for all the research variables are presented in Table 2.

Table 2. Descriptive Statistics and Correlation Matrix Analysis for all Research Variables

Latent Variables	Mean	S.D.	The Correlation Coefficients of Latent Variables										
			X_1	X_2	X_3	X_4	X_5	X_6	X_7	X_8	Y_1	Y_2	Y_3
X_1	3.598	0.535	1.000										
X_2	3.606	0.488	.353**	1.000									
X_3	3.542	0.536	.612**	.283**	1.000								
X_4	3.229	0.553	.432**	.289**	.548**	1.000							
X_5	3.549	0.652	.557**	.170**	.601**	.456**	1.000						
X_6	3.602	0.636	.586**	.191**	.579**	.425**	.660**	1.000					
X_7	3.562	0.475	.549**	.312**	.646**	.415**	.502**	.667**	1.000				
X_8	3.590	0.563	.405**	.237**	.392**	.332**	.413**	.449**	.567**	1.000			
Y_1	3.510	0.606	.458**	.253**	.608**	.427**	.498**	.470**	.579**	.475**	1.000		
Y_2	3.533	0.577	.471**	.276**	.627**	.484**	.532**	.525**	.619**	.519**	.894**	1.000	
Y_3	3.504	0.572	.295**	.279**	.297**	.323**	.381**	.307**	.295**	.347**	.567**	.579**	1.000

X_1:Network X_2:Norm X_3:Belief X_4:Trust
X_5:Size of O.C. X_6:Diversity of Members X_7:Ancillary Resource X_8:Role of Member
Y_1:Information Sharing Y_2:Knowledge Contribution Y_3:Sense of Community

With regard to the validity, all the variances reached a notable level and thus, each dimension really shows an acceptable level of validity.

4.3 Data Analysis

Data were analyzed using AMOS 6.0. The polychoric correlation matrix of the 11 first-order constructs was entered into an Unweighted Least Squares analysis (x^2 =72.745; df=44; P=0.083; GFI=0.94; AGFI=0.95; CFI=0.93; RMR=0.027; RMSEA=0.017). The results show good fit with the model. The Fornell and Larcker(1981) criterion for discriminant validity was met in all cases for the measurement and the structural model. Thus, an adequate level of fit in both the measurement model and the structural model can be assumed. Figure 2 shows the test results regarding the structural model, including the structural equation coefficients, the t values, and the explained variance of endogenous constructs. All tests of hypotheses are significant, so the theoretical model may be accepted as consistent with the data.

The indices as shown in Figure 2 further indicate that the elements of social capital (network, norms, beliefs and trust) and the nature of the online community (size, diversity, ancillary resources and roles of peripheral and central members) are all positively related to the helping behaviors of OC members. According to these results, the hypotheses testing are as following:

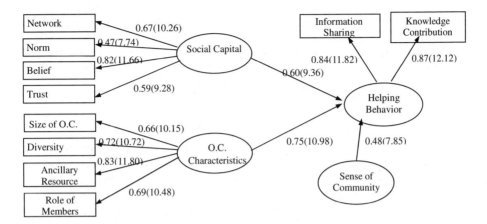

Fig. 2. Empirical results

First, it is suggested that H_{1a} to H_{1d} are supported, indicating that the four dimensions of social capital positively affect members' helping behavior. The results for the path correlations reached 0.67, 0.47, 0.82, 0.59, as relates to: beliefs, network, trust and norms. Second, Hypothesis H_{2a} to H_{2d} are supported thereby indicating that the four dimensions of social capital positively affect members' helping behavior. The path correlations reach 0.66, 0.72, 0.83, 0.69, relating, respectively to, auxiliary resources, diversities, the interaction between periphery and core members and the number of members. Third, the results of the relationship between the dimensions of members' helping behavior in the online community indicated these dimensions positively affect members' helping behavior, with correlations reaching 0.84 and 0.87. Finally, Hypothesis H_3 is supported, indicating the sense of community has a positive moderating effect on the impact of helping behaviors, with correlation value of 0.48. Thus, it can be argued that helping behaviors involve the notion of willingness, and therefore, the concept of the sense of community plays a preponderant role in it.

5 Discussion and Managerial Implications

The aim of this paper is to develop a theoretical model that enables us to examine the antecedents and consequences of members' helping behavior in the online community. The phenomenon of helping behavior among members may become a major source and channel for information in the decision making process for the purchase of products. Therefore, a major finding derived from the empirical application is that community members are capable and willing to contribute to virtual co-development.

Social capital is not only an important resource in a physical community; it is also a major factor in constructing an online community. Positive gains are derived from mutually beneficial coordination among community members in achieving common goals and solving problems. However, from the perspective of the four dimensions of social capital (given here in order of intensity as: belief, network, trust and norms), there are significant differences when compared with Grootaert & Bastelaert's (2002) findings. In their study they asserted that norms were the most significant variable in a physical community. From this we may infer that to an online community member, having the same goals and prospects during the process of interacting and establishing good interpersonal relationships with others is far more useful than having to operate with norms that are imposed.

The inherent size, diversity, auxiliary resources and the interaction of both periphery and core members within the network of communities will act as a positive influence to improve helping behaviors. This result contrasts with previous research results proposed by Butler et al. (2002). Thus, the online community members will put greater emphasis on whether the community is equipped with assistance from computer databases and if there is network technology available for members to enjoy a convenient, interactive environment, which can effectively facilitate information sharing and knowledge contribution.

The mutual cooperation behaviors of online community members can have a positive influence on information sharing and knowledge contribution. These results are the same as those of Clemmensen (2005). Thus, we now know that cooperation will have a positive influence on the interaction among online communities no matter what the relevant answers and solutions or on-line discussion or debate are. A wider discussion within the network environment can positively effect knowledge creation and exchange.

6 Further Research Directions

The research results can form the basis for some suggestion for practice by respectively focusing on "The Owners of Online Communities and Websites" and "The Firms". We hope that both the owners of online communities and websites and firms can refer to our proposal when they are creating their marketing strategies. If the leaders of online communities are vested with sufficient capability, knowledge and willingness, they will be able to promote the activities of information exchange and knowledge sharing among network members. In this manner they should be able to attract the enthusiastic views and visits by website on-lookers. Thus, the owners of online communities should dream up ways to promote the sharing of experience and opinions with interactions among group leaders, information sharing participants and people posting queries so that the communication depth and frequency among members can be enhanced and the loyalty (coherence) of members will be also intensified.

Finally, the exchange of opinions across various network roles is featured by promoting the consciousness of consumers. The external information resource controlled by non-marketing people possibly has a diffusing or inhibiting influence on advertising and personnel promotion. Thus, the business administrators must make use of this new marketing channel.

References

1. Batson, C.D.: Altruism and Prosocial Behavior. The handbook of social psychology, 4th edn., pp. 282–316. McGraw-Hill, New York (1998)
2. Blanchard, A.L., Markus, M.L.: The Experienced "Sense" of a Virtual Community: Characteristics and Processes. The Data Base for Advanced Information Systems 35(1) (2004)
3. Burnett, G., Buerkle, H.: Information Exchange in Virtual Communities: a Comparative Study. Journal of Computer Mediated Communication 9(2) (2004) (retrieved October 1, 2004), http://www.ascusc.org/jcmc/vol9/issue2/burnett.html
4. Butler, B., Sproull, L., Kiesler, S., Kraut, R.: Community Effort in Online Groups: Who does the Work and Why? In: Weisband, S., Atwater, L. (eds.) Leadership at a Distance (2004) (accessed January 2),
   http://opensource.mit.edu/papers/butler.pdf
5. Cook, K.S.: Networks, Norms, and Trust: The Social Psychology of Social Capital. Social Psychology Quarterly 68(1), 4–14 (2005)
6. Cross, R., Parker, A., Prusak, L., Borgatti, S.P.: Knowing What We Know: Supporting Knowledge Creation and Sharing in Social Networks. Organizational Dynamics 30(2), 100–120 (2001)
7. Davenport, T.H., Prusak, L.: Working Knowledge: How Organizations Manage What They Know. Harvard Business School Press (1998)
8. Dholakia, U.M., Bagozzi, R., Pearo, L.K.: A Social Influence Model of Consumer Participation in Network- and Small-Group-Based Virtual Communities. International Journal of Research in Marketing 21(3), 241–263 (2004)
9. Franke, N., Shah, S.: How Community Matters for User Innovation: the "Open Source" of Sports Innovation, Sloan Working Paper # 4164 (2001)
10. Galston, W.A.: How does the Internet affect community? In: Kamarck, E.C., Nye Jr., J.S. (eds.) Democracy.com? Governance in a networked world, pp. 45–69. Hollis Publishing, Hollis (1999)
11. Hall, H., Graham, D.: Creation and recreation: motivating collaboration to generate knowledge capital in online communities. International Journal of Information Management (24), 235–246 (2004)
12. Hann, D., Glowacki-Dudka, M., Conceicao-Runlee, S.: 147 Practical Tips for Teaching Online Groups: Essentials of Web-Based Education. Atwood Publishing, Madison (2000)
13. Harris, R.: Evaluating Internet Research Sources (1997),
    http://www.sccu.edu/faculty/R-Harris/evalu8it.htm
14. Hopkins, L., Thomas, J., Meredyth, D., Ewing, S.: Social Capital and Community Building through an Electronic Network. Australian Journal of Social Issues 39(4), 369–379 (2004)
15. Horrigan, J.B.: Online Communities: Networks That Nurture Long-Distance Relationships and Local Ties, Pew Internet and American Life Project (2001),
    http://pewinternet.org
16. Hutchins, E.: The Technology of Team Navigation. In: Galegher, J.R., Kraut, R.E., Egido, L. (eds.) Intellectual Teamwork: Social and Technological Foundations of Co-operative (1990)
17. von Iwan, W., Katja, R., Thorsten, T.: The Creation of Social and Intellectual Capital in Virtual Communities of Practice. In: The Fifth European Conference on Organizational Knowledge, Learning and Capabilities, Conference Paper in Proceedings (2004)
18. Jarvenpaa, S.L., Staples, S.: Exploring Perceptions of Organizational Ownership of Information and Expertise. Journal of Management Information Systems, 151–183 (2001)

19. Jonassen, D., Peck, K., Wilson, B.: Learning with Technology - A Constructivist Perspective. Prentice Hall, Inc., Upper Saddle River (1999)
20. Kearsley, G.: Online education: Learning and teaching in cyberspace. Wadsworth/Thomson Learning, Belmont (2000)
21. Koh, J., Kim, Y.: Sense of Virtual Community: A Conceptual Framework and Empirical Validation. International Journal of Electronic Commerce 8(2), 75–93 (2003)
22. Kollock, P.: The Economies of Online Cooperation: Gifts and Public Goods in Cyberspace, Communities in Cyberspace. In: Smith, M.A., Kollock, P. (eds.). Routledge, New York (1999)
23. Lesser, E., Everest, K.: Using Communities of Practice to Manage Intellectual Capital. Ivey Business Journal, 37–41 (2001)
24. Mathwick, C., Klebba, J.: The Nature and Value of Virtual Community Participation. In: Proceedings of the Summer AMA Educators Conference (2003)
25. Nonaka, I.: A Dynamic Theory of Organization Knowledge Creation. Organization Science 5(1), 14–37 (1994)
26. Pelled, L.H., Eisenhardt, K.M., Xin, K.R.: Exploring the Black Box: An Analysis of Work Group Diversity, Conflict, and Performance. Administrative Science Quarterly 44(1), 1–28 (1999)
27. Raban, D.R., Rafaeli, S.: Investigating Ownership and the Willingness to Share Information Online. Computers in Human Behavior 23, 2367–2382 (2007)
28. Raymond E. S.: The Hacker Milieu as Gift Culture (2003),
    http://futurepositive.synearth.net/stories/storyReader$223
29. Rodon, J., Cataldo, C., Christiaanse, E.: Antecedents of Knowledge Sharing and Effectiveness of Online Professional Communities: A Conceptual Model, accepted in the EURAM 2005, TUM Business School, Munich, Germany, May 4-7 (2005)
30. Ruuskanen, P.: Social Capital and Innovations in Small and Medium Sized Enterprises. In: DRUID Summer Conference, Elsinore, Denmark, Theme F. (2004)
31. Smith, M.A.: Tools for Navigating Large Social Cyberspaces. Communications of the ACM 45(4), 51–55 (2002)
32. Subramani, M.R., Peddibhotla, N.: Determinants of Helping Behaviors in Online Groups: A Conceptual Model. In: Academy of Management Conference, New Orleans, LA (2004)
33. Tiwana, A., Bush, A.: A Social Exchange Architecture for Distributed Web Communities. Journal of Knowledge Management 5(3), 242–248 (2001)
34. Uzzi, B., Lancaster, R.: Relational Embeddedness and Learning: The Case of Bank Loan Managers and Their Clients. Management Science 49(4), 383–399 (2003)
35. Wang, Y., Fesenmaier, D.R.: Modeling Participation in an Online Travel Community. Journal of Travel Research 42, 261–270 (2004)
36. Wasko, M.M., Faraj, S.: It is What One Does': Why people participate and help others in electronic communities of practice. Journal of Strategic Information Systems 9, 155–173 (2000)
37. Wenger, E.C., Snyder, W.M.: Communities of Practice: The Organizational Frontier. Harvard Business Review 78(1), 139–145 (2000)
38. Wind, Y., Mahajan, V.: Convergence Marketing. Journal of Interactive Marketing 16(2), 64–79 (2002)
39. Yli-Renko, H., Autio, E., Sapienza, H.J.: Social Capital, Knowledge Acquisition and Knowledge Exploitation in Young Technology-Based Firms. Strategic Management Journal 22, 587–613 (2001)
40. Zhang, W., Storck, J.: Peripheral Members in Online Communities. In: Proceedings of the Americas Conference on Information Systems, Boston, MA (2001)

Mobile Forensics: An Introduction from Indian Law Enforcement Perspective

Rizwan Ahmed and Rajiv V. Dharaskar

P.G. Department of Computer Science and Engineering, G.H. Raisoni College of
Engineering, Hingna Road, Nagpur- 440016 (MS) India
rizwanmailbox@gmail.com

Abstract. Mobile phone proliferation in our societies is on the increase. Advances in semiconductor technologies related to mobile phones and the increase of computing power of mobile phones led to an increase of functionality of mobile phones while keeping the size of such devices small enough to fit in a pocket. This led mobile phones to become portable data carriers. This in turn increased the potential for data stored on mobile phone handsets to be used as evidence in civil or criminal cases. This paper examines the nature of some of the newer pieces of information that can become potential evidence on mobile phones. It also highlights some of the weaknesses of mobile forensic toolkits and procedures. Finally, the paper shows the need for more in depth examination of mobile phone evidence.

Keywords: Mobile Forensics, Mobile Phone evidence, Mobile Forensics toolkits, Digital device Forensics.

1 Introduction

Mobile phone proliferation is on the increase with the worldwide cellular subscriber base reaching 4 billion by the year end of 2008 [1]. In India alone, there are 364 million mobile phone subscribers [33] which are growing at a rapid pace. While mobile phones outsell personal computers three to one, mobile phone forensics still lags behind computer forensics [2]. Data acquired from mobile phones continues to be used as evidence in criminal, civil and even high profile cases [3]. However, validated frameworks and techniques to acquire mobile phone data are virtually non-existent.

1.1 The Need for Mobile Phone Handset Forensics

The following section of the paper will discuss the need for mobile forensics by highlighting the following:

- Use of mobile phones to store and transmit personal and corporate information
- Use of mobile phones in online transactions
- Law enforcement, criminals and mobile phone devices

S.K. Prasad et al. (Eds.): ICISTM 2009, CCIS 31, pp. 173–184, 2009.

1.2   Use of Mobile Phones to Store and Transmit Personal and Corporate Information

Mobile phones applications are being developed in a rapid pace. Word processors, spreadsheets, and database-based applications have already been ported to mobile phone devices [4]. The mobile phone's ability to store, view and print electronic documents transformed these devices into mobile offices. The ability to send and receive Short Message Service (SMS) messages also transformed mobiles into a message centre. In India alone, nearly 1.5 billion (1,492,400,769) text messages (SMS) were sent per week between January and May, 2008, the Mobile Data Association (MDA) said. In India, more than 10 million (10,734,555) pictures and video messaging (MMS) were sent per week — a year on year growth of 30 percent [1]. Furthermore, technologies such as "push e-mail" and always-on connections added convenience and powerful communications capabilities to mobile devices. Push e-mail provided users with instant email notification and download capability, where when a new e-mail arrives; it is instantly and actively transferred by the mail server to the email client, in this case, the mobile phone. This in turn made the mobile phone an email storage and transfer tool. Roughly 40% of all Internet users worldwide currently have mobile Internet access. The number of mobile Internet users will reach 546 million in 2008, nearly twice as many as in 2006, and is forecast to surpass 1.5 billion worldwide in 2012 [5].

1.3   Use of Mobile Phones in Online Transactions

Wireless Application Protocol (WAP) enabled the use of mobile phones in online transactions. Technologies such as digital wallets (E-Wallet) added convenience to online transactions using a mobile phone. Further enhancements in connectivity and security of mobile devices and networks enabled mobile phones to be used securely to conduct transactions such as stock trading, online shopping, mobile banking [5] and hotel reservations and check-in [6] and flight reservations and confirmation [7]. As part of development of mobile systems [6, 7], the novel idea of mobile forensics came to our mind and so this research paper is a milestone to achieve the same objectives.

1.4   Law Enforcement, Criminals and Mobile Phone Devices

The gap between law enforcement and organised crime is still considerable when it comes to the utilisation of mobile phone technologies. Mobile phones and pagers were used in the early 1980s by criminal organisations as a tool to evade capture as well as a means to facilitate everyday operations. Ironically, while it took decades to convince legitimate businesses that mobile connectivity can improve their operations, just about every person involved at any level of crime already knew in the early 1980s that mobile phones can provide a substantial return on investment [8].

On the other hand, law enforcement and digital forensics still lag behind when it comes to dealing with digital evidence obtained from mobile devices. This is partly due to some of the following reasons [9]:

- The mobility aspect of the device requires specialized interfaces, storage media and hardware

- The file system residing in volatile memory versus stand alone hard disk drives
- Hibernation behaviour in which processes are suspended when the device is powered off or idle but at the same time, remaining active
- The diverse variety of embedded operating systems in use today
- The short product cycles for new devices and their respective operating systems

These differences make it important to distinguish between mobile phone and computer forensics.

2 Computer Forensics v/s Mobile Phone Handset Forensics

The following sections of the paper compare computer and mobile forensics in the following aspects:

- Reproducibility of evidence in the case of dead forensic analysis
- Connectivity options and their impact on dead and live forensic analysis
- Operating Systems (OS) and File Systems (FS)
- Hardware
- Forensic Tools and Toolkits Available

2.1 Reproducibility of Evidence in the Case of Dead Forensic Analysis

Digital investigations can involve dead and/or live analysis techniques. In dead forensic analysis, the target device is powered off and an image of the entire hard disk is made. A one-way-hash function is then used to compute a value for both, the entire contents of the original hard disk and the forensically acquired image of the entire hard disk. If the two values match, it means that the image acquired represents a bitwise copy of the entire hard disk. After that, the acquired image is analysed in a lab using a trusted OS and sound forensic applications. This process is referred to as offline forensic analysis or offline forensic inspection.

One of the key differences between traditional computer forensics and mobile phone forensics is the reproducibility of evidence in the case of dead forensic analysis. This is due to the nature of mobile phone devices being constantly active and updating information on their memory. One of the causes of that is the device clock on mobile phones which constantly changes and by doing so alters the data on the memory of that device. This causes the data on the mobile device to continuously change and therefore causing the forensic hash produced from it to generate a different value every time the function is run on the device's memory [9]. This means that it will be impossible to attain a bit-wise copy over the entire contents of a mobile phone's memory.

2.2 Connectivity Options and Their Impact on Dead and Live Forensic Analysis

Live forensic analysis in this context refers to online analysis verses offline analysis. Online analysis means that the system is not taken offline neither physically nor logically [10]. Connectivity options refer to the ways in which a system or device is connected to the outside world be it a wired or wireless connection. Even though built-in connectivity options for computers are limited when compared to the increasingly developing connectivity options on mobile phone devices, connectivity options are addressed in both live and dead computer forensics. On the other hand, live analysis is not even heard of yet when it comes to mobile phone handset forensics.

2.3 Operating Systems and File Systems

Computer forensic investigators are very familiar with computer operating systems and are comfortable working with computer file systems but they are still not as familiar with working with the wide range of mobile OS and FS varieties. One of the main issues facing mobile forensics is the availability of proprietary OS versions in the market. Some of these OS versions are developed by well known manufacturers such as Nokia and Samsung while some are developed by little known Chinese, Korean and other regional manufacturers. Mobile phone operating systems are generally closed source with the exception of Linux based mobile phones. This makes developing forensics tools and testing them an onus task. Moreover, mobile phone manufacturers, OS developers and even forensic tool developers are reluctant to release information about the inner workings of their codes as they regard their source code as a trade secret.

Another issue with mobile OS and FS when compared to computers is the states of operation. While computers can be clearly switched on or off, the same can not be said about some mobile phone devices. This is especially true for mobile phones stemming from a PDA heritage where the device remains active even when it is turned off. Therefore, back-to-back dead forensic acquisitions of the same device will generate different hash values each time it is acquired even though the device is turned off [11].

A key difference between computers and mobile phones is the data storage medium. Volatile memory is used to store user data in mobile phones while computers use non-volatile hard disk drives as a storage medium. In mobile phones, this means that if the mobile phone is disconnected from a power source and the internal battery is depleted, user data can be lost. On the contrary, with non-volatile drives, even if the power source is disconnected, user data is still saved on the hard disk surface and faces no risk of deletion due to the lack of a power source. From a forensics point of view, evidence on the mobile phone device can be lost if power is not maintained on it. This means that investigators must insure that the mobile device will have a power supply attached to it to make sure data on the device is maintained.

One of the drawbacks currently facing mobile OS and FS forensic development is the extremely short OS release cycles. Symbian, a well known developer of mobile phone operating systems is a prime example of the short life cycle of each of its OS releases. Symbian produces a major release every twelve months or less with minor releases coming in between those major releases [12]. This short release cycle makes timely development, testing and release of forensic tools and updates that deal with the newer OS releases difficult to achieve.

2.4 Hardware

Mobile phones are portable devices that are made for a specific function rather than computers which are made for a more general application. Therefore, mobile phone hardware architecture is built with mobility, extended battery life, simple functionality and light weightiness in mind. This makes the general characteristics of a mobile phone very different from a computer in the way it stores the OS, how its processor behaves and how it handles its internal and external memory.

The hardware architecture of a typical mobile phone usually consists of a microprocessor, main board, Read Only Memory (ROM), Random Access Memory (RAM), a radio module or antenna , a digital signal processor, a display unit, a microphone and speaker, an input interface device (i.e., keypad, keyboard, or touch screen) and a battery. The OS usually resides in ROM while RAM is generally used to store other data such as user data and general user modifiable settings. The ROM may be re-flashed and updated by the user of the phone by downloading a file from a web site and executing it on a personal computer that is connected to the phone device.

This general architecture does not apply to all models of mobile phones as mobile phones are very diverse in hardware architecture and OS varieties [13]. Some mobile devices might contain additional devices and modules such as a digital camera, Global Positioning device (GPS), wireless and network modules, and even a small hard disk. Manufacturers highly customize operating systems to suit their hardware devices and the feature sets they want to support on them [14]. This means that a certain version of an OS on a certain manufacturer's phone model does not mean that the same version of the same OS on a different manufacturer's hardware will be exactly the same. This is true also for on the same manufacturer's phones with different hardware architectures. Moreover, ROM updates are not only OS specific but are also hardware specific. Also, some phone providers add functionality and customization options to their ROMs which mean that the same version phone of a phone purchased from two different providers might not be exactly the same.

Proprietary hardware is another issue facing mobile phone forensics. Support for such devices is not available from mobile forensics tools. About 16% of mobile phones in the market today come from proprietary manufacturers and are not supported by forensic tools [15]. Moreover, some manufacturers produce mobile phones that have no interfaces that are accessible through a computer. This makes forensically acquiring those mobile phones harder to achieve if not impossible.

The wide array of connection socket and cable types for connecting a mobile phone to a computer makes identifying the right cable for the right phone model an onus task for the forensic investigator. Phone chargers also come in different shapes, sizes and socket types and make identifying the right charger for the right model a hard task for the investigator. Short product cycles also contribute to the difficulty in dealing with mobile phones forensically. Support for newer models by forensic tools is usually slow. The following section discusses in more detail some of the mobile forensic tools and their features and drawbacks when compared to computer based forensic tools.

2.5 Forensic Tools and Toolkits Available

Early mobile phones did not have the capacity to store large amounts of information so law enforcement officers did not need to access mobile phone handsets to get information on a suspect. The focus was more on phone records from the telecommunications companies. Nowadays, mobile phones have large storage capacity and a wide array of applications and connectivity options besides connectivity with the telecommunications provider. Mobile phone forensic tools and toolkits are still immature in dealing with these advances in mobile phone technology. Mobile forensic toolkits are developed by third party companies and the toolkits are not independently verified or tested for forensic soundness. The developers of the toolkits admit to using both, manufacturer supplied and self developed commands and access methods to gain data access to memory on mobile devices [16]. The tools often limit themselves to one or more phone manufacturer handsets with a limited number of devices supported. Some of the tools are also limited when it comes to connectivity options when it comes to acquisition of data from the handset. For example, some tools are limited to wired connections as opposed to Infrared (IrDA) and Bluetooth access to data on mobile devices. Moreover, while some toolkits provide acquisition capabilities, they do not provide examination or reporting facilities [17]. Moreover, direct access to data on the mobile phone is not achievable. Phone software and/or hardware must be used to acquire data from the mobile phone's memory as shown in Figure 1:

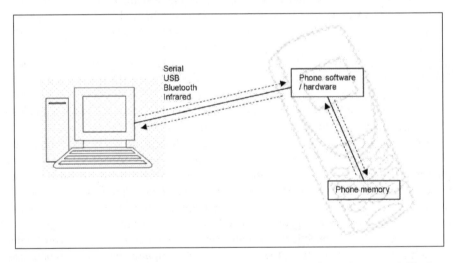

Fig. 1. Indirect Access to Data in Mobile Phone Memory via Software and Hardware Commands and Methods [16]

This inherent difference between computer forensics and mobile phone forensics effects how data acquired from mobile phones is perceived. To make this data trustable, independent evaluation of mobile forensic tools has to become an integral part of their development. The only currently available tools evaluation document for mobile phone forensics is published by the National Institute of Standards and Technology (NIST) in the United States [9]. The document evaluated eight mobile phone forensic toolkits. It covered a range of devices from basic to smart phones. It showed that none of forensic toolkits supported all the mobile phone devices covered in the document. The document however limited its scope to a set of scenarios with a definite set of prescribed activities that were used to gauge the capabilities of each of the eight toolkits evaluated. The document also tested the toolkits in one set of conditions which was a virtual machine installed on a windows machine. This insured toolkit segregation and ruled out the possibility of conflicts amongst the tools [13].

3   Mobile Phone Data as Evidence

This section of the paper will highlight some forensic definitions, principles and best practice guidelines and how they address mobile phone forensics issues. It will also discuss some of the forensic guides that cover mobile phone forensics and mention their shortcomings.

3.1   Definition of Digital Evidence

According to the Scientific Working Group on Digital Evidence (SWGDE), Digital Evidence [18] is "information of probative value that is stored or transmitted in binary form". Therefore, according to this definition, evidence is not only limited to that found on computers but may also extend to include evidence on digital devices such as telecommunication or electronic multimedia devices. Furthermore, digital evidence is not only limited to traditional computer crimes such as hacking and intrusion, but also extends to include every crime category in which digital evidence can be found [19]. However, the Australian Standards HB171 document titled "Guidelines for the Management of IT Evidence" refers to IT Evidence as: "any information, whether subject to human intervention or otherwise, that has been extracted from a computer. IT evidence must be in a human readable form or able to be interpreted by persons who are skilled in the representation of such information with the assistance of a computer program". This definition is lacking as it does not address evidence on digital devices other than a computer [19]. The latter definition shows that not all digital evidence definitions or procedures related to them are updated to address mobile phone evidence. Even the Information Technology Act 2000 (No. 21 of 2000) is not updated to include information about mobile phone evidence [30]. This fact again can be clearly highlighted in view of two big criminal cases [31, 32] in India which involved mobile phone evidence. The following section of the paper will cover some of these definitions and procedures and highlight their shortcomings.

3.2   Principles of Electronic Evidence

According to the United Kingdom's Association of Chief Police Officers (ACPO) Good Practice Guide for Computer based Electronic Evidence, Four principles are involved with Computer-Based Electronic Evidence. ACPO's guide regards computer based electronic evidence as no different from documentary evidence and as such is subject to the same rules and laws that apply to documentary evidence [20]. The ACPO guide also recognized that not all electronic evidence can fall into the scope of its guide and gave an example of mobile phone evidence as evidence that might not follow the guide. It also mentioned that not following the guide does not necessarily mean that the evidence collected is not considered as viable evidence.

However, Principle 1 of the ACPO guide can not be complied with when it comes to mobile phone forensics. This is because mobile phone storage is continually changing and that may happen automatically without interference from the mobile user [11]. Thus, the goal with mobile phone acquisition should be to affect the contents of the storage of the mobile as less as possible and adhere to the second and third principles that focus more on the competence of the specialist and the generation of a detailed audit trail [11]. In adhering with Principle 2, the specialist must be competent enough to understand both the internals of both hardware and software of the specific mobile device they are dealing with as well as have an expert knowledge of the tools they are using to acquire evidence from the device.

More than one tool is recommended to be used when acquiring evidence from mobile phone as some tools do not return error messages when they fail in a particular task [11]. When it comes to adhering with Principle 3, providing a thorough record of all processes used to obtain the evidence in a way that can be duplicated by an independent third party is essential in order for the evidence gathered to be admissible in court.

When it comes to the recovery of digital Evidence, "The Guidelines for Best Practice in the Forensic Examination of Digital Technology" publication by the International Organization on Computer Evidence (IOCE) considers the following as the General Principles Applying to the Recovery of Digital Evidence [21]:

- The general rules of evidence should be applied to all digital evidence.
- Upon seizing digital evidence, actions taken should not change that evidence.
- When it is necessary for a person to access original digital evidence that person should be suitably trained for the purpose.
- All activity relating to the seizure, access, storage or transfer of digital evidence must be fully documented, preserved and available for review.
- An individual is responsible for all actions taken with respect to digital evidence whilst the digital evidence is in their possession.

As with the ACPO principles, principle B can not be strictly applied to evidence recovered from Smartphone devices because of their dynamic nature. Furthermore, mobile phone acquisition tools that claim to be forensically sound do not directly access the phone's memory but rather use commands provided by the phone's software and/or hardware interfaces for memory access and thus rely on the forensic soundness of such software or hardware access methods [16]. Therefore, when using such tools, the ability to extract that information in a manner that will not significantly change the mobile phone's memory is not verifiable.

3.3 Mobile Phone Evidence Guides

There are a number of guides that briefly mention potential evidence on mobile phone devices. In this section, some of these guides will be highlighted and their shortcomings explained. The Best Practices for Seizing Electronic Evidence published by the

United States Secret Service (USSS) referred to mobile phones as "Wireless Tele-phones" under the "Other Electronic Storage Devices" heading [22]. The National Institute of Justice (NIJ), which is under the United States Department of Justice lists mobile phones under the heading of "Telephones" in their "Electronic Crime Scene Investigation: A guide for First Responders" publication [23]. Both of the guides do not provide sufficient details on how to forensically approach smart phones. This might be in part because these guides are outdated. Both guides however mention that mobile phones might have some potential evidence on them. The extent of the coverage is very limited and does not address smart phone storage capabilities and applications on them. The USSS document also lists a set of rules on whether to turn on or off the device [22]:

- If the device is "ON", do NOT turn it "OFF".
  - Turning it "OFF" could activate lockout feature.
  - Write down all information on display (photograph if possible).
  - Power down prior to transport (take any power supply cords present).
- If the device is "OFF", leave it "OFF".
  - Turning it on could alter evidence on device (same as computers).
  - Upon seizure get it to an expert as soon as possible or contact local service provider.
  - If an expert is unavailable, USE A DIFFERENT TELEPHONE and contact 1-800-LAWBUST (a 24 x 7 service provided by the cellular telephone industry).
  - Make every effort to locate any instruction manuals pertaining to the device.

On the other hand, the NIJ guide for first responders lists the following as potential evidence [23]: Appointment calendars/information., password, caller identification information, phone book, electronic serial number, text messages, e-mail, voice mail, memos, and web browsers. The guide however failed to mention that mobile devices could have external storage attached to them even though it mentioned that other equipment such as fax machines may contain such external storage devices. It did however emphasize that miscellaneous electronic items such as cellular phone cables and cloning equipment may contain information of evidentiary value.

Both guides fail to mention that mobile phones could have electronic documents, handwriting information, or location information on them. The guides also fail to mention that phone based applications such as Symbian, Mobile Linux and Windows Mobile applications could have evidential significances. Both, Symbian and Windows Mobile based phones were found to execute malicious code such as Trojans and vi-ruses especially ones transferred via Bluetooth technology [16, 24]. Non malicious applications on mobile phones could also be considered as evidence as they might be used to conduct illegal activities or can have log files or data that can be considered as evidence. Therefore all phone applications and data related to them should be consid-ered as potential evidence. This includes logs relating Bluetooth, Infrared (IrDA), Wi-Max and Wi-Fi communications and Internet related data such as instant messaging data and browser history data. Java applications should also be considered as evidence as many mobile phone operating systems support a version of Java [16].

When it comes to handling instructions for mobile phones, the United Kingdom's Association of Chief Police Officers (ACPO) Good Practice Guide for Computer based Electronic Evidence lists the following instructions [25]:

- Handling of mobile phones:
  - Any interaction with the handset on a mobile phone could result in loss of evidence and it is important not to interrogate the handset or SIM.
  - Before handling, decide if any other evidence is required from the phone (such as DNA/fingerprints/drugs/accelerants). If evidence in addition to electronic data is required, follow the general handling procedures for that evidence type laid out in the Scenes of Crime Handbook or contact the scenes of crime officer.
  - General advice is to switch the handset OFF due to the potential for loss of data if the battery fails or new network traffic overwrites call logs or recoverable deleted areas (e.g. SMS); there is also potential for sabotage. However, investigating officers (OIC) may require the phone to remain on for monitoring purposes while live enquiries continue. If this is the case, ensure the unit is kept charged and not tampered with. In all events, power down the unit prior to transport.

Note that the on/off rules here initially conflict with the USSS guide but both guides agree to turn off the device before transport. The ACPO guide contains flowcharts when it comes to seizure of electronic evidence and PDAs which may not be applied to mobile phone devices. The charts are included in the Appendix section as a reference only. An updated chart for examining mobile phones by NSLEC in the U.K. contains references to the appropriate action to be taken when seizing a mobile phone and whether it was turned on or off when it was seized [26]. The chart is in no way all-inclusive as it refers to only three types of evidence from mobile phones and they are SMS messages, voicemail and address book/call history details. The guidelines and procedures need to be continually updated to cater for future trends in mobile phones. Some of these trends are mentioned in the next section.

4   Concluding Remarks and Future Work

With increased connectivity options and higher storage capacities and processing power, abuse of mobile phones can become more main stream. Mobile phones outsell personal computers and with digital crime rates rising, the mobile phone may be the next avenue for abuse for digital crime. Mobile phones with their increased connectivity options may become a source of viruses that infect computers and spread on the internet. Virus writers typically look for operating systems that are widely used. This is because they want their attacks to have the most impact. When it comes to mobile phones and their operating systems, there seems to be certain operating systems that are dominating the market which makes them a prime candidate for attacks. According to recent studies, phone virus and malware infection rates are expected to increase with newer smart phones [28, 29].

Mobile phone technology is evolving at a rapid pace. Digital forensics relating to mobile devices seems to be at a stand still or evolving slowly. For mobile phone forensics to catch up with release cycles of mobile phones, more comprehensive and in depth framework for evaluating mobile forensic toolkits should be developed and data on appropriate tools and techniques for each type of phone should be made available a timely manner. The authors are developing an open source and generic mobile forensics framework called MFL3G to aid the forensic experts to carry out investigations on vide variety of devices with an ease [34, 35]. The Indian Laws should be modified to accommodate the digital evidence, specially the mobile phone based evidence in the Court of Laws with the perspectives as elaborated in the paper.

References

1. Doran, P.: MDA, 2008- the year of mobile customers (2008),
   http://www.themda.org/documents/PressReleases/General/
   _MDA_future_of_mobile_press_release_Nov07.pdf
2. Canalys. Smart mobile device shipments hit 118 million in 2007, up 53% on 2006 (2007),
   http://www.canalys.com/pr/2008/r2008021.htm
3. Aljazeera. Phone Dealers in al-Hariri Probe Net (2005),
   http://english.aljazeera.net/archive/2005/09/
   200841014558113928.html
4. Westtek. ClearVue Suite (2008), http://www.westtek.com/smartphone/
5. Manfrediz, A.: IDC Press Release. IDC Finds More of the World's Population Connecting to the Internet in New Ways and Embracing Web 2.0 Activities (2008),
   http://www.idc.com/getdoc.jsp?containerId=prUS21303808
6. FoneKey (2008), http://www.FoneKey.net,
   http://www.youtube.com/watch?v=qW8MdpZFKUY,
   http://www.youtube.com/watch?v=BqJiNvQ3xp8,
   http://www.youtube.com/watch?v=9eAKvCKanH0
7. Ducell (2008), http://www.DuCell.org
8. Mock, D.: Wireless Advances the Criminal Enterprise (2002),
   http://www.thefeaturearchives.com/topic/Technology/
   Wireless_Advances_the_Criminal_Enterprise.html
9. Ayers, R., Jansen, W., Cilleros, N., Daniellou, R.: Cell Phone Forensic Tools: An Overview and Analysis (2007),
   http://csrc.nist.gov/publications/nistir/nistir-7250.pdf
10. Carrier, B.D.: Risks of Live Digital Forensic Analysis. Communications of the ACM 49(2), 56–61 (2006),
    http://portal.acm.org/citation.cfm?id=1113034.1113069&coll=G
    UIDE&dl=GUIDE
11. Jansen, W., Ayers, R.: Guidelines on PDA Forensics (2004),
    http://csrc.nist.gov/publications/nistir/
    nistir-7100-PDAForensics.pdf
12. Symbian History (2008),
    http://www.symbian.com/about/overview/history/history.html
13. Jansen, W., Ayers, R.: Guidelines on Cell Phone Forensics (2006),
    http://csrc.nist.gov/publications/nistpubs/800-101/
    SP800-101.pdf
14. Zheng, P., Ni, L.M.: The Rise of the Smart Phone. IEEE Distributed Systems Online, 7(3), art. no. 0603-o3003 (2006)
15. Espiner, T.: Mobile Phone Forensics 'Hole' Reported (2006),
    http://news.zdnet.co.uk/hardware/0,1000000091,39277347,00.htm

16. McCarthy, P.: Forensic Analysis of Mobile Phones. Unpublished Bachelor of Computer and Information Science (Honours) Degree, University of South Australia, Adelaide (2005)
17. Jansen, W.: Mobile Device Forensic Software Tools. In: Techno Forensics 2005, Gaithersburg, MD, USA (2005)
18. SWGDE. SWGDE and SWGIT Digital & Multimedia Evidence Glossary (2006), http://www.swgde.org/documents/swgde2005/SWGDE%20and%20SWGIT%20Combined%20Master%20Glossary%20of%20Terms%20-July%2020.pdf
19. Ghosh, A.: Guidelines for the Management of IT Evidence (2004), http://unpan1.un.org/intradoc/groups/public/documents/APCITY/UNPAN016411.pdf
20. ACPO. Good Practice Guide for Computer based Electronic Evidence (2003), http://www.acpo.police.uk/asp/policies/Data/gpg_computer_based_evidence_v3.pdf
21. IOCE. Best Practice Guidelines for Examination of Digital Evidence (2002), http://www.ioce.org/2002/Guidelines%20for%20Best%20Practices%20in%20Examination%20of%20Digital%20Evid.pdf
22. USSS. Best Practices for Seizing Electronic Evidence (2006), http://www.ustreas.gov/usss/electronic_evidence.shtml
23. NIJ. Electronic Crime Scene Investigation: A Guide for First Responders (2001), http://www.ncjrs.gov/pdffiles1/nij/187736.pdf
24. Keizer, G.: First Mobile Phone Java Trojan on the Loose (2006), http://www.crn.com.au/story.aspx?CIID=35467&r=rstory
25. CCIPS. Searching and Seizing Computers and Related Electronic Evidence Issues (2002), http://www.usdoj.gov/criminal/cybercrime/searching.html
26. Mellars, B.: Forensic Examination of Mobile Phones. Digital Investigation: The International Journal of Digital Forensics & Incident Response 1(4), 266–272 (2004)
27. Becker, D.: Toshiba Reports Battery Breakthrough (2005), http://news.com.com/2061-10786_3-5649141.html?tag=nl
28. Long, M.: Airborne Viruses: Real Threat or Just Hype (2005), http://www.newsfactor.com/story.xhtml?story_id=12100002P4HM
29. McAfee Mobile Security Report (2008), http://www.mcafee.com/mobile
30. The Information Technology Act 2000, India (2000), http://www.legalserviceindia.com/cyber/itact.html
31. Yahoo News India. The Arushi Murder Case: CBI says it has found the evidence (2008), http://in.news.yahoo.com/32/20080731/1053/tnl-aarushi-case-cbi-says-it-has-found-e_1.html
32. Helplinelaw, Pramod Mahajan Murder Trial: SMS cannot be valid evidence, says defence (2007), http://news.helplinelaw.com/1207/echo12.php
33. Telecom Paper (2008), http://www.telecompaper.com/news/article.aspx?cid=647427
34. Ahmed, R., Dharaskar, R.V.: MFL3G: Mobile Forensics Library for digital analysis and reporting of mobile devices for collecting digital evidence. In: 2nd Asia Pacific Mobile Learning & Edutainment Conference 2008, PWTC Kuala Lumpur, Malaysia (2008)
35. Ahmed, R., Dharaskar, R.V.: MFL3G: An Open Source Mobile Forensics Library For Digital Analysis And Reporting Of Mobile Devices For Collecting Digital Evidence, An Overview From Windows Mobile OS Perspective. In: International Conference on Advanced Computing Technologies (ICACT 2008), GRIET, Hyderabad, India (2008)

A Study to Identify the Critical Success Factors for ERP Implementation in an Indian SME: A Case Based Approach

Parijat Upadhyay[1] and Pranab K. Dan[2]

[1] International School of Business and Media, EN-22, Sector V, Saltlake Electronic Complex, Kolkata, India
parijat.upadhyay@gmail.com
[2] West Bangal University of Technology, Kolkata, India
dan1pk@hotmail.com

Abstract. To achieve synergy across product lines, businesses are implementing a set of standard business applications and consistent data definitions across all business units. ERP packages are extremely useful in integrating a global company and provide a "common language" throughout the company. Companies are not only implementing a standardized application but is also moving to a common architecture and infrastructure. For many companies, a standardized software rollout is a good time to do some consolidation of their IT infrastructure across various locations. Companies are also finding that the ERP solutions help them get rid of their legacy systems, most of which may not be compliant with the modern day business requirements.

Keywords: Business process, vendor support, SME.

1 Introduction

Organizations of any magnitude have implemented or in the process of implementing Enterprise Resource Planning (ERP) systems to remain competitive in business.ERP are essentially commercial software packages that enable the integration of transaction-oriented data and business processes throughout an organization (Markus and Tanis, 2000). More and more organizations across the globe have chosen to build their IT infrastructure around this class of off-the-shelf applications and at the same time there has been and should be a greater appreciation for the challenges involved in implementing these complex technologies. Although ERP systems can bring competitive advantage to organizations, the high failure rate in implementing such systems is a major concern (Davenport, 1998). Enterprise Resource Planning (ERP) implementation in business these days can be considered to be the price for running a business.ERP systems are complex and expensive, and the decision to install an ERP system necessitates a choice of mechanisms for determining whether ERP is needed

S.K. Prasad et al. (Eds.): ICISTM 2009, CCIS 31, pp. 185–196, 2009.

and, once implemented, whether it is successful. Numerous cases can be cited where in organizations which have spent high amount on implementation have not been able to derive any benefit at all out the implementation.

A number of publications have highlighted the failures and the frustrations that enterprises go through in implementing ERP systems. Some notable failures are Dell, Fox-Meyer and Hershley. A Gartner group study was carried out in 1300 European and American companies and found that 32% of ERP projects were delivered late and thus unable to reap the true benefits of the implementation. Allied Waste Industries, Inc. decided to abort a $130 million system built around SAP R/3, while Waste Management, Inc., called off an installation with the same vendor after spending about $45 million of an expected $250 million on the project.

As cases of ERP failures have increased, many studies have been initiated to identify the factors that could lead to ERP implementation success. In this study we try to identify the factors affecting ERP implementation in an Indian small and medium scale enterprise (SME).

2 The Concept of ERP

ERP is a standard software package built on the framework based on best industry practices that provides integrated transaction processing and access to information that can span multiple organizational units and multiple business functions of an organisation. These functions may include financial and accounting, human resources, supply chain, customer services among others. The organization generally decides to implement a package for a business process which may be the most problematic for it. At the heart of the ERP system is a single central database. This database collects data from, and feeds data into, modular applications supporting virtually all of a company's business activities – across functions, across business units and across the world, as shown in Figure 1. When new information is entered in one place, related information is then automatically updated (Davenport, 1998).Most companies expect ERP to reduce their operating costs, increase process efficiency, improve customer responsiveness and provide integrated decision information. They also want to standardize processes and learn the best practices embedded in ERP systems to ensure quality and predictability in their global business interests by reducing cycle times from order to delivery (Ross, 1999). Hence, ERP implementation and business process reengineering (BPR) activities usually are closely connected.

ERP systems differ from in-house or custom development systems in the following ways:

(1) the user may have affect changes to business processes and procedures;

(2) the user may need customisations to match their business process; and

(3) the user usually becomes dependent on the ERP vendor for assistance and updates (Somer and Nelson, 2003; Wu and Wang, 2005).

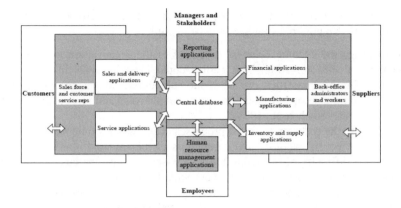

Fig. 1. Anatomy of an ERP (Source: Adapted from Devenport)

3   The ERP Implementation Context

ERP deployment in an organization can be viewed as one kind of outsourcing – some parts of the business depend on external organizations and the process of introducing the final system depends on at least three stakeholders: ERP package developers, ERP consultants and ERP-system users.

The ERP environment differs from those of traditional Information System (IS) management. (Mabert et al.,2003). Thus, any instrument developed for measuring the success of IS implementation may not necessarily apply to an ERP environment. An ERP satisfaction construct may need to consider ERP-specific factors, such as the quality of the project team, ultimate-user – project team interaction, ERP system product, and ultimate-user – ERP system interaction. Therefore, the factors of product information,cost, support and service, user knowledge and involvement can be a good starting point when considering suitable constructs.The competences required to manage properly the organizational change determined by an ERP system implementation is still a debated issue. A review of literature stressed the importance of change and project management competences as critical success factors for ERP implementation (Davenport, 2000;Mandal and Gunasekaran, 2003; Motwani et al., 2002), hereby indirectly raising the issue of small to medium-sized enterprise's (SME's) lack of organizational preparation. Such a situation is mainly caused by the low extent of formalization of people's roles and responsibilities that is expressed by with their continuous re-shuffle (Dutta and Evrard, 1999). This structural condition makes the identification of ERP implementation's main figures, such as the process owner and the key user (Davenport, 2000), extremely difficult to achieve. Beside this, SMEs generally suffer from a widespread lack of culture, as to the concept of business process: it is not by chance that the reinforcement of the concept of business process is often claimed among the critical success factors in ERP implementation (Beretta, 2002). In particular, the business process concept helps promoting co-operation and convergence of efforts among managers (i.e. managerial integration), versus the internal competition induced by the functionally-oriented organizational models which is typical of SMEs.

When the features of the software application do not correctly fit the business requirements two possible strategies can be identified:

(1) Change the existing business processes of the organisation to fit the software with minimal customization: This choice could mean changes in long-established ways of doing business (that often provide competitive advantage), and could shake up or even eliminate or introduce important people roles and responsibilities (Dewett and Jones, 2001; Koch et al., 1999).

(2) Modify the software to fit the processes: This choice would slow down the project, could affect the stability and correctness of the software application and could increase the difficulty of managing future releases, because the customizations could need to be torn apart and rewritten to work with the newer version.

Although ERP vendors are concentrating on the customization process needed to match the ERP system modules with the actual features of existing processes in a number of different industries, several studies show that configuring and implementing ERP systems is a complex and expensive task.

Several aspects related to this twofold approach towards ERP adoption and implementation become even more critical, for their known specificities, within SMEs Although the effective use of business information is a strategic goal for companies of any size, nowadays most of the ERP systems available on the market are too expensive for the financial capabilities of smaller companies.SMEs differ from large companies in important ways affecting their information-seeking practices (Lang et al., 1997). These differences include the:

- lack of (or substantially less sophisticated) information system management (Kagan et al., 1990);
- frequent concentration of information-gathering responsibilities into one or two individuals, rather than the specialization of scanning activities among top executives;
- lower levels of resource available for information-gathering; and
- quantity and quality of available environmental information.

Chan (1999) asserts that many SMEs either do not have sufficient resources or are not willing to commit a huge fraction of their resources due to the long implementation times and high fees associated with ERP implementation. The resource scarcity, the lack of strategic planning of information systems (IS) , the limited expertise in IT and also the opportunity to adopt a process-oriented view of the business are among the factors that strongly influence, either positively or negatively, ERP adoption by SMEs. Thus it is necessary to find out alternative solutions providing the ERP capabilities at an affordable price, including implementation costs.

Lured by seer magnitude of the SME sector in any country and particularly in a country like India, many ERP vendors have been moving their attention toward SMEs by offering simplified and cheaper solutions from both the organizational and technological points of view,pre-configured systems based on best-practices at a fraction of the cost originally required and promising implementation times of 60 days. In spite of such promises, there is not a general agreement on the effectiveness of such systems. As a result, the current ERP systems adoption rate in SMEs is still low.

Such a scenario raises some serious questions: Are SMEs informational needs different from that of large companies? Can certain factors be identified that can be considered critical in context to SME's on account of their peculiarities?

Through a detailed literature review, a set of indicators are identified as variables which could influence the ERP adoption process. These indicators have been tested on the field through an empirical study.

4 Identification of Critical Success Factors (CSFs)

The different CSFs were identified after doing an extensive literature study and from the questionnaire survey of organizations that have gone through the implementation process. The content validity of these constructs was tentatively established by extensive review with top executives and other stakeholders. Some items were removed from the construct if their removal results in an increase in the reliability estimates, however care was taken to ensure the validity of the measures is not threatened by the removal of a key conceptual element.

4.1 Methodology

Based on the literature review, certain factors were identified which were highlighted in the majority of paper that the researcher could access. Some of the journals that were accessed by the researchers have been mentioned in the appendix. In addition to, the preceding journals, the following databases were searched: Emerald, Proquest Computing, Proquest European Business, Web of Science and J Stor.

The CSFs identified can be listed as follows:

- The support of higher authorities throughout the entire implementation phase.
- The implementation required minimum customization and BPR.
- Cost of the package
- The competency level of implementation team
- The proper project management.
- Effective communication amongst the team members during the process.
- The clearly defined goals and objectives.
- Acceptance of changes brought about by the implementation of the new system.
- The users were subjected to adequate training and education regarding usage of the new system.
- Adequate user involvement and participation during the implementation phase.
- Adequate vendor support throughout the implementation phase and after.
- The participation and contribution of external consultant.
- Compatibility of the new system with the existing technical infrastructure.
- The composition and leadership of the project team.
- Proper selection of package.
- Scope of implementation.

- There are adequate features of scalability in the package chosen and implemented.
- User friendliness of the implementation process..

A questionnaire was designed comprising of the factors that were identified and mentioned above. Data was collected from people who have been associated with the implementation process. Some of them were not very much aware of the certain specific concepts and hence session was conducted to explain them the meaning of the questions contained in the questionnaire. Data were finally analyzed with SPSS v13.Reliability study was done to establish the reliability of the data set and subsequently factor analysis was done and the results are analysed.

5 About the SME

Amik Group has been a major player in Poultry Feed and day-old chicks market in Eastern India for many years. In order to become a strong food processing company by capitalizing on its core competence in poultry sector, the company thought that the best way to grow and gain is to integrate the operations by going into broiler bird farming and chicken processing. In view of this, the Group aims to achieve its future core growth by providing more value added chicken products through higher forward and backward integration of its various poultry related activities. This way quality product can be made available to end consumer at affordable cost along with good profit margin to the Group. To support this, they will also create a wide spread distribution and marketing network. As mentioned above, to offer more products to customers and to capitalize on the proposed marketing and distribution network, fruit processing products and layer eggs will also be incorporated in our business model. The Group will also expand its geographical presence in India and will also start exports.

The major focus of their management is into these functions: Process Customer Orders, Handle Inquires and Complaints, Collect Customer Data, Manage Product Pricing, Perform Invoicing/Billing, Manage Customer Credit, Manage Receivables, Plan Capacity, Plan Production Requirements, Schedule Production, Order Materials and Supplies, Manage Inbound Logistics, Manage raw materials and Return Materials to Vendor

Problem Faced by the existing System:

- Higer Delivery time.
- Prospect –sale Cycle is higher.
- Price justification problem are occurring.
- Expense incurred due to telephone calls. Because in each day each marketing man confirm their customer requirement over phone.
- Expense incurred due to annual maintenance contact.

The company has identified the following key driver for the ERP project:

- Reduction in delivery times.
- Reduction in post- sale cycle.

- Reviewing of pricing strategy on a real time basis to remain competitive in business.
- Amik Group evolved as an organization over the last few years and has not restructured. Through the ERP implementation, it is envisaged that they will reengineer the organization to current management paradigms.
- Information flow and roles of individuals need to be restructured.

6 Survey Methodology

Invitations to participate in the survey requested responses from managers and users of client organizations who had worked on a completed ERP implementation, and who were familiar with the selection of the consultant and the management of the implementation project. At the beginning of the survey, the respondents indicated if their ERP system was operational or they are currently in the implementation phase.

6.1 Data Collection and Data Analysis

Overall, 72 responses to the survey were obtained for analysis. The responses were also received from those users who are working in branch offices of the organization. The pilot surveys were excluded from analysis.

Chi-square tests were performed to match the distributions of some of the demographic variables in this study to the same ones from other similar studies (Somers and Nelson, 2003; Hong and Kim, 2002; Aladwani, 2002; Sena, 2001). Annual revenue of the firm (from two different studies), firm size (as total employee numbers), and respondents' functional area were the three demographic variables compared to the same corresponding ones from the similar papers. The results did not show differences in the distributions of the variables in both the samples. An unpaired t-test was also performed with the current employment duration variable, and no difference between the means was found at a statistically significant level.

Early respondents were statistically compared to late ones (considered as surrogates for non-respondents) to detect any systematic differences between the two groups and thus test for response bias.

A p-value on a median test comparing the two groups on the demographic variable exceeded .05. Finally, none of the t-tests on the items were significant at the p <.05 level. Thus, all of the comparisons failed to indicate any significant difference between the two groups, thus suggesting the absence of non-response bias.

With the objective of establishing the reliability of the data collected and that of the study, Cronbach's alpha was calculated. The value was found to be 0.842 and since any value greater than .80 is considered reliable, this data set can also be considered to be reliable.

Factor analysis is performed on the explanatory variables with the primary objective of determining minimum number of factors that will account for maximum variance in the data. Here, principle components method with varimax rotation has been applied. The table named communalities explains the proportion of each variable's variance that can be explained by the factors.

Factor analysis is performed on the explanatory variables with the primary objective of determining minimum number of factors that will account for maximum variance in the data. Here, principle components method with varimax rotation has been applied. Fig.2 named communalities explains the proportion of each variable's variance that can be explained by the factors.

Factors	Extraction
Project Champion Higher Authority	.832
Project Sponsor Higher Authority	.903
Customization & BPR	.855
Cost of Package	*.941*
Proper Project mgmt	.910
Effective Communication among team members during project	.862
Clearly defined goals &objectives	*.968*
Experience of similar implementation	*.919*
Training required for the users	*.952*
Participation and involvement of users	.854
Vendor support for implementation	*.918*
Significant participation by external consultant	*.918*
System compatibility with the existing technical infrastructure	.907
Satisfactory composition & leadership of Project team	.897
Package is appropriate	.741
Appropriate scope of implementation	.742
Adequate scalability features are present in the package	.646
Package is user friendly	.883
Support from Higher Authority	.822

Fig. 2. Communalities (Extraction Method: Principal Component Analysis)

Fig.3 Total Variance gives the eigen values in decreasing order of magnitude as we go from factor 1 to 19. Eigen value for a factor indicates the total variance attributed to that factor. The total variance for all the factors is 19, which is equal to the number of variables. We will consider factors having eigen value greater than one. Thus we have extracted 6 factors and the first two factors are the major factors out of six factors contributing 47% of the total variance. This implies that ERP implementation is necessary for six major reasons.

Total Variance Explained

Component	Initial Eigenvalues			Extraction Sums of Squared Loadings			Rotation Sums of Squared Loadings		
	Total	% of Variance	Cumulative %	Total	% of Variance	Cumulative %	Total	% of Variance	Cumulative %
1	5.124	26.968	26.968	5.124	26.968	26.968	3.999	21.048	21.048
2	3.851	20.269	47.237	3.851	20.269	47.237	3.402	17.905	38.954
3	2.632	13.855	61.091	2.632	13.855	61.091	3.081	16.217	55.170
4	2.299	12.098	73.189	2.299	12.098	73.189	2.514	13.233	68.403
5	1.422	7.482	80.671	1.422	7.482	80.671	1.820	9.580	77.983
6	1.143	6.016	86.688	1.143	6.016	86.688	1.654	8.704	86.688
7	.773	4.067	90.754						
8	.562	2.959	93.713						
9	.494	2.600	96.313						
10	.279	1.468	97.782						
11	.235	1.235	99.017						
12	.117	.615	99.631						
13	.048	.252	99.883						
14	.022	.117	100.000						
15	2.78E-016	1.46E-015	100.000						
16	-3.2E-017	-1.70E-016	100.000						
17	-9.9E-017	-5.19E-016	100.000						
18	-2.2E-016	-1.15E-015	100.000						
19	-3.8E-016	-1.99E-015	100.000						

Extraction Method: Principal Component Analysis.

Fig. 3. Total variance

The component matrix contained the unrotated factor loadings, i.e., the correlations between the variable and the factor. A coefficient with a large absolute value implies that the factor and the variable are highly correlated. But this unrotated factor matrix hardly results in factors that can be interpreted as the factors are correlated with many variables.

The study shows that, factor 1 is highly correlated with at least 12 variables. If several factors have high correlations with the same variable, then it is difficult to interpret the factors. The rotation keeps the communality and total variance unchanged, only the variance explained by the individual factors are redistributed.

In this study, orthogonal rotation has been applied which is also called varimax procedure. The rotation minimizes the number variables having high correlations on a factor, thus increasing the interpretability of the factors. Since the rotation is orthogonal, the resulting factors will be uncorrelated. The study revels that factor 1 has high correlations with the variables like clearly defined goals and objectives, similar experience of successful ERP implementation, system compatibility with the existing technical infrastructure, adequate scalability features are present in the package, and package is user friendly.

Similarly, factor 2 has high coefficients for cost, participation and involvement of users, vendor support for implementation; package is appropriate, appropriate scope of implementation. Factor 3 is constituted of higher authority has been project champion; the project has been managed properly, satisfactory composition and leadership of the project team and a supportive higher authority. Component four consists of higher authority has been project sponsor, customization and BPR, effective communication among the team members during the project. Factors 5 and 6 consist of single variables. Factor 5 consists of training required for the users and factor 6 is significant participation and contribution by the external consultant.

7 Recommendation and Limitations

Small to Mid-size Enterprises face a dichotomy of business drivers impacting ERP strategies. A recent research report published by Aberdeen(2008) finds SME's tends to focus on growth strategies (49%) and customer service (48%), sometimes balancing between the two, but more often focusing exclusively on one or the other. As companies grow in size and improve performance, they are 37% more likely to have invested in an ERP system that will grow with them. As mid-size companies grow, they must learn to operate in a distributed environment and often experience a proliferation of ERP and other enterprise applications. The study, which has been discussed in this paper would attempt to give some insight to those organizations in the manufacturing and distribution community and essentially SMB in nature in Indian context. Such SME's desire to reduce costs, improve accuracy of inventory and schedules, and develop customer responsiveness through successful ERP implementations. Given the fact that collection of data for such studies from any small and medium scale enterprise is a difficult proposition because of the inherent peculiarities of such organizations still, the sample size considered for the study may seem inadequate. Efforts are being made to identify more such types of organizations and include those data into this study.

References

1. Aladwani, A.: An integrated performance model of information systems projects. Journal of Management Information Systems 19(1) (2002)
2. Bailey, J., Pearson, S.: Development of a tool for measuring and analyzing computer user satisfaction. Management Science 25(5), 530–545 (1983)
3. Bingi, P., Sharma, M.K., Godla, J.K.: Critical issues affecting an ERP implementation. Information Systems Management 16(3), 7–14 (1999)
4. Boudreau, M.C., Robey, D.: Organizational transition to enterprise resource planning systems: theoretical choices for process research. In: Proceedings of the 20th International Conference on Information Systems, Charlotte, CA, pp. 291–299 (1999)

5. Cerveny, R., Scott, L.W.: A survey of MRP implementation. Production & Inventory Management Journal 30(3), 31–34 (1989)
6. Chan, A., Scot, D., Lam, E.: Framework of success criteria for design build projects. Journal of Management in Engineering, 120–128 (July 2002)
7. Doll, W.T., Torkzaden, G.: The measurement of end-user computing satisfaction. MIS Quarterly 12(2), 259–271 (1988)
8. Davenport, T.: Putting the enterprise into the enterprising system. Harvard Business Review 76(4) (1998)
9. Gibson, N., Holland, C.P., Light, B.: Enterprise resource planning: a business approach to system development. In: Proceedings of the 32nd Hawaii International Conference on System Sciences, pp. 1–9 (1999)
10. Guimaraes, T., Yoon, Y., O'Neal, Q.: Success factors for manufacturing expert system development. Computers & Industrial Engineering 28(3), 545–559 (1995)
11. Holsapple, C.W., Wang, Y.-M., Wu, J.-H.: Empirically testing user characteristics and fitness factors in ERP success. International Journal of Human-Computer Interaction 19(3), 323–342 (2005)
12. Hong, K., Kim, Y.: The critical success factors for ERP implementation: An organizational fit perspective. Information and Management 40(1), 25–40 (2002)
13. Luo, W., Strong, D.M.: A framework for evaluating ERP implementation choices. IEEE Transactions on Engineering Management 51(3), 322–333 (2004)
14. Mabert, V.A., Soni, A., Venkataramanan, M.A.: Enterprise resource planning: managing the implementation process. European Journal of Operational Research 146, 302–314 (2003)
15. Markus, M.L., Tanis, C., Fenema, P.C.V.: Multisite ERP implementations. Communications of the ACM 43(4), 42–46 (2000)
16. Melone, N.P.: A theoretical assessment of the user-satisfaction construct in information systems research. Management Sciences 36(1), 76–91 (1990)
17. Nunnally, J.C.: Psychometric Theory, 2nd edn. McGraw-Hill, New York (1978)
18. Petroni, A.: Critical factors of MRP implementation in small and medium-sized firms. International Journal of Operations & Management 22(3), 329–348 (2002)
19. Raymond, L.: Validating and applying user satisfaction as a measure of success in small organizations. Information & Management 12, 173–179 (1987)
20. Ross, J.W.: Surprising facts about implementing ERP, IT Pro, (July/August 1999)
21. Sengupta, K., Zviran, M.: Measuring user satisfaction in an outsourcing environment. IEEE Transactions on Engineering Management 44(4), 414–421 (1997)
22. Somers, T., Nelson, K.: The impact of strategy and integration mechanisms on enterprising system value: Empirical evidence from manufacturing firms. European Journal of Operational Research 146(2), 315–338 (2003)

Appendix

List of Journals

1. Information & Management.
2. Journal of Management Information Systems.
3. MIS Quarterly.
4. Information Systems Research.
5. Decision Sciences.

6. Management Science.
7. IEEE Journals.
8. Communications of the ACM.
9. Information Systems Management.
10. European Journal of Information Systems.
11. Business Process Management Journal.
12. Information Systems Management.

Hardware-Software Co-design of QRD-RLS Algorithm with Microblaze Soft Core Processor

Nupur Lodha, Nivesh Rai, Rahul Dubey, and Hrishikesh Venkataraman

Dhirubhai Ambani Institute of Information and Communication Technology
Gandhinagar, Gujarat, India
{nupur_lodha,nivesh_r,rahul_dubey}@daiict.ac.in,
hrishikesh@ieee.org
http://www.daiict.ac.in

Abstract. This paper presents the implementation of QR Decomposition based Recursive Least Square (QRD-RLS) algorithm on Field Programmable Gate Arrays (FPGA). The design is based on hardware-software co-design. The hardware part consists of a custom peripheral that solves the part of the algorithm with higher computational costs and the software part consists of an embedded soft core processor that manages the control functions and rest of the algorithm. The use of Givens Rotation and Systolic Arrays make this architecture suitable for FPGA implementation. Moreover, the speed and flexibility of FPGAs render them viable for such computationally intensive application. The system has been implemented on Xilinx Spartan 3E FPGA with Microblaze soft core processor using Embedded Development Kit (EDK). The paper also presents the implementation results and their analysis.

Keywords: Hardware-Software Co-Design, QRD-RLS Algorithm, FPGA, Givens Rotation, Systolic Array.

1 Introduction

Adaptive signal processing plays an important role in broadband wireless communications with very high signal transmission rates. The process of calculating adaptive weights for such environments can be computationally very demanding task, as bandwidth of the order of megahertz is required. A processor that can estimate the parameters related to the communication channels on a real time basis is necessary in such applications. QR based Recursive Least Square (RLS) is a well established technique for solving the Least Mean Squares (LMS) problem by calculating adaptive weights and is extensively used in applications like Adaptive Beamforming, Multiple-Input-Multiple-Output (MIMO) and Software Defined Radio (SDR) [1][2].

There is a considerable research devoted to algorithms and VLSI architectures for RLS filtering [3-6], with the aim of reducing computational complexity. Much of this work has been concentrated on calculating the inverse of a matrix, in a more stable and less computational manner rather than simple matrix inversion

S.K. Prasad et al. (Eds.): ICISTM 2009, CCIS 31, pp. 197–207, 2009.

[1]. The standard RLS algorithm recursively updates the weights using the matrix inversion lemma. A commonly used alternative solution performs a set of orthogonal rotations on the incoming data thereby transforming the over specified rectangular data matrix into an upper triangular form. The weights are then obtained by back substitution. This method is known as QR decomposition based RLS. It enables the matrix to be retriangularised when new inputs are present, without the need to perform triangularisation from scratch [1]. Good numerical performance is achieved by performing the algorithm using Givens Rotation. An efficient implementation of RLS algorithm achieves much faster convergence than the LMS algorithm; however its complexity increases in proportion to the square of the number of parameters to be estimated [7]. To circumvent this problem our architecture is based on a pipelining technique referred to as systolic array.

The research presented in [1] is part of a project to design a single chip adaptive beamforming system. The core of the project is a QR array processor implementing the Enhanced-Squared Givens Rotation algorithm (E-SGR). The work in [3] demonstrates the implementation of Application Specific Integrated Processors (ASIP), specifically targeting QR matrix decomposition.

Conventionally, architectures address different ends of the performance spectrum. At one end there is a general purpose processor which provides flexibility but implements software algorithms in the order of milliseconds. At the other end exists a fixed dedicated custom hardware which can execute algorithms in the order of nanoseconds, but is highly inflexible. Often what is required is the balance between the high performance obtained from hardware, and the flexibility of software. Reconfigurable logic devices such as FPGAs are most suitable for such applications because they provide speed and are upgradeable. Hence they reduce the risk of depending on evolving standards of industries and are cost effective solutions. Moreover, FPGAs with embedded processors are flexible by nature and allow reconfiguration of logic with optimum use of resources. So programmable logic and reconfiguration is now seen as a suitable solution for several wireless applications.

The hardware software co-design term refers to the integration of hardware and software components at the design and development stages [8]. It deals with the problem of designing heterogeneous systems. One of the goals of co-design is to shorten the time-to-market while reducing the design effort and costs of the designed products. The evolution of hardware-software co-design concepts has also accelerated the developments of systems on single chip. It is based on an architecture that implements an embedded processor and one or more dedicated hardware coprocessors to improve the system efficiency. This approach takes advantage of both the flexibility of processors and power and speed of a dedicated hardware.

In this work, we have adopted hardware software co-design methodology for implementing QRD-RLS, an adaptive filter algorithm. This algorithm is partitioned into two parts. The part involving matrix computations is made as a custom peripheral as it is computationally more expensive. Back substitution and other control functionalities are executed on the processor. Interfacing between the processor and the peripheral is done using interfacing buses. Thus, such a design will enable us to make a complete system on programmable chip (SoPC). Also, it will enable us to achieve an optimum trade off between speed and flexibility.

The rest of the paper is organized as follows. Section 2 mentions the key points of QRD-RLS algorithm. Section 3 presents the architecture of the algorithm including Coordinate Rotation Digital Computer (CORDIC) algorithm and systolic arrays. Implementation methodology and results are presented in section 4 and 5 respectively. Section 6 gives the conclusion.

2   QRD-RLS Algorithm

2.1   QR Decomposition

QR Decomposition is an elementary operation which decomposes a matrix into an orthogonal and a triangular matrix [7]. QR decomposition of a real square matrix U is as $U = Q \times R$, where Q is an orthogonal matrix ($Q^T Q = I$) and R is an upper triangular matrix. There are different methods to compute QR decomposition like Gram-Schmidt ortho-normalization, Householder reflections and the Givens rotations.

2.2   Givens Rotations

Givens Rotation is used to find an operator which rotates each vector through a fixed angle and this operator can be represented as a matrix [7]. Givens method is useful when adaptive algorithms like RLS are considered. The reason being that in order to obtain the triangular form, the entries in the matrix are zeroed in a particular manner. This leads to a simple implementation which combined with its numerical stability results in an efficient algorithm.

2.3   QR Decomposition Based RLS

The least squares approach attempts to find the set of coefficients w_n that minimizes the sum of squares of error. Equation (1) refers to the least square solution,

$$U\mathbf{w} = \mathbf{d} + \mathbf{e}. \tag{1}$$

where U is an input matrix ($m \times n$ with $m > n$), $\mathbf{d}$ is a known desired sequence and $\mathbf{w}$ is the coefficients vector to be computed such that the error vector $\mathbf{e}$ is minimized. Direct computation of coefficient vector $\mathbf{w}$ involves matrix inversion, which is undesirable for hardware implementations. QR decomposition based least square methods avoid such explicit matrix inversions and are more robust for hardware implementation.

The QRD is used to solve the least square problem as follows. Equation (2) shows the QR decomposition of the compound matrix [U|d] as,

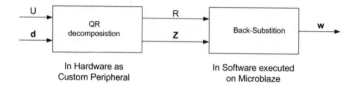

Fig. 1. QR Decomposition based RLS

$$[U|\mathbf{d}] = [Q|\mathbf{q}] \cdot \left[\begin{array}{c|c} R & \mathbf{z} \\ \hline 0 & \varsigma \end{array}\right] \qquad (2)$$

where $[Q|\mathbf{q}]$ is orthogonal and $\left[\begin{array}{c|c} R & \mathbf{z} \\ \hline 0 & \varsigma \end{array}\right]$ is triangular matrix and $\mathbf{z} = Q^T\,\mathbf{d}$. Once the QRD is available, $\mathbf{w}$ is easily available through the process of back-substitution in (3)

$$R\mathbf{w} = \mathbf{z}. \qquad (3)$$

Givens Rotation is applied to compound U matrix to calculate QRD. This gives QRD solution at time k. Now, in order to calculate QRD at time $k+1$ computations do not start from the beginning. QRD is recursively computed at time $k+1$ from time k by updating it as in (4),

$$\left[\begin{array}{c|c} R(k) & \mathbf{z}(k) \\ \hline \mathbf{u}_{k+1}^T & \mathbf{d}_{k+1} \end{array}\right] = Q(k+1) \cdot \left[\begin{array}{c|c} R(k+1) & \mathbf{z}(k+1) \\ \hline 0 & * \end{array}\right] \qquad (4)$$

where $*$ is a don't care entry, left hand side matrix is a *"triangular-plus-one-row-matrix"* and right hand is an orthogonal times triangular matrix. The next value of coefficient $\mathbf{w}$ follows from the back substitution in (5)

$$R(k+1)\mathbf{w}(k+1) = \mathbf{z}(k+1). \qquad (5)$$

This constitutes the complete QRD-RLS algorithm [7]. The above procedure avoids calculating the inner product $U^T U$ and hence it is numerically more stable. Fig. 1 shows QR decomposition based RLS algorithm.

3   QRD-RLS Architecture

3.1   CORDIC in QRD

CORDIC describes a method to perform a number of functions including trigonometric, hyperbolic and logarithmic functions. The algorithm is iterative and uses only additions, subtractions and shift operations [9]. This makes it more suitable for hardware implementations. CORDIC cells are used to calculate and apply the unitary transformation represented by Givens rotation.

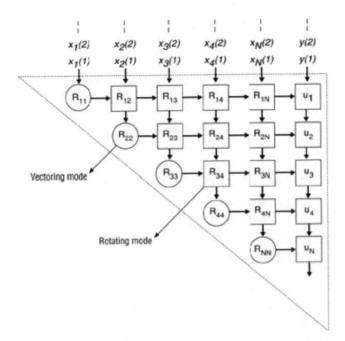

Fig. 2. Systolic Array Architecture

3.2 Systolic Arrays

Systolic arrays have two types of processing nodes: Boundary cells and Internal cells. Boundary cells are used to calculate the Givens rotation that is applied across a particular row in a matrix. The unitary transform which is calculated by the boundary cells, is taken as an output and applied to the remainder of the row containing internal cells [5-6].

Fig. 2 shows the use of systolic array architecture for performing the QR decomposition of the input matrix X. The rows of matrix X are fed as inputs to the array from the top along with the corresponding element of the vector **u**, which is the desired sequence. The data is entered in a time skewed manner. The calculations for a particular decomposed matrix R propagate through the array. The R and **u** values held in each of the cells once all the inputs have been passed through the matrix are the outputs from QR-decomposition. These values are subsequently used to derive the coefficient vector **w**. Each of the cells in the array can be implemented as a CORDIC block.

Direct mapping of the CORDIC blocks to the systolic array in fully parallelized mode consumes a significant number of logic blocks but at the same time yields very fast performance. The resources required to implement the array can be reduced by trading throughput for resource consumption via mixed and discrete mapping schemes, which map multiple nodes on a single instantiation of hardware [1][4].

3.3   Back Substitution

The back substitution procedure primarily involves multiplication and division operations. The Microblaze embedded soft core processor can be configured with its optional functionalities like hardware multiplier and division unit. This feature makes it ideal to implement back substitution. The CORDIC block performs the QR decomposition and stores the R and z values in registers accessible to the Microblaze processor, which then calculates coefficient values and stores the results back into memory.

4   Implementation Methodology

4.1   Profiling and Hardware-Software Partitioning

The complete algorithm has been programmed in C. It was executed on a 2 GHz Intel Dual Core high performance processor. As the profiler showed, matrix decomposition part took almost 94% of the total time, indicating that it is the most expensive operation in terms of computational cost, while back-substitution represented remaining 6% of the time. So, we chose to implement matrix decomposition in hardware by means of a custom peripheral. Back-substitution was executed in software on Microblaze.

4.2   Hardware Flow

Hardware platform describes the flexible, embedded processing subsystem which is created according to the demands of the application. The hardware platform consists of one or more processors and peripherals connected to the processor buses [10]. Fig. 3 shows the block diagram of the complete system configured for our design.

Microblaze: Microblaze is 32 bit Reduced Instruction Set Computer (RISC) architecture. A "Harvard" style bus architecture is used which includes 32 bit general purpose registers and separate instructions and data buses. It features a five stage instruction pipeline. As it is a soft core processor, the functional units incorporated into its architecture can be customized as per the needs of the application. Thus, the barrel shifter unit, hardware divider unit, data cache and instruction cache can be optionally instantiated along with the processor. Extra peripherals like UART, Ethernet controllers or other IP cores can be configured using EDK [10].

Fast Simplex Link (FSL) Bus: FSL is a unidirectional point-to-point communication channel bus used to perform fast communication between any two design elements on the FPGA when implementing an interface to the FSL bus. Microblaze can be configured with up to 16 FSL interfaces, each consisting of one input and one output port. FSL provides mechanism for unshared and non-arbitrated communication mechanism. This can be used for fast transfer of data words between master and slave implementing the FSL interface [10]. One FSL link was configured for communicating between Microblaze and custom peripheral.

4.3   Software Flow

A software platform is a collection of software drivers, libraries and, optionally, the operating system on which to build an application. The Library Generator tool configures libraries, device drivers, file systems and interrupt handlers for the embedded processor system by taking Microprocessor Software Specification (MSS) file as an input [11].

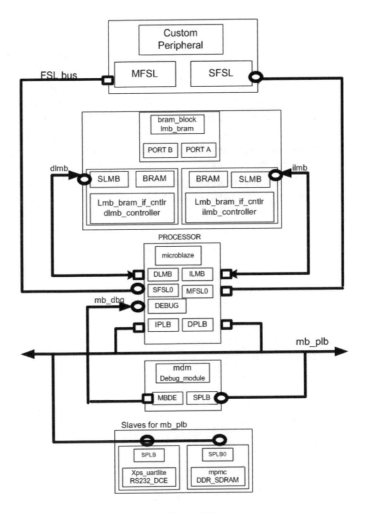

Fig. 3. Block Diagram

5   Implementation Results

5.1   MATLAB Results

QRD-RLS algorithm was implemented in Matlab, in order to verify its functionality. The desired data vector was obtained by reading a wave file. The input matrix was obtained by filtering this data through a channel and adding white Gaussian noise. The outputs obtained were the coefficient vectors and the error vector in modeling the channel.

Fig. 4 shows the error curve corresponding to the number of samples. It shows that as the number of samples increase, the error value decreases and converges to a very small value of 4.9×10^{-6}.

5.2   Verification of Our Design

An example scenario of 3×3 matrix ($M = 3, N = 3$) was considered because of the ease in implementing the algorithm, targeting the given Spartan 3E FPGA device. 16 bit real inputs to CORDIC blocks were considered. Fixed point representation was used to represent real numbers. The input data was the same which was used for Matlab implementation. Table I shows the comparison between the output coefficient values obtained from our design and the values obtained from Matlab for two updates of the coefficients. Update implies when all cells in the systolic array are updated with their new R and $\mathbf{z}$ values.

We see that the coefficient values obtained from our design i.e. Hardware Software (H/W-S/W) Co-design, do not match exactly with Matlab values. Table 1 also shows the error computed between both results. There is a small error value

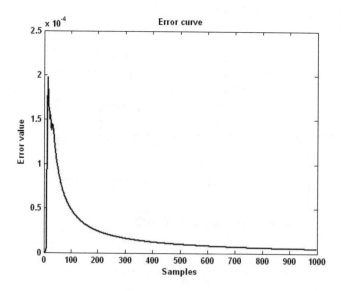

Fig. 4. Error Curve

Table 1. Comparison between Matlab and H/W-S/W Co-design Results

Weight Coefficients	MATLAB Results	H/W-S/W Co-design Results	Error
w1 (update1)	0.58e-2	0.65e-2	-6.071e-4
w2 (update1)	0.275e-1	0.269e-1	6.767e-4
w1 (update2)	-0.89e-2	-0.88e-2	-1.245e-4
w2 (update2)	0.255e-1	.0254e-1	1.371e-4

because our design is tailor made to work for four decimal places, while Matlab is a simulation software which uses IEEE double precision. Such a precision is very difficult to obtain in hardware due to resource constraints.

5.3 Analysis of Our Design

The custom peripheral was described in a high level description language like Verilog. The design was synthesized on a Xilinx Spartan-3E $XC3S500E - 4FG320$ FPGA. CORDIC blocks were used with the speed of 170 MHz.

Two mapping schemes were used:- Direct mapping and Discrete mapping. In Direct mapping, where each cell of the systolic array was mapped to a CORDIC block, 5 CORDIC blocks were required for 3×3 matrix. The update time obtained was $387ns$. Update time refers to the time required before all the cells in the systolic array are updated with their R and $\mathbf{z}$ values. But it led to huge consumption of resources, amounting to 5262 slices. This number even exceeded the number of slices available in targeted FPGA device (4656). So discrete mapping was adopted. In this scheme, two CORDIC blocks are used; one for translation operations and the other for rotation operations. The resource consumption reduced to 1689 slices, which is less than 40% of the available resources. The update time obtained was $454ns$, which is slightly higher than direct mapping.

Table 2 gives the resource estimates of *'Pure Hardware'* approach compared with our *'Hardware-Software Co-design'* approach for two matrix sizes of 3×3 and 4×4.

It can be seen that for $N = M = 3$ the resource estimates of "H/W-S/W Co-design" approach and "Pure Hardware" approach differ by only a small

Table 2. Resource Estimates of Complete Design

Matrix size	Method	Slices	LUT's	Flip Flops
3x3	Pure Hardware	3416	4588	6259
	H/W-S/W Co-design	3791	5247	5685
4x4	Pure Hardware	3844	5322	7473
	H/W-S/W Co-design	3821	5286	6190

number. For $N = M = 4$, the resource estimates of our approach are less than "Pure Hardware" approach. Thus our approach is scalable to larger matrix sizes with only a minor increase in resources, while there is a significant increase in resources for "Pure Hardware" approach. This is because in "Pure Hardware" approach back substitution part also takes resources, but this does not hold true for "H/W-S/W Co-design" approach where back substitution is executed in software. The software part of the design also lends flexibility, which is an important requirement for continually evolving standards of design cycle.

The back substitution part of the algorithm was coded in C language and it was executed on Microblaze soft core processor with clock frequency of 50 MHz. For $N = 3$, the Microblaze processor took 64 clock cycles or 1.28 μs which is acceptable for many applications. Though it is slower than a pure hardware approach, the presence of Microblaze processor lends flexibility to the whole system. Moreover, as matrix size increases, the execution time of custom peripheral which contains complex logic increases considerably as compared to software time, thereby allowing Microblaze to implement other data and control functions on the FPGA.

This QRD-RLS architecture achieves a throughput of 1.68 μs or 0.59M updates per second.

The design is easily extendable to other matrix sizes of $N \times M$ by changing the control unit. Also, there is a trade off between area of the design and throughput by using different mapping schemes. The abundant resources of newer and bigger FPGA families support the realization of a fully parallel hardware design, should the throughput requirements of the target application demand extremely high performance.

6   Conclusion

A novel design methodology of Hardware Software Co-design has been proposed in this paper. QRD-RLS algorithm has been implemented using Xilinx Spartan 3E FPGA with embedded Microblaze soft core processor. The use of systolic array and CORDIC architecture makes the algorithm suitable for hardware implementation. This QRD-RLS architecture achieves a throughput of 1.68 μs or 0.59M updates per second. The interfacing of the peripheral with Microblaze embedded processor provides an element of flexibility, which cannot be achieved in pure hardware approach. The flexibility will allow the system to configure adaptively for varying wireless conditions and external requirements. Moreover, it facilitates to create a SoPC. The independent custom unit which does the decomposition of the matrix can be used in any application depending upon different conditions. For example, the same hardware can be used in beamforming, MIMO, SDR or any such application where decomposition of matrix is required.

References

1. Lightbody, G., Walke, R., Woods, R., McCanny, J.: Linear QR Architecture for a Single Chip Adaptive Beamformer. Journal of VLSI Signal Processing Systems 24, 67–81 (2000)
2. Guo, Z., Edman, F., Nilsson, P.: On VLSI Implementations of MIMO Detectors for Future Wireless Communications. In: IST-MAGNET Workshop, Shanghai, China (2004)
3. Eilert, J., Wu, D., Liu, D.: Efficient Complex Matrix Inversion for MIMO Software Defined Radio. In: IEEE International Symposium on Circuits and Systems, Washington, pp. 2610–2613 (2007)
4. Gao, L., Parhi, K.K.: Hierarchical Pipelining and Folding of QRD-RLS Adaptive Filters and its Application to Digital Beamforming. IEEE Transactions on Circuits and Systems-II: Analog and Digital Signal Processing 47 (2000)
5. Walke, R.L., Smith, R.W.M., Lightbody, G.: Architectures for Adaptive Weight Calculation on ASIC and FPGA. In: Conference Record of the Thirty-Third Asilomar Conference on Signals, Systems, and Computers, California, vol. 2, pp. 1375–1380 (1999)
6. Yokoyama, Y., Kim, M., Arai, H.: Implementation of Systolic RLS Adaptive Array Using FPGA and its Performance Evaluation. In: 2006 IEEE Vehicular Technology Conference (VTC 2006 Fall), Montreal, Canada, vol. 64, pp. 1–5 (2006)
7. Haykin, S.: Adaptive Filter Theory, 4th edn., pp. 513–521. Prentice Hall, Englewood Cliffs (2001)
8. Gupta, R.K., Micheli, G.D.: Hardware-Software Cosynthesis for Digital Systems. Design and Test of Computers 10, 29–41 (1993)
9. Andraka, R.: A survey of CORDIC algorithms for FPGA based computers. In: 1998 ACM/SIGDA sixth International Symposium on FPGAs, Monterey, pp. 191–200 (1998)
10. Xilinx Inc., Microblaze Processor Reference Guide (2004), http://www.xilinx.com
11. Xilinx Inc., Embedded System Tools Reference Manual (2004), http://www.xilinx.com

Classification of Palmprint Using Principal Line

Munaga V.N.K. Prasad[1], M.K. Pramod Kumar[1], and Kuldeep Sharma[2]

[1] Institute for Development and Research in Banking Technology,
Hyderabad, India
[2] Dept. Of Mathematics, Indian Institute of Technology Delhi
mvnkprasad@idrbt.ac.in, mkpramod2005@gmail.com,
kuldeepmail2005@gmail.com

Abstract. In this paper, a new classification scheme for palmprint is proposed. Palmprint is one of the reliable physiological characteristics that can be used to authenticate an individual. Palmprint classification provides an important indexing mechanism in a very large palmprint database. Here, the palmprint database is initially categorized into two groups, right hand group and left hand group. Then, each group is further classified based on the distance traveled by principal line i.e. Heart Line During pre processing, a rectangular Region of Interest (ROI) in which only heart line is present, is extracted. Further, ROI is divided into 6 regions and depending upon the regions in which the heart line traverses the palmprint is classified accordingly. Consequently, our scheme allows 64 categories for each group forming a total number of 128 possible categories. The technique proposed in this paper includes only 15 such categories and it classifies not more than 20.96% of the images into a single category.

Keywords: Palmprint, ROI, Heart line, Palmprint Classification.

1 Introduction

Reliable authorization and authentication is becoming necessary for many everyday applications, be it boarding an aircraft or performing a financial transaction. Authentication of a person becomes a challenging task when it has to be automated with high accuracy and hence with low probability of break–ins and reliable non-repudiation. In this paper, we propose a new technique for classification of palmprint to address this major challenge of authentication. The use of biometrics in user authentication these days is increasingly prevalent [1, 2]. Many biometrics techniques such as fingerprints, iris, face, and voice pre exist [3, 8]. Although there are numerous distinguishing traits used for personal identification, the present approach focuses on using palmprints to effectively authenticate a person by classification of palmprint. This technique besides from being effective is also implemented at a lower cost. Palmprint is considered as the biometric because of its user friendliness, environmental flexibility, and discriminating ability.

Although fingerprint system [4] is reliable and is in wide spread use, the complexity lies in the extraction of the minutiae from unclear fingerprints [11] and furthermore,

S.K. Prasad et al. (Eds.): ICISTM 2009, CCIS 31, pp. 208–219, 2009.

high-resolution images are required. Iris and retina recognition [7] provide very high accuracy but suffer from high costs of input devices or intrusion into users. Hence in this paper, palmprint has been used as the biometric considered for classification and thus ensure that authentication is effective. In this connection, palm has several features to be extracted like principal lines, wrinkles, ridges, singular points [5], texture and other minutiae.

A palmprint image with principal lines and wrinkles represented is shown in Fig.1.

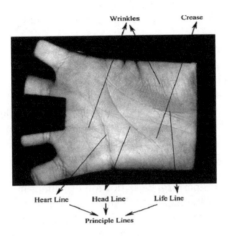

Fig. 1. Palmprint Image

There are usually three principal lines made by flexing the hand and wrist in the palm [6], namely heart line, headline and lifeline. In this paper, Heart line is chosen for classification purpose since there is sufficient variability for discrimination and also because the heart line lies independently in the palm almost in parallel orientation with horizontal axis. After a careful investigation of palms, we found that for some palms there exists gaps in the heart line and in rare cases some palms do not have a heart line at all. In general the heart line lies in the area between the fingers and the point where the lifeline and headline come closer.

There are basically two stages in the Palmprint Recognition:

1. Palmprint Registration 2. Palmprint Verification

Palmprint Registration is a process of registering new images into the database. The block diagram for the image registration is shown in the Fig.2.

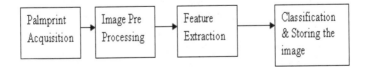

Fig. 2. Block Diagram for Palmprint Registration

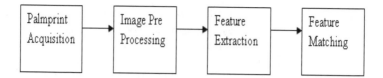

Fig. 3. Block Diagram for Palmprint Verification

Palmprint Verification is a process to verify whether the captured palmprint belongs to the same individual or not. Block diagram for Palmprint verification is shown in the Fig.3.

After feature extraction, the class to which the palmprint belongs is determined and the palmprint is compared with the images stored under the same class. In the existing systems [9, 10], firstly, the user's palmprint is captured by the system. Then, it is compared to palmprint in the database and a matching result is produced. This process got constrained with few limitations. The comparison of captured palmprint with every single palmprint in the database demands excessive amount of resources. Furthermore, it is also time consuming. Even if one comparison of palmprint takes up a few milliseconds, in this context we are referring thousands of images. Such is the computational complexity of those techniques.

Thus, in this paper we perform classification of palmprint using a heart line for authentication. The rest of the paper is organized as follows: Section 2 presents related work in the field of palmprint classification. Section 3 presents preprocessing, segmentation of ROI and proposed algorithm for classification. In Section 4 experimental results are described. Conclusions are presented in section 5.

2   Related Work

The palmprint image is captured and it is pre processed. The processed image is then used to extract the features. Duta et.al. extracted some points on palm lines from off-line palmprint images for verification [11]. Han et al. used Sobel and morphological operations to extract line-like features from the palmprints [12]. Zhang et al. used 2D-Gabor filters to extract the texture features from low-resolution palmprint images and employed these features to implement a highly accurate online palmprint recognition system [13]. Chih-Lung et al. used finger web points as datum points to extract region of interest (ROI) [16]. Every individual's palmprint is stored in the database. The authentication methods require the input palmprint to be matched with the large set of palmprints stored in the database. The search time should be very small of the order less than 1sec. In order to reduce the search time and space complexity, the database is classified according to the principal line of the palmprints. Wang et al. [14] classified CCD camera based palmprints, but were not able to categorize the palms efficiently as 78.12% of palms belong to one category and very few come in other categories. In his approach the matching process may still have to search through 78.12% of the original database samples before finding a match. In another approach Negi et al divided the ROI into 8 groups and classified the groups into 257 categories [15]. They considered the alignment of the heart line in both hands to be same. But in the current work, we considered different methodologies for right and left hand palm-prints. In this paper a Region of Interest (ROI) is extracted in which only heart line is present. So, here ROI is divided into six regions leading to 64 (2^6) categories for each group. Thus, for left and right hand palmprint groups the number of categories is 128 (64+64). By following this methodology, the classification rate is much improved in comparison to Negi et al [15]. The accuracy of the system is also much improved than Negi et al and Wang et al as the right hand and left hand palms are considered separately and accordingly the alignment of the heart line is considered differently for both the groups.

3 Proposed Classification Procedure

The method for classification of the palmprint images is to preprocess the palmprint images, extract the heart line and find the distance it travels along the palm as well as the presence of heart line parts. So, the images captured are to be segmented to extract the Region of Interest (ROI). The technique used here rotates the palmprint image captured to align the heart line with the horizontal axis as much as possible. The starting point of the bottom line is found out by scanning the image from the bottom left most pixels. The distance to all the other border pixels is calculated and distance distribution diagram is plotted as shown in Fig.5. The local minima in the distance distribution diagram are found out, which are nothing but the finger web locations. After rotation the co-ordinates of the finger web location are changed, so again the finger web locations are calculated for the rotated image, which was a limitation in earlier approaches. In the proposed technique a rectangular region containing only heart line is extracted from the rotated image and the median filter is used to remove noise from the ROI. In median filter [10], each output pixel is set to an average of the pixel values in the neighborhood of the corresponding input pixel with median filtering; the value of an output pixel is determined by the median of the neighborhood pixels, rather than the mean. So, this filter removes the outliers without reducing the sharpness of the image.

3.1 Steps to Find the Finger Web Points

In this paper a method [16] is adopted which uses finger webs as the datum points to develop a approximate ROI to which changes are made to overcome the limitations of the existing method. Gray scale images are first converted to Binary with gray value 0 or 1.The following processes is performed to locate finger web locations using binary palmprint images.

1. Boundary tracing 8-connected pixels algorithm is applied on the binary image to find the boundary of palmprint image, as shown in Fig.4. P_s is the starting point and W_m is middle point of the base of the palm. Tracing is done in counter clockwise direction starting and ending at P_s, and the boundary pixels are collected in Boundary Pixel Vector (BPV).
2. Euclidean distance is calculated between BPV and W_m using Eq.1

$$D_E(i) = \sqrt{(X_{wm} - X_b(i))^2 + (Y_{wm} - Y_b(i))^2} \tag{1}$$

where (X_{wm}, Y_{wm}) are the X and Y coordinates of W_m. ($X_b(i)$, $Y_b(i)$) are the coordinates of the ith border pixel stored in BPV, and $D_E(i)$ is the Euclidian distance between W_m and i^{th} border pixel.

Fig. 4. Boundary Pixels of Palm image

3. A distance distribution diagram is constructed using the vector D_E as shown in Fig. 5. The constructed diagram pattern is similar to geometric shape of the palm. In the diagram, four local maxima and three local minima can be seen, which resembles the four finger tips (local maxima) and three-finger webs (local minima) i.e., valley between fingers excluding thumb finger.

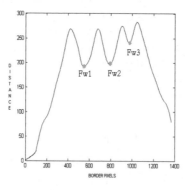

Fig. 5. Distance Distribution Diagram

4. Here the proposed approach first segregates the entire palmprint database into two groups i.e. Right hand Image Group and Left hand Image Group and then classification algorithm is applied separately and is explained in section 3.4. The first finger web point (FW1) and third finger web point (FW3) as in Fig. 5 is taken and slope of the line joining these two points is calculated using Eq.2

$$\tan \alpha = Y/X \qquad (2)$$

where Y=y1-y3, X=x1-x3, where (x1, y1) and (x3, y3) are the coordinates of FW1 and FW3 finger web points and α is slope of the line joining FW1 and FW3.

Depending upon the value of $\tan \alpha$ the entire palmprint database is divided into two groups i.e. Right-hand images group and Left hand images group.

If $\tan \alpha >= 0$ then image is grouped into Right Hand Group.
Else $\tan \alpha < 0$ then image is grouped into Left Hand Group.

In this context the finger web point FW1 is at higher position than finger web point FW3 for Right hand images as shown in Fig 6 and the angle formed between finger web points α is acute with respect to horizontal axis for right hand images.

The finger web point FW1 is at lower position than finger web point FW3 shown in Fig. 7 and the angle between these finger web points α is obtuse with respect to horizontal axis for left hand images group. Once a palm database is grouped, the region of interest (ROI) is extracted separately.

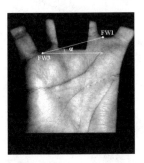

Fig. 6. Right Hand Palm Image

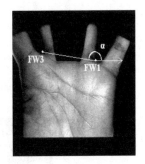

Fig. 7. Left Hand palm Image

3.2 ROI Extraction

A rectangular region containing only heart line is to be extracted in order to classify the palmprint database. The following steps are performed to extract the ROI.

Step 1. The image is rotated at an angle α to align the straight line joining FW1 and FW3 with the horizontal axis.

Step 2. After rotation again the procedure of finding the finger web points (Section 3.1) is applied to get finger web points of the rotated image as the co-ordinates of the finger web point's changes after rotation. The finger web points after rotation are named as FR1, FR2 and FR3 (from right).

Step 3. Vertical distance D1 is calculated between FR2 and FR3 and the same distance is projected vertically downward from FR3 and from that point half of the horizontal distance ds/2 between FR2 and FR3 is projected horizontally towards the left edge of the palm to get the point LT1. Similarly RT1 is marked on the other side of the palm as shown in Fig.8.

Step 4. Mid points of FR1, FR2 and FR3 are M1 and M2 and mid point of FR2 and M2 is taken as M. It is seen that the distance (D) between LT1 and M, if projected vertically from the point LT1, the heart lines always lies in between the points LT1 and LT2. Similarly RT1 and RT2 is also marked on other side of the palm as shown in Fig. 9.

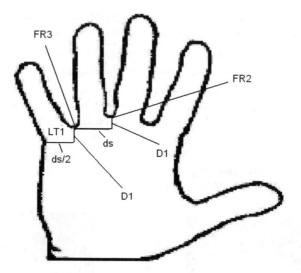

Fig. 8. Palmprint with Left Edge point LT1

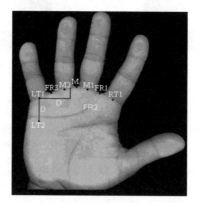

Fig. 9. Palmprint with Left Edge point LT2

3.2.1 ROI Extraction for Left Hand

i. For left hand group images, the heart line originates in between the points RT1 and RT2, and the images are rotated in anti clockwise direction with an angle $180-\alpha$ to align the heart line horizontal as much as possible as shown in Fig 10.

ii. The Steps 2, 3 and 4 of ROI extraction explained in section 3.2 are applied to get the ROI part for left hand group images as shown in Fig 10.

3.2.2 ROI Extraction for Right Hand

i. For Right Hand group images, the heart line originates in between the points LT1 and LT2 and the images are rotated in clockwise direction with an angle of α as shown in Fig.11.

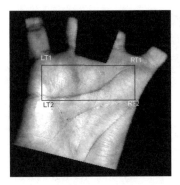 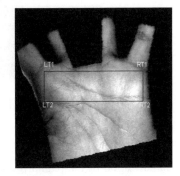

Fig. 10. Left Hand Palmprint Rotated Anti Clock wise and ROI marked

Fig. 11. Right Hand Palmprint Rotated Clock wise and ROI marked

ii. The steps 2, 3 and 4 of ROI extraction explained in section 3.2 are applied to get the ROI for Right Hand Group images as shown in Fig.11.

Step 5. Using the points LT1, LT2, RT1 and RT2 a rectangular ROI containing only heart line is extracted. The extracted ROI's of left hand and right hand palmprints are shown in Fig.12 & 15.

3.3 Feature Extraction

The ROI contains the heart line, wrinkles and ridges. In order to remove the wrinkles and ridges Adaptive Median Filter is applied to smoothen the image. After adaptive median filter is applied there is possibility for the presence of fine wrinkles and ridges, which are removed completely from ROI using "Canny edge detection" technique [17]. The images are further smoothened by Gaussian filter that is embedded in Canny algorithm. The Canny edge detector an optimal edge detector with lower and upper thresholds is used to extract prominent heart line. If the magnitude of an edge is below the lower threshold, it is set to zero (made a no edge) and if the magnitude is above the higher threshold, it is made an edge. If the magnitude is between the 2 thresholds, then it is set to zero unless there is a path from this pixel to a pixel with a gradient above higher threshold. ROI of the palm print after applying the canny edge detection is shown in Fig. 13 & 16.

Fig. 12. ROI of Left Hand Palmprint

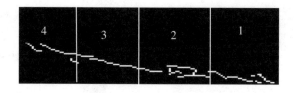

Fig. 13. ROI of Left Hand Palmprint Divided into 4 Regions

3.4 Classification

The classification of the palmprint images is separately done for right and left hand group images depending upon the heart line traversing in the ROI. The process is as follows:

1. The ROI is first divided into 4 regions of equal size as shown in Fig. 13 &16. The 1st and 3rd regions are further divided into two halves. It is made so because, in most of the palms the starting and the ending points of the heart line varies mainly in 1st and 3rd regions. So a group of 6 regions are formed where 1st, 2nd, 4th and 5th are of equal size and 3rd, 6th region is of equal size as shown in Fig.14 & 17.
2. For left hand group the numbering of regions is done from right to left as the heart line starts from right side of the palm as shown in Fig.14.
3. For right hand group the numbering of regions is done from left to right as heart line starts from left side of the palm as shown in Fig.17.
4. The palm image then can be classified by grouping together the region numbers in a sorted order in which the heart line is present. For example consider the heart line traversal of the left hand palmprint which is classified into 123456 class as shown in Fig.14 and the right hand palmprint is classified into 1234 class as shown in Fig.17.
5. The total number of classes for each group is based on the number of regions n. As the number of regions are 6, the number of classes can be $2^n = 2^6 = 64$. Hence total number of classes will be 128 for right and left hand groups.

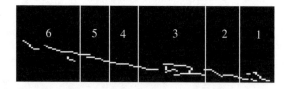

Fig. 14. ROI of Left Hand Palmprint Divided into 6 Regions

Fig. 15. ROI of Right Hand Palmprint

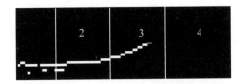

Fig. 16. ROI of Right Hand Palmprint Divided into 4 Regions

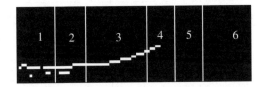

Fig. 17. ROI of Right Hand Palmprint Divided into 6 Regions

4   Experimental Results

The experiment has been done on a system of 2.40 GHz CPU and 256 MB of RAM. The proposed classification algorithm is coded in MATLAB7 and executed on Windows 2000 platform. The proposed classification algorithm is tested on a pegged palmprint database containing 4992 indexed grayscale palmprint images. The major source of our database is Hong Kong Polytechnic University Database [18]. Prior to Classification, the database is grouped into two major groups i.e. right hand group and left hand group based upon the angle formed between the finger web points (FW1, FW3).The ROI is then divided into 6 regions, so there are 64 different categories(2^6) for each group forming a total of 128 categories for both the groups. Now the classification algorithm is applied on right hand group and left hand group separately and the corresponding palm image is classified into one of the 64 classes for each group. Out of 4992 palm images, 2890 images are grouped into right hand group and 2102 into left hand group. The distribution of different classes obtained for right hand group and left hand group are shown in Table 1 and Table 2 respectively. In the right hand palmprint group database 13 different classes are found. They are 1234, 123456, 345, 0, 456, 23456, 2345, 1345,3456, 12345, 236, 346 and 13456. Here out of 2890 palm images, highest pecentage of palms i.e 20.96% are found in 123456 class. For the left hand group database out of 2102 palms 18.31% palms are found in class 123456. In left hand group 12 different classes are found they are 236, 2345, 123456, 0, 456, 12345, 1236, 3456, 2346, 13456, 346, 23456.

Total 15 unique different classes are found in both the groups as shown in Table 1 and Table 2. They are 1234, 123456, 345, 0, 456, 23456, 2345, 1345, 3456, 12345, 236, 346, 13456, 1236 and 2346. The distribution of palms in the corresponding classes depends upon heart line and thresholds used in Canny Edge Detection method. The class 0 is for the case where there is no heart line or the line could not be detected by our threshold Canny Edge detector. The upper and lower thresholds used in Canny Edge Detection are 0.29 and 0.59 respectively. The change in the values of the threshold will lead distribution of palm images into different classes.

Table 1. Right Hand Group Database

S.No	Class	Number of palm Images	Percentage
1	1234	179	6.19
2	123456	606	20.96
3	345	234	8.09
4	0	46	1.59
5	456	158	5.47
6	23456	378	13.07
7	2345	120	4.15
8	1345	136	4.705
9	3456	472	16.33
10	12345	162	5.60
11	236	56	1.94
12	346	81	2.802
13	13456	262	9.06

For both the groups it is observed that classes 123456, 3456, 23456 and 13456 were common and also accounted for a large fraction of the classification. It is also observed that certain palms have broken heart line or the lines which are not detected by thresholds. The classes 1345, 236, 346, 13456, 1236 and 2346 have broken heart lines for both the groups.

Table 2. Left Hand Group Database

S.No	Class	Number of palm Images	Percentage
1	236	119	5.66
2	2345	145	6.90
3	123456	385	18.31
4	0	52	2.47
5	456	85	4.04
6	12345	149	7.08
7	1236	132	6.27
8	3456	262	12.46
9	2346	104	4.94
10	13456	196	9.32
11	346	178	8.47
12	23456	295	14.03

Out of 500 images randomly selected from the database 498 images are correctly classified and only 2 images are misclassified that shows a classification accuracy of 99.6%.The accuracy rate is much improved than Negi et al [15] and Wang et al [14] because here the right hand images and left hand images are considered separately and accordingly alignment of heart line is considered to classify the palmprint database.

5 Conclusion

In this paper a new classification technique to classify palmprint database is proposed. Here, the palmprint database is initially categorized into two groups, right hand group and left hand group. A rectangular Region of Interest (ROI) in which only heart line is present is extracted and is divided into 6 regions. Further, based on type of the group and regions in which heart line traverses the palmprint is classified accordingly. Consequently, our scheme allows 64 categories for each group forming a total number of 128 possible categories. By following this technique, there is a considerable improvement in the system accuracy. Moreover, in our approach there is flexibility to increase the number of classes based on different combinations of regions in which heart line traverses. Hence, it can be concluded that this new classification technique classifies the palmprint database efficiently and reduces the search time considerably.

References

[1] Zhang, D.: Automated Bimetrics – Technologies and Systems. Kluwer Academic Publishers, Dordrecht (2000)

[2] Jain, A., Hong, L., Bolle, R.: On-line fingerprint verification. IEEE Trans. On Pattern Analysis Machine Intelligence 19(4), 302–314 (1997)

[3] Miller, B.: Vital signs of identity. IEEE Spectrum 31(2) (1994)

[4] Coetzee, L., Botha, E.C.: Fingerprint recognition in low quality images. Pattern Recognition 10(26), 1441–1460 (1993)

[5] Dass, S.C.: Markov Random Field Models for Directional Field and Singularity Extraction in Fingerprint Images. IEEE Transactions on Image Processing 10(13), 1358–1367 (2004)

[6] Kumar, A., Wong, D.C.M., Shen, H.C.: Personal verification using palmprint and hand geometry biometric. In: Kittler, J., Nixon, M.S. (eds.) AVBPA 2003. LNCS, vol. 2688, pp. 668–678. Springer, Heidelberg (2003)

[7] Ma, L., Tan, T., Wang, Y., Zhang, D.: Personal Identification Based on Iris Texture Analysis. IEEE Trans. on Pattern Analysis and Machine Intelligence 25(12), 1519–1533 (2003)

[8] Jain, A.K., Ross, A., Prabhakar, S.: Arun Ross and Salil Prabhakar, An Introduction to Biometric Recognition. IEEE Trans. on Circuits and Systems for Video Technology 14(1), 4–20 (2004)

[9] Zhang, D., Kong, W.-K., You, J., Wong, M.: Online palmprint identification. IEEE Trans. On Pattern Analysis and Machine Intelligence 25(9), 1041–1051 (2003)

[10] Image Processing, analysis and machine vision. In: Sonaka, M., Vaclav, H., Roger, B. (eds.), 2nd edn. PWS Publishing Company (1999)

[11] Duta, N., Jain, A.K., Mardia, K.V.: Matching of palmprint. Pattern Recognition Letters 23(4), 477–485 (2001)

[12] Han, C.C., Chen, H.L., Lin, C.L., Fan, K.C.: Personal authentication using palmprint features. Pattern Recognition 36(2), 371–381 (2003)

[13] Zhang, D., Kong, W.-K., You, J., Wong, M.: Online palmprint Identification. IEEE Trans. On Pattern Analysis and Machine Intelligence 25(9) (2003)

[14] Wang, K., Wu, B.H.X., Zhang, D.: Palmprint classification using principal lines. Pattern Recognition 37(10), 1987–1998 (2004)

[15] Negi, A., Panigrahi, B., Prasad, M.V.N.K., Das, M.: A Palmprint Classification Scheme using Heart Line Feature Extraction. In: ICIT apos: 2006, vol. 18(21), pp. 180–181 (2006)

[16] Lin, C.-L., Chaung, T.C., Fan, K.-C.: Palmprint Verification using hierarchial decomposition. Pattern Recognition 38(12), 2639–2652 (2005)

[17] Canny, J.F.: A Computational approach to edge detection. IEEE Trans. Pattern Analysis and Machine Intelligence 8(6), 679–698 (1986)

[18] PolyU Palmprint Palmprint Database,
       http://www.comp.polyu.edu.hk/~biometrics/

Workflow Modeling Using Stochastic Activity Networks

Fatemeh Javadi Mottaghi[*] and Mohammad Abdollahi Azgomi

Performance and Dependability Engineering Lab., Department of Computer Engineering,
Iran University of Science and Technology, Tehran, Iran
javadimottaghi_f@comp.iust.ac.ir, azgomi@iust.ac.ir

Abstract. The essence of workflow systems is workflow patterns. The aim is to use an existing powerful formal modeling language with workflow systems. Stochastic activity networks (SANs) are a powerful extension of Petri nets. Having the SAN model of a system, one can verify the functional aspects and evaluate the operational measures, both on a same model. SANs have already been used in a wide range of applications. As a new application area, we have used SANs for modeling workflow systems. The results show that the most important workflow patterns can be modeled in SANs. In addition, the resulting SAN models of workflow systems can be used for model checking and/or performance evaluation purposes using the existing tools. In this paper, we will present the results of this work. For this purpose, we will present the SAN submodels corresponding to the most important workflow patterns. Then, the proposed SAN submodels are used in a case study for workflow modeling, which will also be presented in this paper. Finally, we will present the results of the evaluation of the model using the Möbius modeling tool.

Keywords: Workflow modeling, workflow patterns, Petri nets, stochastic activity networks (SANs).

1 Introduction

Requirements for workflow languages are indicated through workflow patterns [12]. Different aspects such control flow, data, source, and exception management which are needed to be supported by a workflow language or a business process modeling language are investigated using these patterns. The behavior of a model can investigated using workflow models, and errors occurring in the model will be revealed before implementing the model. On the whole we could say that workflow patterns are used for classification of main concepts of system design, business analysis, and the design of architectural and business process of software [7].

Workflow patterns are used to model workflow systems. Several workflow modeling languages are introduced, which have been used in wide range of application. Among these systems we could name *workflow nets* (WFNs), *business process modeling notation* (BPMN), and *business process execution language* (BPEL). WFNs are

---

[*] First author was supported by Mobile Company of Iran (MCI).

S.K. Prasad et al. (Eds.): ICISTM 2009, CCIS 31, pp. 220–231, 2009.

a sub-class of Petri nets. Using WFNs, the business process can be modeled and correctness can be checked [14]. Since the WFNs are based on Petri nets, the resulting models can be every large.

Workflow modeling languages do not fully support the implementation of all workflow patterns. Considering the stochastic activity networks (SANs) [5] characteristics, model and its elements, the workflow patterns can be modeled using SAN model. Modeled system using SANs can be analyzed in functional aspects and evaluated in operational aspects. In this paper, we use SANs for modeling workflow systems. For this purpose, we have introduced submodels of SANs corresponding to the most important workflow patterns. Then, we have used the proposal SAN submodels in a case study for workflow modeling. Finally, we have presented the results of analysis of the model using Möbius tool [9].

The rest of this paper is organized as follows. In section 2, the related works and in section 3, the workflow patterns are discussed. In section 4, the stochastic activity networks (SANs) and their elements are introduced. In section 5, the workflow patterns are modeled as some SAN submodels. In section 6, a sample workflow system is modeled and evaluated using the proposed submodels. Finally, some concluding remarks are mentioned in section 7.

2 Related Works

A workflow process is modeled by workflow management systems (WFMS). Each of workflow systems give a definition for workflow patterns and propose a method for patterns implementation using those definitions, and use that method to model a workflow process. The workflow patterns are used to examine the capabilities of business process modeling languages such as BPMN and UML [7].

The most important workflow modeling systems are introduced in the following:

The Petri net is a formal graphical modeling language. This language is used for description and implementation of concepts and process flow control [11]. There are three specifications that are used to choose the high-level Petri nets as a workflow language which are [15]: (1) formal semantics, (2) state-based, and (3) lack of analysis techniques.

The Petri nets have problems and limitations for workflow processes modeling which are as follows [15]:

1. Incapability of modeling multiple instances.
2. Difficulty in modeling the advanced parallel patterns with high-level Petri Networks.
3. Difficulty in modeling the cancellation patterns with high-level Petri Networks.

The workflow net is a sub-class of Petri nets that can be used to model the process and then check its correctness. The workflow nets can also perform verification of processes [14]. The workflow nets model the lifecycle of a process and process control dimensions. The workflow net besides supporting the existing elements in the Petri nets also supports other elements such as AND-join, AND-split, XOR-split, and OR-join [14]. In the workflow net four structures of continuous, parallel, conditional, and repetition routing are defined [11]. Due to the fact that the workflow nets are based on Petri nets the resulting workflow net models can be very large.

Yet another workflow language (YAWL) is a language based on Petri nets. YAWL is more conceptual and expressive than the Petri net and can directly support most of the workflow patterns. In YAWL some symbols in addition to XOR-join, AND-join, AND-split, and XOR-split are defined [10].

Business process execution language (BPEL) is a language based on XML that is used to define the business processes in web services [21]. Due to using XML, BPEL faces problems in designing the processes [6].

BPMN is a graphical language that can be used to demonstrate a business process by business analyzer or developer. The principle of BPMN is the Petri nets theory. BPMN is a bridge to cover the distance between business process design and process implementation [20].

In software engineering field, the unified modeling language (UML) is a de facto standard language for modeling the objects. Elements in UML cannot be formally defined. Due the fact that formal definition is the basis of verification and evaluation of models, it is not possible to verify the modeled systems using UML [1].

3 Workflow Patterns

The fundamental concepts of workflow in [12] are required to be introduced in order to inspect the workflow patterns.

The principles of workflow patterns are elements of AND-join, AND-split, XOR-join, and XOR-split. These elements and the manner of their mapping to Petri Net are fully explained in [4, 11].

The workflow patterns can be classified into six categories [7, 12]: Basic Control Flow Patterns, Advanced Branching and Synchronization Patterns, Structural Patterns, Patterns Involving Multiple Instances, State Based Patterns, and Cancellation Patterns.

- Basic Control Flow Patterns: These are the basic constructs present in most workflow languages to model sequential, parallel and conditional routing. Patterns of this category are: continuity, parallel, synchronization, exclusive choice, and simple merge [12].
- Advanced Branching and Synchronization Patterns:
- These patterns transcend the basic patterns to allow for more advanced types of splitting and joining behavior [10]. In contrast to basic control flow patterns, the implementation of all of this category's patterns is not supported by all workflow systems. Patterns of this category are: Multi-Choice, Convergent Combination, Multi-Merge, and Discriminator [12].
- Structural Patterns: In programming languages a block structure which clearly identifies entry and exit points is quite natural, but it can be expressed in modeling using related patterns [10]. Patterns of this category are: Arbitrary Cycles and Implicit Termination [12].
- Patterns Involving Multiple Instances: Within the context of a single case sometimes parts of the process need to be instantiated multiple times. Patterns of this category are: multiple instances without synchronization, multiple instances with

prior design time knowledge, multiple instances with prior runtime knowledge, and multiple instances without prior design time knowledge [12].

- State Based Patterns: Typical workflow systems focus only on activities and events and not on states. This limits the expressiveness of the workflow language because it is not possible to have state dependent patterns. Thereby, the state based patterns are introduced [10]. Patterns of this category are: Deferred Choice, Interleaved Parallel Routing.

- Cancellation Patterns: The occurrence of an event (e.g., a customer canceling an order) may lead to the cancellation of activities. In some scenarios such events can even cause the withdrawal of the whole case. [7]. Patterns of this category are: Cancel Activity and Cancel Case [10, 12].

4 An Overview of Stochastic Activity Networks

Stochastic activity networks (SANs) [5, 8] are a stochastic generalization of Petri nets that are defined for modeling and analysis of real-time systems. By having the SAN model of a system, can verify the correctness of that system and evaluate the operational aspects of it.

The stochastic activity networks are more powerful and flexible than most other stochastic extensions of Petri nets such as stochastic Petri net and generalized stochastic Petri net. Concurrency is one of the important aspects of modeling which itself has two aspects of non-determinacy and parallelism. There are three settings in SAN: nondeterministic, probabilistic, and stochastic. In a nondeterministic setting two aspects of concurrency, non-determinacy and parallelism are represented in a nondeterministic manner. In a probabilistic setting, non-determinacy is specified as probabilistically and parallelism is treated non-deterministically. In a stochastic setting non-determinacy and parallelism are modeled as probabilistically. These three settings are described in [5].

The purpose of SAN is twofold. First, it allows for a better a most formal definition, and second it allows for the use of the model for the analysis of both functional and operational aspects of the system such as performance, dependability and performability [5].

SANs have the following elements: Place, timed activity, instantaneous activity input gate and output gate which are introduced as following [5, 8].

- ○ Places show the state of the modeled system.
- ▮ Timed activities represent activities of the modeled system whose durations impact the System's ability to perform.
- ‖ Instantaneous activity: Instantaneous activities represent system activities, which, relative to the performance variable in question, are completed in a negligible amount of time.
- ⇒▷ Input gates are introduced to permit greater flexibility in defining enabling and completion rules.
- ◁⊏ An output gate has a finite set of outputs and one input. To each such output gate is associated a computable function called the output function.

5 Modeling Workflow Patterns in SANs

As mentioned before, the current workflow modeling languages have limitations and problems. These languages do not fully support the implementation of all workflow patterns.

According to SANs model specifications and with regard to its primitives, we have modeled the workflow patterns using SANs. Modeling workflow patterns by SAN has simplified the model design according to capability of input gate and output gate of SANs. Furthermore, using SAN model a modeled system can be analyzed in functional aspects and evaluated in operational aspects.

In this section the concepts and patterns of workflow are defined using primitives of SANs. In the patterns definition, the activities showing a task are shown with timed activities and activities that are only for decision making are modeling instantaneous activities. The input gate is used for defining how activities can enable and how marking is changed, and the output gate is used for choosing the route and also changing marking.

In the following paragraphs, the flow control elements, AND-join, XOR-split and then the discriminator patterns are modeled using SANs.

5.1 AND-Join Submodel

A SAN submodel for AND-join pattern is shown in Fig. 1, which is the convergence of two or more branches into a single subsequent branch, such that the thread of control is passed to the subsequent branch, when all of the input branches have been enabled. Table 1 shows the predicate and functions of the input gates for the Fig. 1.

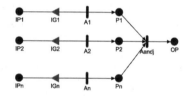

Fig. 1. SAN submodel for AND-Join pattern

Table 1. Gate table for SAN model of Fig. 1

Gates	Enabling Predicate	Function
IGi	IPi->Mark()>0 && Pi->Mark()=0	Pi->Mark()++

Formal Definition: An AND-join submodel is a 6-tuple (IP, IG, A, P, Anx, OP), where:

- IP = {IP1, IP2, ..., IPn}, a set of n input places where $IPi \in P$
- IG = {IG1, IG2, ..., IGn}, a set of n input gates where $IGi \in IG$
- A = {A1, A2, ..., An}, a set of n timed activities where $Ai \in TA$
- P = {P1, P2, ..., Pn}, a set of n places where $Pi \in P$

- $Anx \in TA$, a timed activity
- $OP \in P$, an output place

5.2 XOR-Split Submodel

A SAN submodel for AND-Join is shown in Fig. 2, which is the divergence of a branch into two or more branches. Table 2 shows the predicate and functions of the input gates for the Fig.2 submodel.

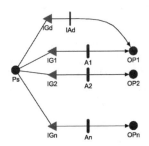

Fig. 2. SAN submodel for XOR-Split pattern

Table 2. Gate table for SAN model of Fig. 2

Gates	Enabling Predicate	Function
IGi	Ps->Mark()>0 && condi	Ps->Mark()--
IGd	Ps->Mark()>0 && (!(cond1) && ...&& !(condn))	Ps->Mark()--

Formal Definition: An XOR-Split submodel is a 6-tuple (P1, IG, IGd, Id, A, OP), where:

- $Ps \in P$, the first place of element
- IG = {IG1, IG2, ..., IGn}, a set of n input gates where $IGi \in IG$
- $IGd \in IG$, an input gate to choose the pre-defined branch
- $Id \in IA$, an instantaneous activity in the pre-defined branch
- A = {A1, A2, ..., An}, a set of n timed activities where $Ai \in TA$
- OP = {OP1, OP2, ..., OPn}, a set of n output places where $OPi \in P$

5.3 Discriminator Submodel

A SAN submodel for discriminator pattern is shown in Fig.3. The discriminator is a point in a workflow process that waits for *m* of the incoming branches to complete before activating the subsequent activity.

Table 3 shows the predicate and functions of the input and output gates for the Fig.3 submodel.

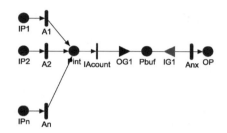

Fig. 3. SAN submodel for discriminator pattern

Table 3. Gate table for SAN model of Fig.4

Gates	Enabling Predicate	Function
IG1	Pbuf->Mark()>0	Pbuf->Mark()--

OG1		counter++
		if (counter==m) Pbuf->Mark()++
		if (counter==n) counter=0

Formal Definition: A discriminator pat submodel is a 9 tuple (IP, A, Pint, IG, Anx, IAcount, OP, OG, Pbuf), where:

- IP = {IP1, IP2, ..., IPn}, a set of n input places where $IPi \in P$
- A = {A1, A2, ..., An}, a set of n timed activities where $Ai \in TA$
- $Pint \in P$, an intermediate place
- IG = {IG1}, an input gate
- $Anx \in TA$, a next timed activity
- $OP \in P$, an output place
- OG = {OG1}, an output gate
- $IAcount \in IA$, an Instantaneous activity
- $Pbuf \in P$, a place

5.4 Other Workflow Patterns

Figures 4 through 11 show some other SAN submodels and their relative gate tables for other important workflow patterns.

6   An Example of Workflow Modeling with SANs

In this section a travel agency workflow model is represented in SANs. First, the travel agency process is modeled by workflow nets (WFNs), and then it is modeled in SANs, using the above proposed SAN submodels. In the travel agency process, first the customer is registered in the agency. Then an employee searches for opportunities which are communicated to the customer. Then the customer will be contacted to find out whether she or he is still interested in the trip of this agency and whether more alternatives are desired. There are three possibilities: (1) the customer is not interested at all, (2) the customer would like to see more alternatives, and (3) the customer selects an opportunity. If the customer selects a trip, the trip is booked. In parallel one or two types of insurance are prepared if they are desired. A customer can take insurance for trip cancellation or/and for baggage loss.

Note that a customer can decide not to take any insurance, just trip cancellation insurance, just baggage loss insurance, or both types of insurance. Two weeks before the start date of the trip the documents are sent to the customer. A trip can be cancelled at any time after completing the booking process (including the insurance) and before the start date. Please note that the customers, who are not insured for trip cancellation, can cancel the trip (but will get no refund).

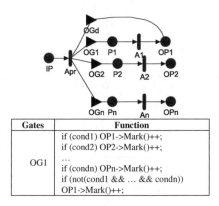

Gates	Function

OG1	if (cond1) OP1->Mark()++;
	if (cond2) OP2->Mark()++;
	...
	if (condn) OPn->Mark()++;
	if (not(cond1 && ... && condn))
	OP1->Mark()++;

Fig. 4. SAN submodel for OR-split

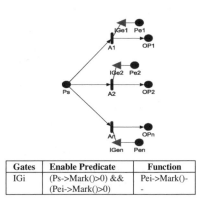

Gates	Enable Predicate	Function
IGi	(Ps->Mark()>0) && (Pei->Mark()>0)	Pei->Mark()- -

Fig. 5. SAN submodel for deferred choice

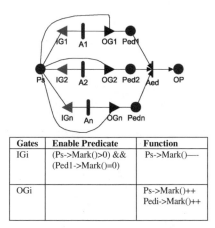

Gates	Enable Predicate	Function
IGi	(Ps->Mark()>0) && (Ped1->Mark()=0)	Ps->Mark()—
OGi		Ps->Mark()++ Pedi->Mark()++

Fig. 6. SAN submodel for interleaved parallel routing

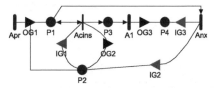

Gates	Enable Predicate	Function
IG1	(P2->Mark()>0) && (x>0)	P2->Mark()-- P1->Mark()--
IG2	(P2->Mark()>0) && (x==0)	P2->Mark()--
IG3	P4->Mark()==numinst	P4->Mark()=0
OG1		P1->Mark()++ P2->Mark()++ x=numinst
OG2		P2->Mark()++ x--
OG3		P4->Mark()++

Fig. 7. SAN submodel for multiple instance with a prior, design time knowledge

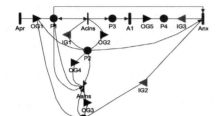

Gates	Enable Predicate	Function
IG1	(P2->Mark()>0) && (x>0)	P2->Mark()-- i++
IG2	(P2->Mark()>0) && (i==numinst)	P2->Mark()--
IG3	(P4->Mark()>0) && (i==numinst)	P4->Mark()--
OG1		P1->Mark()++ P2->Mark()++ P5->Mark()++ x=numinst i=0
OG2		P2->Mark()++ x--
OG3		P5->Mark()++ numinst++
OG4		P2->Mark()++ x++

Fig. 8. SAN submodel for multiple instances without a priori runtime knowledge

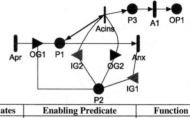

Gates	Enabling Predicate	Function
IG1	(P2->Mark()>0) && (x<0)	P2->Mark()--
IG2	(P2->Mark()>0) && (x==0)	P2->Mark()--
OG1		P1->Mark()++ P2->Mark()++ x=numinst
OG2		P2->Mark()++ x--

Fig. 9. SAN submodel for multiple instances without synchronization

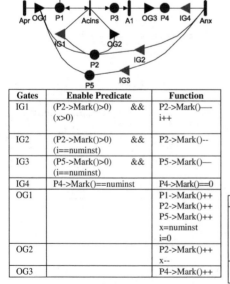

Gates	Enable Predicate	Function
IG1	(P2->Mark()>0) && (x>0)	P2->Mark()— i++
IG2	(P2->Mark()>0) && (i==numinst)	P2->Mark()--
IG3	(P5->Mark()>0) && (i==numinst)	P5->Mark()—
IG4	P4->Mark()==numinst	P4->Mark()==0
OG1		P1->Mark()++ P2->Mark()++ P5->Mark()++ x=numinst i=0
OG2		P2->Mark()++ x--
OG3		P4->Mark()++

Fig. 10. SAN submodel for multiple instances with a priori runtime knowledge

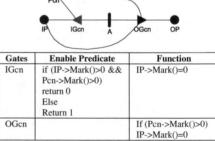

Gates	Enable Predicate	Function
IGcn	if (IP->Mark()>0 && Pcn->Mark()>0) return 0 Else Return 1	IP->Mark()=0
OGcn		If (Pcn->Mark()>0) IP->Mark()=0

Fig. 11. SAN submodel for cancel activity

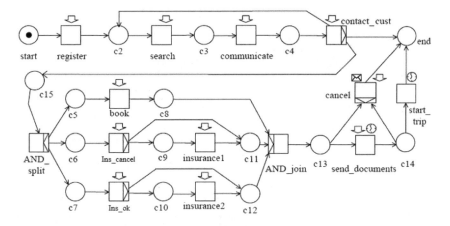

Fig. 12. Travel agency workflow model with workflow net

Based on this informal description, the travel agency workflow model using WFNs is shown in Fig. 12 and the travel agency workflow model using SANs is shown in Fig. 13, which is built by Möbius modeling tool. Table 4 shows the gate table for output gates for Fig. 13.

Variable assignments are shown in Table 5. The model has been evaluated by Möbius tool and the results are given in Table 6. The goal is to measure the throughput and utilization of travel agency system, at a particular time (*one month*).

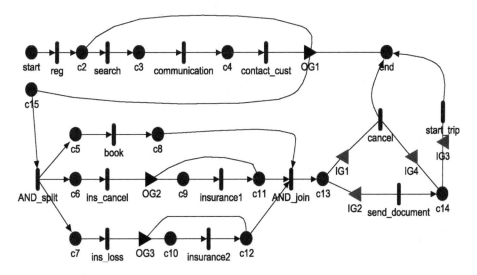

Fig. 13. Travel agency workflow model in SANs using the proposed submodels

Table 4. Gate table for SAN model of Fig. 13

Gates	Enabling Predicate	Function
IG1	(c13->Mark()>0) && (canceltrip=.true.)	c13->Mark()--
IG2	(c13->Mark()>0) && (canceltrip=.false.)	c13->Mark()--
IG3	(c14->Mark()>0) && (canceltrip=.false.)	c14->Mark()--
IG4	(c14->Mark()>0) && (canceltrip=.true.)	c14->Mark()--

OG1		if (cust_notaccepted==true) end->Mark()++
		if (cust_search==true) c2->Mark()++
		if (cust_accepted==true) c15->Mark()++

OG2		if (inscancel==true) c9->Mark()++ if (inscancel==false) c11->Mark()++
OG3		if (insloss==true) c10->Mark()++ if (insloss==false) c12->Mark()++

Table 5. Study variable assignments

Variable	Type	Range Type	Range	Variable	Type	Range Type	Range
canceltrip	bool	Fixed	false	fr_inscancel	double	Fixed	108.0
cust_accepted	bool	Fixed	true	fr_insloss	double	Fixed	108.0
cust_notaccepted	bool	Fixed	false	fr_insurance1	double	Fixed	20.0
cust_search	bool	Fixed	false	fr_insurance2	double	Fixed	80.0
fr_andjoin	double	Fixed	108.0	fr_reg	double	Fixed	100.0
fr_andsplit	double	Fixed	108.0	fr_search	double	Fixed	120.0
fr_book	double	Fixed	108.0	fr_senddoc	double	Fixed	91.9
fr_cancel	double	Fixed	12.624	fr_starttrip	double	Fixed	90.426
fr_com	double	Fixed	120.0	inscancel	bool	Fixed	true
fr_contact	double	Fixed	108.0	insloss	bool	Fixed	true

7   Conclusions

In this paper, the primitive elements of workflow and workflow patterns are modeled using stochastic activity networks (SANs) and a submodel has been given for each of them. The proposed SAN submodels can be used for workflow modeling.

By using the proposed SAN submodels for workflow patterns and considering the two different nondeterministic and stochastic settings of SANs, the behavior of work-flow systems can be modeled and then verified and/or evaluated using the existing tools like SharifSAN [22], SANBuilder [23] or Möbius [9].

According to the SANs capabilities and the workflow patterns presented in this pa-per, we intend to introduce a workflow modeling language based on SANs.

References

[1] Kamandi, A., Abdollahi Azgomi, M., Movaghar, A.: Derivation and Evaluation of OSAN Models from UML Models of Business Processes. In: Proc. of the IPM Int'l. Workshop on Foundations of Soft. Eng.: Theory and Practice (FSEN 2005), Tehran, Iran, pp. 287–305 (2005)

[2] Dehnert, J., Eshuis, R.: Reactive petri nets for workflow modeling. In: van der Aalst, W.M.P., Best, E. (eds.) ICATPN 2003. LNCS, vol. 2679, pp. 296–315. Springer, Heidelberg (2003)

[3] Mercx, H.: Business Process Management Notations within Business Process Management (2006)
[4] Kiepuszewski, B., ter Hofstede, A.H.M., van der Aalst, W.M.P.: Fundamentals of Control Flow in Workflows. BPM Center Report BPM-03-07 (2003),
     http://www.BPMcenter.org
[5] Movaghar, A.: Stochastic Activity Networks: A New Definition and Some Properties. Scientia Iranica 8(4), 303–311 (2001)
[6] Ouyang, C., van der Aalst, W.M.P., Dumas, M., ter Hofstede, A.H.M.: Translating BPMN to BPEL. BPM Center Report BPM-06-02, URL (2006),
     http://www.BPMcenter.org
[7] Russell, N., et al.: Workflow Control-Flow Patterns: A Revised View. BPM Center Report BPM-06-22 (2006), http://www.BPMcenter.org
[8] Sanders, W.H., Meyer, J.F.: Stochastic activity networks: Formal definitions and concepts. In: Brinksma, E., Hermanns, H., Katoen, J.-P. (eds.) EEF School 2000 and FMPA 2000. LNCS, vol. 2090, pp. 315–343. Springer, Heidelberg (2001)
[9] Sanders, W.H.: Möbius: User Manual, University of Illinois (2006)
[10] van der Aalst, W.M.P., ter Hofstede, A.H.M.: YAWL: Yet another Workflow Language. BPM Center Report BPM-05-01 (2005), http://www.BPMcenter.org
[11] van der Aalst, W.M.P.: The Application of Petri Nets to Workflow Management. The Journal of Circuits, Systems and Computers 8(1), 21–66 (1998)
[12] van der Aalst, W.M.P., et al.: Workflow Patterns. BPM Center Report BPM-03-06 (2003), http://www.BPMcenter.org
[13] van der Aalst, W., van Hee, K.: Workflow Management Models, Methods and Systems, Eindhoven University of Technology (2000)
[14] van der Aalst, W.M.P.: Challenges in Business Process Management: Verification of Business Processes Using Petri Nets. BPM Center Report BPM-03-11 (2003),
     http://www.BPMcenter.org
[15] van der Aalst, W.M.P., ter Hofstede, A.H.M.: Workflow Patterns: On the Expressive Power of Petri Net-Based Workflow Languages. In: Proc. of the Fourth Workshop on the Practical Use of Coloured Petri Nets and CPN Tools (CPN 2002), Aarhus, Denmark. DAIMI, vol. 560, pp. 1–20 (2002)
[16] Martin, V., Schahram, D.: A View Based Analysis of Workflow Modeling Languages. In: Proc. of the 14th Euromicro Int'l. Conf. on Parallel, Distributed, and Network-Based Processing (2006)
[17] Stephen, A.: Process Modeling Notations and Workflow Patterns,
     http://www.omg.org/bp-corner/pmn.htm
[18] Business Process Execution Language (BPEL),
     http://www.computerworld.com/comments/node/102580
[19] Workflow Pattern, http://www.workflowpatterns.com
[20] Introduction to BPMN, http://www.bpmn.org/Documents
[21] An Introduction to BPEL,
     http://www.developer.com/services/article.php/3609381
[22] Abdollahi Azgomi, M., Movaghar, A.: A Modeling Tool for A New Definition of Stochastic Activity Networks. Iranian J. of Science and Technology, Transaction B (Technology) 29(B1), 79–92 (2005)
[23] Abdollahi Azgomi, M., Movaghar, A.: A Modelling Tool for Hierarchical Stochastic Activity Networks. Simulation Modelling Practice and Theory 13(6), 505–524 (2005)

AIDSLK: An Anomaly Based Intrusion Detection System in Linux Kernel

Negar Almassian[1], Reza Azmi[2], and Sarah Berenji[3]

[1] Sharif University of Technology, School of Science and Engineering-International Campus
– Kish Island, Iran
almassian@kish.sharif.edu
[2] Azzahra University, Computer Department
azmi@alzahra.ac.ir
[3] Sarah.berenji@gmail.com

Abstract. The growth of intelligent attacks has prompted the designers to envision the intrusion detection as a built-in process in operating systems. This paper investigates a novel anomaly-based intrusion detection mechanism which utilizes the manner of interactions between users and kernel processes. An adequate feature list has been prepared for distinction between normal and anomalous behavior. The method used is introducing a new component to Linux kernel as a wrapper module with necessary hook function to log initial data for preparing desired features list. SVM neural network was applied to classify and recognize input vectors. The sequence of delayed input vectors of features was appended to examine the effectiveness of the system call consecution. The evaluation method for the Intelligent Intrusion Detection system was simulation method and improvement in some metrics such as accuracy, training time and testing time in comparison with the other similar systems.

Keywords: Intrusion Detection System (IDS), Kernel module, Support Vector Machine (SVM), Virtual File System (VFS).

1 Introduction

As the computer systems are used widely in our daily activities and the complexity of these systems are increasing all the times, the computer systems are becoming increasingly vulnerable to different attacks. The existence of flaws either in operating systems level or application programs level can be taken as an opportunity by these attacks. The final goal of attackers is to abuse the traditional security mechanism of the system and executes some specific actions to compromise security goals; therefore, in order to increase the security level, these flaws should be eliminated.

Security is based on four concepts of Confidentiality, Integrity, Availability and Authenticity [14]. Intrusion detection does not attempt to prevent attack occurrence. It has the duty of alerting the system administrator in the time of system violation. As such it is a reactive rather than proactive [3].

S.K. Prasad et al. (Eds.): ICISTM 2009, CCIS 31, pp. 232–243, 2009.
© Springer-Verlag Berlin Heidelberg 2009

The two main intrusion detection techniques which are widely used are rule or signature-based misuse detection and anomaly detection. Anomaly detection techniques mainly focus on defining normal activity patterns and any current activity that deviates from these known normal patterns is detected as an intrusion. In contrast, misuse detection techniques attempt to build and memorize a model of attack signatures, when a current signature matches with one of these predefined signatures it is classified as an intrusion. Both of these techniques have two main drawbacks: false positives which are due to misclassification of normal behavior as an attack and false negatives in which a novel attack is ignored and classified as a normal behavior.

Our approach to detecting irregularities in the behavior of system calls is to redirect recalled system calls to trace them and associate parameters in a log file. In this level of behavior monitoring we can ignore the application based flaws such as miss configuration or crash of an application. In this method, in order to collect a proper dataset, a Linux kernel module was developed and the feature list was directly collected from kernel which is a novel method of preparing dataset with a wide attributes of system calls. To have deeper insight to the kernel and to have more accuracy, our attention was concentrated on the file system of the Linux kernel.

The machine learning method used for IDS is support vector machine. Support vector machine computes the maximal margin for separating data points. Only those patterns closed to the margin can affect the computations of that margin and other points do not have any impact on the final result. These closest patterns to the margin are called support vectors.

In this paper, an investigation and an evaluation are carried out on the performance of SVM; this process is done by making a distinction between normal and abnormal behavior of the file system of the Linux kernel and tracing the invoked system calls during all system activities. The behavior model of normal activities was randomly selected by a random generator of normal system calls of the Linux file system. In order to simulate an attack, the two famous samples of User to Root and Denial of Service exploit codes were executed.

Using one-class SVM experiment with the feature list of normal recalled system calls as the train set, improved the robustness of the anomaly based IDS. To optimize the performance of the SVM kernels, the grid search method described in [5] was applied. With combination of normal and anomalous feature list we evaluated a SVM binary-feature classifier with the use of linear, polynomial and radial basis function kernels. Furthermore, a sequence of delayed vectors of invoked system calls with their parameters was used for implementing one of the SVM classifier experiments to investigate the impact of this issue on the performance of the system. The libsvm-2.85 software is used to implement all of the experiments. The result indicated the improvement of the accuracy with a very small performance overhead.

Following the mentioned concepts above, section 2 deals with related works to the Anomaly Based Intrusion Detection Systems using SVM'. Section 3 discusses the architecture, design and implementation of our proposed Intrusion Detection System and proposed kernel module. Section 4 represents proposed method for constructing the feature lists and model selection. Experiments results and performance evaluation of proposed kernel based IDS are discussed in section 5. In section 6, conclusions, contributions and future scope of our work is presented.

2 Related Works

Support Vector Machines were first introduced as a machine learning method by Cortes and Vapnik (1995) [4] to observe a two-class training set and specify a maximum-margin separating hyperplane between the datapoints of two classes in order to find the optimal hyperplane that can be generalized well to unseen data. Joachims (1998) [8] and Dumais et al. (1998) [6], performed the text categorization task of the Reuters database; they compared different machine learning methods. After that Liu et al. (2002) [10] and LeCun et al. (1995) [9], by using SVM in the field of handwritten digits recognition, achieved the best results. They also used SVM in active learning. The random sampling was used to improve the training of SVM [15] and many people worked on sampling to speed up the training and improving the classification accuracy such as [1], [13] and [20].

Support Vector Machines have been applied in many intrusion detection systems such as Upadhyaya et al. (2001) worked on "An analytical framework for reasoning about intrusions" [16], Kayacik et al. (2003) in the"Self-organizing maps (SOM) and support vector machine" [2], Wang et al (2003) in [18] used "one class SVM" based on one set of examples belonging to a particular class and no negative examples rather than using both of them, Laskov et al. (2004) worked on "Visualization of anomaly detection using prediction sensitivity" [12] which is an anomaly based intrusion detection system.

3 Design and Implementation of Proposed Kernel Based IDS

This work proposes an Intelligent Hosed-Based off-line Intrusion Detection System using Anomaly Detection approach. The proposed IDS was designed by the present researchers to implement two aims: firstly, to detect novel attacks by using one-class SVM with training the system just with attribute list of normal behavior of the system and any deviation from this behavior is detected as a new attack. Secondly we have added some anomalous behavior of the previous known attacks to prevent the repetition of that attack.

Figure1 shows the whole architecture of our proposed IDS, called AIDSLK (Anomaly based Intrusion Detection System in Linux Kernel). AIDSLK consists of two main primary parts, preparation on the left side and classification on the right side of the figure.

To prepare an adequate feature list, a new component was introduced to Linux kernel as a wrapper module with necessary hook function to log initial data for preparing desired features list. The other part, as a regular pattern recognition system, consists of a learner and classifier subsystem and training model to update classifier parameters. As it was mentioned before we used SVM classifier to classify and recognize input vectors.

In the classification part, three different classifiers were applied, one-class SVM, binary feature classifier, and lastly the sequence of delayed vectors as an input for the binary classifier.

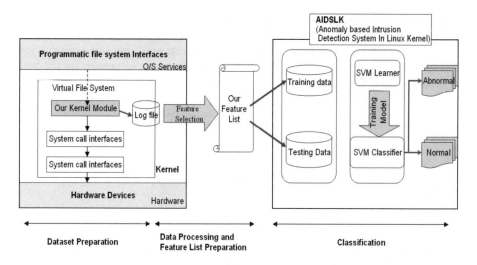

Fig. 1. Whole architecture of our proposed IDS

We implemented a module to change the way a system call works and also wrote our own function to implement the functionality we had expected, this function calls printk to log the parameter of the file system of Linux kernel's system calls such as open(), close(), write(), chmod(), chown(), link(), unlink(), access(), etc. In other word, it calls the original system call function with the same parameters to actually run the normal procedure. Also in each system call invocation the associated uid, gid and pid will be logged and written to the log file by printk. In this module the pointer which points at sys_call_table is changed to point to our function. After the executionn of our function, it is necessary to call the original function, because we might turn off the system later and it is not safe to leave the system in an unstable state, it is important for module_exit to restore the table to its original state.

In order to simulate the normal behavior of the user, a list of normal system calls of the file system was prepared in which each number represents a specific system call and a random generator was written that in each execution generates 10 random numbers which each number is correlated to a normal behavior of the user (such as create and copy a file or folder, link and unlink a file, change the file access or owner, etc.), then the selected system calls will be executed. In order to simulate an attack to the system we have collected the two famous samples of User to Root and Denial of Service exploit codes from which intruders execute them to take advantage of the existing vulnerabilities of the system to enter the system and reach the privileged access or unauthorized data access. Our module logged the invoked system calls during normal and abnormal behavior and provided us with some log files for our attribute list.

4   Feature Selection

The next step of dataset collection is the feature selection of raw attribute list. We used two different scripts for feature selection:

Implementing Expanded Feature List: We select the following fields as our dataset features from our log files. Some of these fields always have value in all the system calls, such as system call number, upid, uid, gid and reiteration, but some of them just have value for the specific system calls such as flags that have value only for the open system call (figure 2).

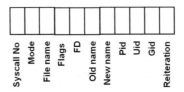

Fig. 2. Initial data collection

In our initial feature list outputs, two fields of mode and flag had heavy weight in comparison to the other fields, so they were most determining features on the training process; therefore, we had to dispread the weight of training in all fields evenly.

The flag field was broken into its binary form. Therefore, we added 14 binary fields instead of one flag field in our dataset. Each of them can be 0 or 1. For example, read only flag has 0x0001 Hex value which is the lowest bit and its equal binary value is 0000,0000,0000,0001. So we assigned the last field of flag in the dataset, 16th field, to this number. When the 16th field of dataset set to 1, it means the read only flag is set.

SVM requires that each data instance be represented as a vector of real numbers. So the categorical attributes such as file name, new name and old name, have to be converted into numeric data. We preferred using m-numbers to represent an m-category attribute for the File Path in these three fields. Only one of the m numbers is equals to one at the same time, and others are zero. So our first dataset is like figure 3.

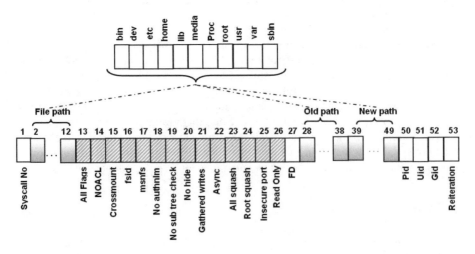

Fig. 3. Final dataset collection

The Implementation of Appending Consecutive Records: We tried another experiment and developed a method for tracing our system call behavior using a fixed-length window. Conceptually, we took a small fixed size window and slid it over each trace, recording which calls precede the current call within the sliding window.

In order to setup this experiment, we assumed every 4 sequenced records as one vector and concatenated them to make a new data set. We also prepared the 5 and 6 consecutive delayed instances to analyze the effect of increasing the length of consecutive sample of system calls attribute list in classification accuracy.

5 Evaluation

We have implemented three experiments, including one-class SVM, Binary classification and sequence of delayed samples. Each of our experiments had two phases, namely, "Normal classification phase" with default parameters and an "Optimization phase". The first experiment as a One-class SVM classifier with 11 normal attributes, was set up to identify the unknown attacks to implement an Anomaly Based Intrusion Detection System. The second Experiment was set up to prevent known attack repetition with 53 normal and anomalous attributes. The third experiment was set up to examine the effect of changing the system architecture of the system with sequence of 4, 5, and 6 consequent sample of feature list as an input instance which have 212, 265 and 318 features. The data set for our experiments contained almost 50000 records. This data set consisted of 30000 lines of attack attribute vectors, and 20000 records of normal data. For preparing training and testing datasets we used a random subset selection script which exists in the libsvm library; and for the optimization phases we used the grid search tool to find the best parameter for the classification; this script is also included in the libsvm library. All the intrusion detection models were trained and tested with the same set of data. The normal data belongs to class1 with label class1; attack class belongs to class2 with label class of -1.

5.1 Kernels Variety Experiment

In this experiment one-class SVM classifier was constructed. The trained dataset was included just the normal attributes and test dataset was partitioned into two classes of "Normal" and "Attack" patterns. The objective was to identify the deviated patterns from normal trained patterns. As mentioned before, various types of kernel function were tried to find the best kernel and the results showed that RBF kernel function had the best performance for our goal. 10% of whole datasets was randomly selected as the training dataset and used to obtain the best result. All the classification processes performed in this work were repeated for five times and all the results presented here are the average of these five repetitions. Table 1 shows the result of our experiment for 1000, 2000, 3000, 4000 and 5000 training patterns; also you can see the rate of accuracy growth by increasing the number of trained data up to 10% of the whole amount of data in Figure 4.

Table 1. Performance of the One-class SVM classification

# of train samples	#of test samples	One-class SVM		
		Train time (s)	Test time (s)	Accuracy (%)
1000	48375	0.66	8.3	72.714
2000	47375	1.918	13.04	72.72
3000	46375	3.304	17.68	76.156
4000	45375	4.938	22.81	74.962
5000	44375	7.026	28.982	77.314

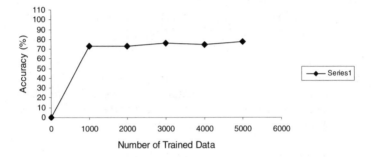

Fig. 4. Effect of increasing the number of trained data on One-class classification accuracy

As you can see in Figure 4 with increasing 1000 training samples in each step the accuracy increased gradually and in the range of 10% of the samples we achieved an acceptable accuracy. In the next experiment, a binary classifier is applied to improve the accuracy of the system.

5.2 C-SVC Binary-Classification Experiment

According to the result of the first experiment we achieved an acceptable level of accuracy, but we decided to improve the results. For this reason we had to repeat model selection. To achieve best model of classification we found out that there were different types of SVM classification algorithms including: C-SVC, nu-SVC, one-class SVM, epsilon-SVR, nu-SVR, and the libsvm software supported them, so we tried another algorithm called C-SVC which used binary-feature classification method. In this kind of SVM classification the train process needs two classes of attributes to train and then predict the test dataset classes. In this case, we added the attack instances feature list to the normal features, so unlike the first experiment, we trained our machine with two classes of positive and negative attributes.

In the next step, the best kernel function was selected. As we mentioned before the Kernel option defines the feature space in which the training set examples will be classified. We ran our experiment on the range of 500 input instances and increased the number of samples 100 by 100. You can see in Table 2 the results of these experiments.

Table 2. Kernel Function's performance comparison for binary classification

# of train	# of Test	Sigmoid			Linear			RBF		
		Train Time (s)	Test Time (s)	Accu-racy (%)	Train Time (s)	Test Time (s)	Accu-racy (%)	Train Time (s)	Test Time (s)	Ac-cura-cy (%)
100	49275	0.05	0.82	58.65	24.45	0.61	59.00	0.27	1.81	92.42
200	49175	0.07	1.29	57.26	88.85	1.43	62.29	0.12	2.08	95.86
300	49075	0.38	2.92	58.63	207.78	1.32	59.58	0.04	2.50	96.43
400	48975	0.28	2.96	58.62	300.75	1.82	62.69	0.06	3.11	97.21
500	48875	0.34	2.83	58.64	222.45	1.54	62.85	0.06	3.17	97.87

Our results showed that the RBF kernel option often performs well on most data sets. The test and trained time is much better than the others where its accuracy is significantly higher than sigmoid and linear. We therefore used the RBF kernel for our next experiments. The sigmoid kernel gives the least accuracy for classification. In case of linear and sigmoid kernel functions both experiments result in almost the same accuracy whereas the train time is increased considerably in the sigmoid kernel.

The graphical view of the different kernel functions accuracy is demonstrated in Figure 5.

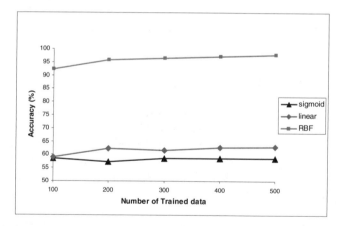

Fig. 5. Kernel Functions' classification accuracy comparison in binary classification

According to the graph there is only a small difference in the accuracy for the sigmoid and linear kernel, but there is a significant difference for RBF kernel against two others.

In this step we ran another experiment to compare the results of the one-class classification and Binary classification in a fair situation. We selected the train inputs in the same range of data (10%) and ran a C-SVC classification with RBF kernel and with the libsvm default parameters C and γ. We saw a considerable improvement in the accuracy. We improved the average accuracy from 77.31% to 99.69% and in the security world 18 per cent increase in the accuracy is a very noticeable success. The results of the experiments are illustrated in Table 3. In Figure 6 there is a comparison between the results of the One-class SVM with the optimized binary-class SVM (C-SVC) using the best parameter of C and γ.

Table 3. One-class SVM and C-SVC comparison

# of train	#of test	One-class SVM			Binary class SVM		
		Train time	Test time	Accuracy	Train time	Test time	Accuracy
1000	48375	0.66	8.3	72.714	0.196	4.68	98.386
2000	47375	1.918	13.04	72.72	0.536	6.248	99.006
3000	46375	3.304	17.68	76.156	0.84	7.75	99.536
4000	45375	4.938	22.81	74.962	1.12	8.26	99.502
5000	44375	7.026	28.982	77.314	1.56	8.794	99.694

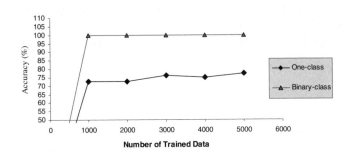

Fig. 6. One-class and Binary classification comparison

Then we optimized the RBF kernel parameters to achieve a better result. We have used a script integrated with libsvm software which is used a grid search to find the best C and γ. Here we could increase the average accuracy to excellent value of 99.88%.

The average result of the optimization experiment in range of 1000 to 5000 is summarized in Table 4 and Figure 7.

Table 4. Binary classification performance comparison between default and optimized parameters

# of train	#of test	Default Binary Classification			Optimized Classification		
		Train time	Test time	Accuracy	Train time	Test time	Accuracy
1000	48375	0.196	4.68	98.386	0.65	3.87	99.890
2000	47375	0.536	6.248	99.006	0.58	1.81	99.930
3000	46375	0.84	7.75	99.536	0.047	1.78	99.900
4000	45375	1.12	8.26	99.502	0.51	3.15	99.820
5000	44375	1.56	8.794	99.698	1.72	10.77	99.860

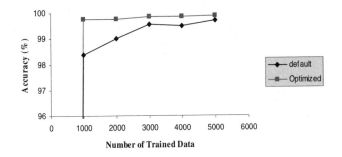

Fig. 7. Binary classification accuracy comparisons with default and optimized parameters

5.3 Sequential Instances Experiment

The idea of the last experiment was formed when we wanted to examine if the random selection of instances had negative effect on the classification or not. To check this fact we decided to consider some consecutives instances of system call attributes as a vector of training input.

In the previous experiments with random selection of input vectors we trained our system using just a vector with function $y(x) \equiv f(X(n))$ and reached reasonable results, in this experiment we sought to study the impact of a set of delayed sample of vectors with function $y(x) \equiv f(X(n)X(n-1)X(n-2)...)$. We tried different sizes of fixed length window including 4, 5 and 6 delayed instances. The dataset we have selected is the same as the other experiments but because of appending the lines, the number of test records was decreased; the number of trained data, however, was the same as before [1000-5000]. The result of this experiment is demonstrated in Table 5 and Table 6.

Table 5. Accuracy comparison between 1 sample of input and 4 consecutives samples

# of pattern	# of train	#of test	1 sample			4 consequences sample		
			Train time	Test time	Accuracy	Train time	Test time	Accuracy
12381	1000	11381	0.65	3.87	99.890	0.326	2.214	98.522
12381	2000	10381	0.58	1.81	99.930	1.34	3.43	99.803
12381	3000	4.498	0.047	1.78	99.900	2.118	3.276	99.338
12381	4000	8381	0.51	3.15	99.820	2.31	4.056	99.578
12381	5000	7381	1.72	10.77	99.860	3.056	2.942	99.890

Table 5 indicates that appending 4 samples of input has almost the same accuracy, and the small decline in the result is because of the confusion of the very similar and duplicated features. With a certain number of inputs for example 100 samples, we increased the length of the input vector by increasing the number of appended instances the accuracy considerably dropped. Table 6 clearly shows this negative effect of feature expansion.

Table 6. Performance comparison of the different window size input vectors

Length of sequence	Accuracy (%)	Best C	Best G
1	92.416	1.85	2.69E-04
4	89.646	1.85	5.25547E-05
5	83.952	2.00	3.51758E-05
6	81.014	1.9	4.25E-01

6 Conclusion

This paper dealt with the use of support vector machines for Intelligent Anomaly Based Intrusion Detection System. We confirmed the impressive performance of Binary classification using SVM with RBF kernel function, also we emphasized on robustness and generalization of one-class SVM classification. This result shows that if we equally spread the weight of training attributes for each vector, we can successfully train our system with random selected instances. Moreover, we expanded the feature list efficiently for the first time to 4 consecutive instances for the binary classification tasks. The RBF kernel was applied and the result was acceptable. With increasing the length of sequence the result had a significant reduction because of the confusion resulted from the similarity and closeness of attributes in the consecutive vectors.

With the use of only 10% of dataset as the train data and a binary-classification with RBF kernel, we achieved 99.88% accuracy; this result shows that the SVM is a powerful intelligent classifier with great ability of classification with a small subset of training data and very small overhead impact on the normal system performance.

The major contributions of this work are: first, a new technique for dataset collection, a novel feature list structure was designed for the training and testing process, also the method of spreading Flag and File Path fields is a novel contribution in this work. The second is to use the sequence of the delayed system calls pattern schema as input vectors SVM.

In the future, we will implement our IDS with other different classifiers such as K-nearest neighbors and naive bayes classifier and compare the results with the SVM classification. We suggest research on a hybrid model of intrusion detection as the future work. In this proposed, model the data sets are first passed through the one-class SVM classifier and check the attributes with previous known attacks and then go to the binary-classifier.

References

1. Balcazar, J., Dai, Y., Watanabe, O.: A random sampling technique for training support vector machines for primal-form maximal-margin classifiers, algorithmic learning theory. In: Abe, N., Khardon, R., Zeugmann, T. (eds.) ALT 2001. LNCS, vol. 2225, p. 119. Springer, Heidelberg (2001)
2. Kayacik, H.G., Zincir-Heywood, A.N., Heywood, M.I.: On the capability of an SOM based intrusion detection system, vol. 3, pp. 1808–1813. IEEE, Los Alamitos (2003)

3. Chavan, S., Shah, K., Dave, N., Mukherjee, S.: Adaptive Neuro-Fuzzy Intrusion Detection Systems. In: Proceedings of the International Conference on Information Technology. IEEE, Los Alamitos (2004)
4. Cortes, C., Vapnik, V.: Support-vector networks. Machine Learning 20(3), 273–297 (1995)
5. Cunningham, R., Lippmann, R.: Improving Intrusion Detection performance using Keyword selection and Neural Networks. MIT Lincoln Laboratory (2002)
6. Dumais, S., Platt, J., Heckerman, D., Sahami, M.: Inductive learning algorithms and representations for text categorization. In: Proceedings of the Seventh International Conference on Machine Learning. ACM Press, New York (1998)
7. Forrest, S., Hofmeyr, S., Somayaji, A., Longstaff, T.A.: A Sense of Self for Unix Processes. In: Proceedings of the 1996 IEEE Symposium on Computer Security and Privacy. IEEE Computer Society Press, Los Alamitos (1996)
8. Joachims, T.: Text categorization with support vector machines: learning with many relevant features. In: Nédellec, C., Rouveirol, C. (eds.) ECML 1998. LNCS, vol. 1398, pp. 137–142. Springer, Heidelberg (1998)
9. LeCun, Y., Jackel, L.D., Bottou, L., Brunot, A., Cortes, C., Denker, J.S., Drucker, H., Guyon, I., Muller, U.A., Sackinger, E., Simard, P., Vapnik, V.: Comparison of learnin algorithms for handwritten digit recognition. In: Fogelman, F., Gallinari, P. (eds.) International Conference on Artificial Neural Networks, Paris, EC2 & Cie, pp. 53–60 (1995)
10. Liu, C., Nakashima, K., Sako, H., Fujisawa, H.: Handwritten digit recognition using state-of-the-art techniques. In: FHR 2002, pp. 320–325 (2002)
11. Marin, J., Ragsdale, D., Surdu, J.: A Hybrid Approach to the Profile Creation and Intrusion Detection. In: Proceedings of the DARPA Information Survivability Conference and Exposition – DISCEX 2001 (June 2001)
12. Pavel, P., Laskov, C., Schäfer, Kotenko, I.: Visualization of anomaly detection using prediction sensitivity. In: Proc. DIMVA, pp. 71–82 (2004)
13. Shih, L., Rennie, Y.D.M., Chang, Y., Karger, D.R.: Text bundling: statistics-based data reduction. In: Proceedings of the 20th International Conference on Machine Learning (ICML), Washington, DC, pp. 696–703 (2003)
14. Tanenbaum, A.S.: Modern Operating Systems. Prentice Hall, Englewood Cliffs (1992)
15. Tufis, D., Popescu, C., Rosu, R.: Automatic classification of documents by random sampling. Proc. Romanian Acad., Ser. 1(2), 117–127 (2000)
16. Upadhyaya, S., Chinchani, R., Kwiat, K.: An analytical framework for reasoning about intrusions. In: Proceedings of the IEEE Symposium on Reliable Distributed Systems, New Orleans, LA, pp. 99–108 (2001)
17. Vasudevan, S.: Immune Based Event-incident Model For Intrusion Detection Systems: A Nature Inspired Approach To Secure Computing, M.S, Thesis, Kent State University, USA (2007)
18. Wang, K., Stolfo, S.J.: One class training for masquerade detection. In: Proceedings of the 3rd IEEE Conference, Data Mining Workshop on Data Mining for Computer Security, Florida (2003)
19. Wei, Ch., Chang, Ch., Lin, J.Ch.: A Practical Guide to Support Vector Classification (July 2007)
20. Yu, H., Yang, J., Han, J.: Classifying large data sets using SVM with hierarchical clusters. In: Proceedings of the SIGKDD 2003, Washington, DC, pp. 306–315 (2003)

New Quantization Technique in Semi-fragile Digital Watermarking for Image Authentication

Raghu Gantasala and Munaga V.N.K. Prasad

Institute for Development and Research in Banking Technology,
Hyderabad, India
gantasalahcuhyd@yahoo.co.in, mvnkprasad@idrbt.ac.in

Abstract. The Internet has been widely used for the distribution, commercialization and transmission of digital files such as images, audio and video. The growth of network multimedia systems has magnified the need for image copyright protection. In this paper we proposed new method for semi-fragile digital watermarking scheme for image authentication. The watermark is embedded in the discrete wavelet domain of the image by quantizing the corresponding wavelet coefficients. Using the proposed method the image distortion is decreased compared to the other techniques and the quantization parameter is a small value. It also robust against attacks including EZW compression, JPEG compression and JPEG 2000 compression algorithms.

Keywords: Digital watermarking, authentication, security, wavelet transformation, semi-fragile watermark.

1 Introduction

Internet and the World Wide Web have gained great popularity, it has become commonplace to make collections of images, stored on Internet attached servers, accessible to vast number of other through the internet. Since multimedia technologies have been becoming increasing sophisticated in the rapidly growing Internet applications, data security including copyright protection and data integrity detection has raised tremendous concerns [13-15]. Some medical applications offer images taken from equipment, such as computed terminology and ultrasonography to be processed in distributed environment [16]. Digital watermarking is a solution for restraining the falsification and illegal distribution of digital multimedia files. An imperceptible signal "mark" is embedded into the host image, which uniquely identifies the ownership. After embedding the watermark, there should no perceptual degradation. These watermarks should not be removable by unauthorized person and should be robust against intentional and unintentional attacks. As discussed in [25], attacks can be classified into four categories: 1) removal attacks; 2) geometrical attacks; 3) cryptographic attacks; and 4) protocol attacks. The robustness of the current watermarking methods has been examined with respect to removal attacks or geometrical attacks or both. In particular, removal attacks contain operations including filtering, compression, and noise adding, that more or less degrade the quality of the media data. Among the currently known attacks [25], the collusion attack [26-28], which is a

S.K. Prasad et al. (Eds.): ICISTM 2009, CCIS 31, pp. 244–255, 2009.

removal attack, and the copy attack [29, 30], which is a protocol attack, are typical examples of attacks that can achieve the aforementioned goal.

There are two domain-based watermarking techniques: 1.Spatial domain [8-12] and 2. Frequency domain [1-7]. In the spatial domain insert watermark into a host image by changing the grey levels of certain pixels. In spatial domain [8-12], one can simply insert a watermark into a host image by changing grey levels of some pixels in the host image, but the inserted information could be easily detected by computer analysis. Transform domain watermarking techniques are more robust in comparison to spatial domain methods. In the frequency domain insert watermark into frequency coefficients of the image transformed by discrete fourier transform (DFT), discrete cosine transform (DCT), or discrete wavelet transform (DWT). Among the transform domain watermarking techniques discrete wavelet transform (DWT) based water-marking techniques are gaining more popularity because of superior modelling of human visual system (HVS) [31]. In general, the frequency-domain watermarking would be robust since the embedded watermarks are spread out all over the spatial extent of an image [5]. The watermarks embedded into location of large absolute values of the transformed image, the watermarking technique would become more robust.

Robust watermarking is mainly aimed at copyright protection [17-19]. On the other hand, the fragile watermark is extremely sensitive to any modification of the digital content, and is easily corrupted by little difficulty [20, 21]. This feature makes it use-ful for authenticity verification. A semi-fragile watermark combines the properties of fragile and robust watermarks [22]. Semi-fragile watermark can differentiate between localized tampering and information preserving and lossy transformations [23, 24]. The primary applications of semi-fragile watermarks involve tamper detection and image authentication, the overall requirements of a semi-fragile watermarking system resembles those of fragile watermarking systems. In general, watermarking technique generates the watermark signal by using a pseudo-random number sequence [5, 32], which makes the watermark unable to reflect the character of the owner or the origi-nal media. Some researchers use a binary image [33], a grey image [34, 35], or a color image for a watermark, such as the copyright symbol, which makes the watermark more meaningful, but causes the problem of insecurity. Another type of watermark is based on the original media. It is adaptive to the original host and reflects its charac-teristics [36, 37].

In this paper, a novel watermark-embedding algorithm is presented in DWT do-main. First, the original image (grey) is transformed into wavelet coefficients up to specified level (L). The wavelet coefficients are selected by user specified key for embedding the watermark. In this, binary watermark is embedded in to wavelet coef-ficients by quantizing the corresponding wavelet coefficients. To show the validity of the proposed method, the watermarked images are tested for different types of attacks and results are compared with the existing methods. The significant advantage of our proposed method over the methods of Deepa Kundur [1] and Rongsheng XIE [2] is that the watermark can be added to each wavelet coefficient with maximum strength without any perceptual degradation. This is because the quantization parameter for embedding the watermark is fixed to a small value in each level instead of increasing the quantization parameter in each level as in the method of Deepa Kundur [1].

The rest of this paper is organized as follows. In Section 2, presents watermark embedding, quantization process and tamper assessing methods. Proposed method for watermark embedding is presented in section 3. The experimental results are shown in Section 4. Finally, the conclusions are stated in section 5.

2 Review of Deepa Kundur and Rongsheng XIE Techniques [1,2]

Deepa Kundur et al.'s [1] embed the watermark by quantizing the coefficients to a prespecified degree, which provides the flexibility to make the tamper-proofing technique as sensitive to change in the signal as desired. Two basic approaches i.e. Watermark embedding and tamper-proofing proposed in [1].

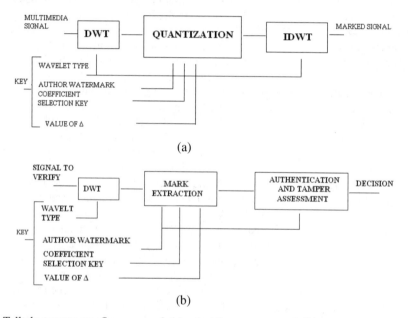

Fig. 1. Telltale tamper-proofing approach.(a)embedding process and (b) tamper assessment process

2.1 Watermark Embedding

In watermark embedding comprises three major steps; during first step it computes the detailed wavelet coefficients up to user specified level. Then watermark is embedded into corresponding wavelet coefficients (selected by user-defined key) through an appropriate quantization procedure. Finally inversing the wavelet transform gives the watermarked image. The general scenario is shown in Fig.1 (a). A validation key comprised of the author's watermark, a coefficient selection key, the quantization parameter Δ, and possibly the specific mother wavelet function are necessary for embedding and extracting the mark. Watermark extraction on a given image is performed as shown in Fig. 1(b). The Lth level discrete wavelet transform (DWT) is applied to the

given image and the coefficient selection key is used to determine the marked coefficients. A quantization function $Q(.)$ is applied to each of the coefficients to extract
the watermark values.

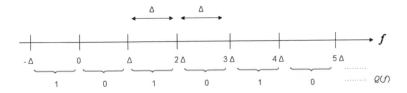

Fig. 2. Quantization Function

2.2 Quantization Process

For an arbitrary wavelet transform, the detail wavelet coefficients are real numbers.
Quantization on the wavelet coefficients in the following manner. Every real number
is assigned a binary number, as shown in Fig.2. The quantization function $Q(f)$,
which maps the real number set to $\{0, 1\}$.

$$Q(f) = \begin{cases} 0, & if \quad r\Delta \le f < (r+1)\Delta \quad for \quad r = 0, \pm 2, \pm 4, ... \\ 1, & if \quad r\Delta \le f < (r+1)\Delta \quad for \quad r = \pm 1, \pm 3, \pm 5, ... \end{cases} \quad (1)$$

Where Δ is a positive real number called the quantization parameter and is shown in
Fig.2. The following assignment rules are used to embed the watermark bit into the
selected wavelet coefficient specified by the user. If $f_{k,l}(m,n) = w(i)$, then no
change in the coefficient is necessary. Otherwise, change $f_{k,l}(m,n)$ so that
$Q(f_{k,l}(m,n)) = w(i)$, using the following assignment:

$$f_{k,l}(m,n) = \begin{cases} f_{k,l}(m,n) + \Delta & if \quad f_{k,l}(m,n) \le 0 \\ f_{k,l}(m,n) - \Delta & if \quad f_{k,l}(m,n) > 0 \end{cases} \quad (2)$$

Rongsheng XIE et al.'s [2] observed that the inherent causation introduces the numerical sensitivity problem in the implementation of (2) in the Deepa Kundur et al.'s
[1] quantization method and changed $f_{k,l}$ (m, n) if Q ($f_{k,l}$ (m, n)) $\ne$ w(i) to satisfy the
condition Q($f_{k,l}$ (m, n)) = w(i). In this method [2], if the user defined random bit
sequence satisfies Q ($f_{k,l}$ (m, n)) = w(i), then there is no need to change the wavelet
coefficients $f_{k,l}$ (m, n). The marked image is just the original one since no distortion
is introduced. In this method no need to apply the DWT, because no watermark is
embedding into the original image. Authentication of the image is possible only if the
extracted watermark is identical to the embedded. If authentication fails then we

employ tamper assessment to determine the credibility of the modified image. To compute the extent of Tamper-proofing, following tamper assessment function (TAF) is used [1].

$$TAF(w, \tilde{w}) = \frac{1}{N_w} \sum_{i=1}^{N_w} w(i) \oplus \tilde{w}(i) \qquad (3)$$

3 Proposed Technique

The proposed method embeds watermark by decomposing the host image and the watermark using wavelet transform. To avoid the numerical sensitivity Deepa Kundur et al. [1] proposed an algorithm in which the changes to the wavelet coefficients guarantee integer changes in the spatial domain. Through which complexity is increases to calculate, whether the coefficient belongs to even or odd class. We propose a function that excludes the procedure (classifying the coefficients even or odd) of changing the coefficient values of the watermarked image without compromising the authenticity and security of the image.

In Deepa Kundur et al technique the quantization parameter sequence $\{\Delta_1, \Delta_2, \Delta_3, \ldots \Delta_n\}$ is increasing order $\Delta_1 < \Delta_2 < \Delta_3 < \ldots \Delta_n$ corresponding to watermark sequence bits $\{w_1, w2, w3 \ldots w_n\}$ is generated by quantization function $Q(.)$. That is watermark bit w_1 is generated from Δ_1, w_2 is generated from Δ_2 and so on. Finally we concluded that image distortion is directly proportional to the quantization parameter Δ. The Quantization parameter is increasing level by level. Due to increasing of quantization parameter watermarked image will be distorted [1].

In the description of Rongsheng XIE et al.'s [2] referred Deepa Kundur [1] and change the implementation of quantization function, as if the user defined ID or random bit sequences satisfies $Q(f_{k,l}(m,n)) = w(i)$, then no need to change the wavelet coefficients $f_{k,l}(m,n)$. This method says that, the original image as watermarked image without embedding the watermark [2]. They have given results without embedding the watermark. Second problem in this method [2] is selection of user defined ID (to select the wavelet coefficients) or key that satisfies the quantization function is time consuming process, because every time we have to check whether it satisfies the quantization function and also it is not possible every time, because the key will be selected by the user randomly.

3.1 Watermark Embedding

The proposed quantization function for watermark embedding scheme illustrated in Fig.3. In this technique watermark-embedding process is implemented in the DWT domain, because the DWT can decompose an image into different frequency components (or different frequency sub bands) [6]. Different frequency components have different sensitivities to image compression [7], which makes it much easier to control the watermark vulnerability. The watermark bits are embedded into wavelet coefficients with prespecified degree, which provides the flexibility to make the tamper-proofing technique as sensitive to change in the signal desired. In all sub bands (LL, LH, HL and HH) the watermarks (binary watermark) are embedded to provide more robust to the watermarked image.

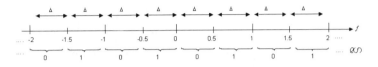

Fig. 3. Proposed Quantization function

3.1.1 Quantization Process

For an arbitrary wavelet transform, the detail coefficients $\{f_{k,l}(m,n)\}$ are real numbers. We perform quantization on the wavelet coefficients in the following manner. Every real number is assigned a binary number, as shown in Fig 3. We denote this function by $Q(.)$ which maps the real number set to $\{0,1\}$. Specifically

$$Q(f) = \begin{cases} 0 & r \le f < (r+\Delta) \ \ for \ \ r = 0,\pm1,\pm2,.... \\ 1 & (r-\Delta) \le f < r \ \ for \ \ r = 0,\pm1,\pm2,.... \end{cases} \tag{4}$$

Where Δ is the quantization parameter is fixed to positive real number 0.5 as shown in Fig.3. The following assignment rules are used to embed the watermark bit (binary watermark either 0 or 1) $w(i)$ into the selected wavelet coefficient $ckey(i)$. We denote the coefficient selected by $ckey(i)$ as $f_{k,l}(m,n)$.

If $Q(f_{k,l}(m,n)) = w(i)$, then no change in the wavelet coefficient is necessary.

Otherwise, change $f_{k,l}(m,n)$ so that we force $Q(f_{k,l}(m,n)) = w(i)$, using the following assignment:

Where Δ is the same parameter as Fig 3 and (4), and = is the assignment operator. In

$$f_{k,l}(m,n) = \begin{cases} f_{k,l}(m,n)+\Delta & if \ \ f_{k,l}(m,n) \le 0 \\ f_{k,l}(m,n)-\Delta & if \ \ f_{k,l}(m,n) > 0 \end{cases} \tag{5}$$

the proposed method (4) quantization parameter Δ fixed to 0.5. Since there are fixed changes in watermark bits $w_1, w_2, w_3,....., w_n$ which are not satisfies the quantization function. In our method the quantization parameter Δ is very less compared to existing algorithms [1,2], so change in wavelet coefficient values are very small in the embedding watermark bits.

3.1.2 Watermark Embedding Algorithm

The wavelet representation of 3-level transformed image is shown in Fig. 4.

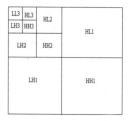

Fig. 4. 3rd -Level wavelet decomposed image representation

Each sub band is represented with {LL, LH, HL, and HH} and $L = \{1, 2, 3...N\}$ gives resolution level of sub band of image size $I \times J$. Let $w(m,n)$ represent the watermark size of $M \times N$. The host image to be watermarked is $f(m,n)$. $w(i), i = 1,..., N_w$ is the watermark. Coefficient selection key $ckey(i), i = 1, 2,...N_w$. The coefficient selection key $ckey(i), i = 1,..., N_w$. Quantization key $qkey(i) \in \{0,1\}$. The algorithm for embedding binary watermark into original image is formulated as follows:

Step 1:   Perform the Lth-level Haar wavelet transform on the host image $f(m,n)$ to produce the 3L detail wavelet coefficient images $f_{k,l}(m,n)$ where $k = h, v, d$ (for horizontal, vertical and diagonal detail coefficient) and $l = 1, 2,..., L$ is the particular detail coefficient resolution level. That is, $f_{k,l}(m,n) = DWT_{Haar}[f(m,n)]$, for $k = h, v, d$ and $l = 1,... L$.

Step 2:   Quantize the detail wavelet coefficients selected by $ckey$. Quantization is applied up to specified level $l = 1, 2,..., L$ for each horizontal, vertical and diagonal coefficient $(k = h, v, d)$.

$if\ ckey(i) = f_{k,l}(m,n)$ for some integer i in

the range 1 to N_w,

$if\ Q(f_{k,l}(m,n)) \neq w(i) \oplus qkey(i),$

$z_{k,l}(m,n)$

$$= \begin{matrix} f_{k,l}(m,n) - \Delta & if\ f_{k,l}(m,n) > 0 \\ f_{k,l}(m,n) + \Delta & if\ f_{k,l}(m,n) \leq 0 \end{matrix}$$

$else$

$z_{k,l}(m,n) = f_{k,l}(m,n)$

end

end

Step 3:   Perform the Lth-level inverse discrete Haar wavelet transform on the marked wavelet coefficients $\{z_{k,l}(m,n)\}$ to produce the marked image $z(m,n)$. That is,

$$z(m,n) = IDWT_{Haar}\left[\{z_{k,l}(m,n)\}\right],$$

for $k = h, v, d$ and $l = 1,...L$.

4   Results

The experiments are conducted on Lena, Baboon, Barbara, Gold hill and Peppers images (512x512) as shown in Fig. 5,Fig. 6 and Fig. 7, with the binary watermark of size is 32x32. Initially, we applied the Haar Wavelet Transform on the original image and embedded the watermark into wavelet   sub bands (LL, LH, HL, and HH) in the 3rd level of the original image. Using a key, the coefficients are selected randomly in each sub band. To find out the performance we tested with different attacks on watermarked images like EZW, JPEG and JPEG 2000 compression algorithms. The tamper assessing values (TAF)   are calculated by the tamper assessing algorithm [1].The TAF values of the different watermarked images showed in Tables 1 and 2. For a given compression ratios the TAF values are in decreasing order from level one to level three. TAF values for a given level (L) increases with respect to compression ratio.

Fig. 5. *(a) Original images* (a) *Lena* (b) Baboon (c) Barbara (d) Gold hill (e) Peppers

Fig. 6. watermarked images   in Level=3, Δ =0.5 (a) Lena PSNR= 55.52 dB (b) Baboon PSNR= 55.53 dB (c) Barbara PSNR= 56.45 dB (d) Gold hill PSNR= 55.73 (e) Peppers PSNR= 55.49 dB

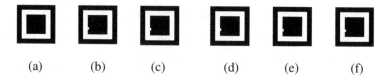

(a)   (b)   (c)   (d)   (e)   (f)

Fig. 7. *(a) Original watermark* (b) – (f) Extracted watermarks from images Lena, Baboon, Barbara, Gold hill, Pepper

Table 1. TAF values of proposed technique for various EZW and JPEG compression ratios on watermarked images in different levels

CR		EZW					JPEG				
		2.5	5	10	15	18	3	5	7	8.5	10
Lena	L1	0.4697	0.5486	0.5652	0.5682	0.5090	0.4126	0.4929	0.5283	0.5557	0.5869
	L2	0.4565	0.5352	0.5645	0.5659	0.5674	0.3975	0.4236	0.4540	0.4727	0.5745
	L3	0.4111	0.4812	0.5095	0.5240	0.5387	0.3716	0.3884	0.4189	0.4526	0.5356
Baboon	L1	0.4971	0.4985	0.5010	0.5914	0.5246	0.4927	0.4980	0.5024	0.5034	0.5045
	L2	0.4875	0.4887	0.4945	0.5018	0.5098	0.4812	0.4849	0.4859	0.4866	0.4895
	L3	0.4531	0.4646	0.4732	0.4831	0.4923	0.4695	0.4738	0.4754	0.4770	0.4881
Barbara	L1	0.6485	0.6516	0.6587	0.6624	0.6641	0.5979	0.6646	0.7004	0.7170	0.7234
	L2	0.4854	0.4893	0.4922	0.4977	0.4999	0.4058	0.45078	0.5088	0.5574	0.5830
	L3	0.4500	0.4524	0.4592	0.4611	0.4630	0.3582	0.3862	0.4600	0.5303	0.5752
Goldhill	L1	0.4663	0.5088	0.5784	0.5922	0.6608	0.4561	0.5378	0.5840	0.6040	0.6228
	L2	0.4140	0.4698	0.5222	0.5315	0.5441	0.3640	0.4355	0.5081	0.5520	0.5750
	L3	0.3906	0.4377	0.4721	0.4970	0.5053	0.3463	0.3783	0.4168	0.4305	0.4557
Pepper	L1	0.5055	0.5156	0.5491	0.5670	0.5700	0.4790	0.5049	0.5300	0.5447	0.5435
	L2	0.4823	0.4978	0.5269	0.5322	0.5465	0.4585	0.4766	0.4939	0.5078	0.5320
	L3	0.4644	0.4766	0.4966	0.5092	0.5104	0.4329	0.4441	0.4639	0.4712	0.4898

Table 2. TAF values of proposed technique for various JPEG2000 compression ratios on watermarked images in different levels

CR		JPEG2000				
		4.5	5	7	10	20
Lena	L1	0.4691	0.4705	0.5962	0.6448	0.6636
	L2	0.4309	0.4424	0.5193	0.5845	0.6274
	L3	0.4036	0.4065	0.4431	0.4065	0.5327
Baboon	L1	0.4893	0.4929	0.4971	0.5014	0.5039
	L2	0.4865	0.4879	0.4897	0.4936	0.4955
	L3	0.4790	0.4798	0.4821	0.4853	0.4867
Barbara	L1	0.7053	0.7086	0.7144	0.7455	0.7680
	L2	0.5017	0.5055	0.5107	0.5490	0.5641
	L3	0.4187	0.4209	0.4534	0.4803	0.4999
Goldhill	L1	0.6035	0.6048	0.6201	0.6477	0.6522
	L2	0.5017	0.5045	0.5378	0.5485	0.5532
	L3	0.4946	0.4958	0.5100	0.5256	0.5370
Pepper	L1	0.4888	0.5037	0.5491	0.5995	0.6035
	L2	0.4795	0.4836	0.5188	0.5508	0.5612
	L3	0.4600	0.4749	0.4986	0.5207	0.5386

5 Conclusions

This paper presents a new method for watrmark embedding algorithm. Changes in the wavelet coefficients reflect quality of the watermarked image. The most important aspect of the proposed method is the changes in the wavelet coefficients, while watermark embedding is very small compared to the existing watermark embedding algorithms. An experimental result shows that the proposed algorithm works with less distortion under image-processing distortions, such as the EZW compression, JPEG compression and JPEG 2000 compression.

References

1. Kundur, D., Hatzinkos, D.: Digital watermarking for Telltale Tamper-Proofing and Authentication. In: Proceedings of the IEEE Special Issue on Identification of Multimedia Information, vol. 87(7), pp. 1167–1180 (1999)
2. R., XIE, R., Wu, K., Li, C., Zhu, S.: An Improved Semi-fragile Digital Watermarking Scheme for Image Authentication. In: IEEE International Workshop on Anti-counterfeiting and Security, vol. 16(18), pp. 262–265 (2007)
3. Shapiro, J.M.: Embedded image coding using zero trees of wavelet coefficients. IEEE Transactions on Signal processing 41(12), 3445–3462 (1993)
4. Berghel, H., O'Gorman, L.: Protecting ownership rights through digital watermarking. IEEE Computer Mag. 29, 101–103 (1996)
5. Cox, I.J., Kilian, J., Leighton, T., Shamoon, T.: Secure spread spectrum watermarking for multimedia. IEEE Trans., Image Process 6(7), 1673–1687 (1997)
6. Chan, P.W., Lyu, M.R., Chin, R.T.: A novel scheme for hybrid digital video watermarking approach, evaluation and experimentation. IEEE Trans. Circuits System. Video Technology 15(12), 1638–1649 (2005)
7. Abdel Aziz, B., Chouinard, J.Y.: On perceptual quality of watermarked images – an experimental approach. In: Kalker, T., Cox, I., Ro, Y.M. (eds.) IWDW 2003. LNCS, vol. 2939, pp. 277–288. Springer, Heidelberg (2004)
8. Nikolaidis, N., Pitas, I.: Robust image watermarking in the spatial domain. Signal Processing 66(3), 385–403 (1998)
9. Celik, M.U., Sharma, G., Saber, E., Tekalp, A.M.: Hierarchical watermarking for secure image authentication with localization. IEEE Trans. Image Process 11(6), 585–595 (2002)
10. Voyatzis, G., Pitas, I.: Application of total automarphisms in image watermarking. In: Proceedings of the IEEE International Conference on Image Processing, Lausanne, Switzerland, vol. 1(19), pp. 237–240 (1996)
11. Mukherjes, D.P., Maitra, S., Acton, S.T.: Spatial domain digital watermarking of multimedia objects for buyer authentication. IEEE Trans. Multimedia 6(1), 1–5 (2004)
12. Wong, P.W.: A public key watermark for image verification and authentication. In: Proceeding of the IEEE International Conference on Image Processing, Chicago, vol. 1(4), pp. 425–429 (1998)
13. Kallel, I.F., Kallel, M., Garcia, E., Bouhlel, M.S.: Fragile Watermarking for medical Image Authentication. In: 2nd International conference on Distributed framework for multimedia applications, pp. 1–6 (2006)
14. Coatrieux, G., Maitre, H., Sankur, B., Rolland, Y., Collorec, R.: Relevance of watermarking in medical imaging. In: Proceedings IEEE EMBS International Conference on Information Technology Applications in Biomedicine, pp. 250–255 (2000)

15. Giakoumaki, Pavlopoulos, S., Koutsouris, D.: Secure and efficient health data management through multiple watermarks on medical images. Journal on medical and biological engineering and computing 44(8), 619–631 (2006)
16. Coelho, Fernanda, B., Barbar, do Salem, J., Carmo, Gustavo, S.B.: The Use of Watermark and Hash Function for the Authentication of Digital Images Mapped through the use of the Wavelet Transform. In: International conference on Internet, web applications and services, pp. 48–49 (2007)
17. Joo, S., Suh, Y., Shin, J., Kikuchi, H., Cho, S.-J.: A New Robust Watermark Embedding into Wavelet DC Components. ETRI Journal 24(5), 401–404 (2002)
18. Liang, T.-s., Rodriguez, J.J.: Robust image watermarking using inversely proportional embedding. In: 4th IEEE Southwest Symposium on Image Analysis and Interpretation, pp. 182–186 (2000)
19. Cox, I.J., Kilian, J., Leighton, T., Shamoon, T.: A Secure, Robust Watermark for Multimedia. In: IEEE International Conference on Image Processing, vol. 3(2), pp. 461–465 (1996)
20. Yuan, H., Zhang, X.-P.: Fragile watermark based on the Gaussian mixture model in the wavelet domain for image authentication. In: International Conference on Image Processing, pp. 505–508 (2003)
21. Zhang, X., Wang, S.: Statistical Fragile Watermarking Capable of Locating Individual Tampered Pixels. IEEE Signal Processing Letters 14(10), 727–730 (2007)
22. Tong, L., Xiangyi, H., Yiqi, D.: Semi-fragile watermarking for image content authentication. In: 7th International Conference on Signal Processing, pp. 2342–2345 (2004)
23. Qin, Q., Wang, W., Chen, S., Chen, D., Fu, W.: Research of digital semi-fragile watermarking of remote sensing image based on wavelet analysis. In: IEEE International Conference on Geoscience and Remote Sensing Symposium, vol. 4(20), pp. 2542–2545 (2004)
24. Zhu, L., Liand, W., Fettweis, A.: Lossy transformation technique applied to the synthesis of multidimensional lossless filters. In: IEEE International Symposium on Circuits and Systems, vol. 3(3), pp. 926–929 (1993)
25. Voloshynovskiy, S., Pereira, S., Iquise, V., Pun, T.: Attack modeling toward a second generation watermarking benchmark. Signal Processing 81(6), 1177–1214 (2001)
26. Kirovski, D., Malvar, H.S., Yacobi, Y.: A dual watermarking and fingerprinting system. IEEE Multimedia 11(3), 59–73 (2004)
27. Su, K., Kundur, D., Hatzinakos, D.: Statistical invisibility for collusion-resistant digital video watermarking. IEEE Trans. on Multimedia 7(1), 43–51 (2005)
28. Swanson, M.D., Zhu, B., Tewfik, A.H.: Multi resolution scene-based video watermarking using perceptual models. IEEE J. Sel. Areas Commun. 16(4), 540–550 (1998)
29. Barr, J., Bradley, B., Hannigan, B.T.: Using digital watermarks with image signatures to mitigate the threat of the copy attack. In: Proc. IEEE Int. Conf. Acoust., Speech, Signal Process., vol. 3, pp. 69–72 (2003)
30. Kutter, M., Voloshynovskiy, S., Herrigel, A.: The watermark copy attack. In: Proc. SPIE: Security and watermarking of multimedia contents II, vol. 3, pp. 97–101 (2000)
31. Meerwald, P., Uhl, A.: A survey of wavelet-domain watermarking algorithms. In: Proc. of SPIE, Electronic Imaging, Security and watermarking of Multimedia Contents III, CA, USA, vol. 4314, pp. 505–516 (2001)
32. Wolfgang, R.B., Delp, E.J.: A watermark for digital images. In: Proceedings of IEEE International Conference on Image Processing, pp. 219–222 (1996)
33. Hsu, C.T., Wu, J.L.: Hidden digital watermarks in images. IEEE Trans. Image Process 1(8), 58–68 (1999)

34. Podilchuk, C.I., Zeng, W.: Image –adaptive watermarking using visual models. IEEE J. Selected Areas Comm. 3(16), 525–539 (1998)
35. Niu, X.-m., Lu, Z.-m., Sum, S.-h.: Digital watermarking of still images with gray-level digital watermarks. IEEE Trans. Consumer Electron 1(46), 137–145 (2000)
36. Craver, S., Memon, N., Yeo, B.L., Yeung, M.: Can invisible watermarks resolve rightful ownerships?, IBM Research Report, TC205209 (1996)
37. Niu, X.-m., Sum, S.-h.: Adaptive gray-level digital watermark. In: Proceedings of International Conference on Signal Processing, vol. 2, pp. 1293–1296 (2000)

A Taxonomy of Frauds and Fraud Detection Techniques

Naeimeh Laleh and Mohammad Abdollahi Azgomi

Department of Computer Engineering,
Iran University of Science and Technology, Tehran, Iran
naimeh_laleh@comp.iust.ac.ir, azgomi@iust.ac.ir

Abstract. Fraud is growing noticeably with the expansion of modern technology and the universal superhighways of communication, resulting in the loss of billions of dollars worldwide each year. Several recent techniques in detecting fraud are constantly evolved and applied to many commerce areas. The goal of this paper is to propose a new taxonomy and a comprehensive review of different types of fraud and data mining techniques of fraud detection. The novelty of this paper is collecting all types of frauds that can detect by data mining techniques and review some real time approaches that have capability to detect frauds in real time.

Keywords: Fraud Detection, Intrusion Detection, Bagging, Stacking, Spike Detection.

1 Introduction

Fraud detection has been implemented using a number of methods. In this paper we describe data mining systems to detect frauds. Fraud detection involves identifying fraud as quickly as possible once it has occurred, which requires the detection module to be accurate and efficient. Fraud detection is becoming a central application area for knowledge discovery in data warehouses, as it poses challenging technical and methodological problems, many of which are still open [44]. Data mining tools for fraud detection are used widely to solve real-world problems in engineering, science, and business. An extensive evaluation of high-end data mining tools for fraud detection has been summarized in [7]. We review and summarize several recent techniques in detecting fraud and all kind of frauds. Also, some of novel approaches that used in fraud detection are reviewed.

2 Types of Fraud

There are several types of fraud including credit card frauds, telecommunication frauds, insurance frauds, internal fraud, computer intrusion, web network fraud and customs frauds. Figure (1) shows our proposed taxonomy of different kinds of frauds.

2.1 Web Network Fraud

Web network fraud is divided into two categories: web advertising and online auction fraud. Web servers and web-based applications are popular attack targets. Web

S.K. Prasad et al. (Eds.): ICISTM 2009, CCIS 31, pp. 256–267, 2009.

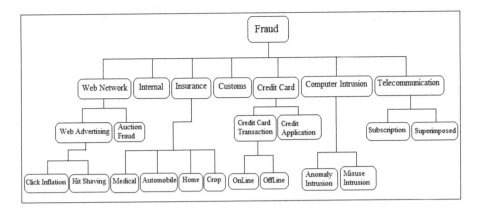

Fig. 1. Taxonomy of fraud types

servers are usually accessible through corporate firewalls, and web-based applications are often developed without following a sound security methodology.

- **Web Advertising Network Fraud.** Web advertising network fraud is a fraud in the setting of Internet advertising commissioners, who represent the middle persons between Internet publishers, and Internet advertisers [8]. In this case publisher falsely increases the number of clicks their sites generate. This phenomenon is referred to as click inflation [31]. The advertising commissioner should be able to tell whether the clicks generated at the publisher's side are authentic, or are generated by an automated script running on some machines on the publisher's end, to claim more traffic, and thus, more revenue. A solution that effectively identifies duplicates in click streams to detect click inflation fraud is developed in [29, 30]. Hit shaving is another type of fraud performed by an advertiser, who does not pay commission on some of the traffic received from a publisher [29, 31]. For detecting Hit Shaving fraud a new notion for association rules between pairs of elements in a data stream is developed in [8]. Forward and backward association rules were defined, and the Streaming-Rules algorithm is devised. Streaming-Rules reports association rules with tight guarantees on errors, using minimal space, and it can handle very fast streams, since limited processing is done per element [8].
- **Online Auction Fraud.** Online auctions have been thriving as a business over the past decade. People from all over the world trade goods worth millions of dollars every day using these virtual marketplaces. EBay1, the world's largest auction site, reported third quarter revenue of $1,449 billion, with over 212 million registered users [45]. Despite the prevalence of auction frauds, auctions sites have not come up with systematic approaches to expose fraudsters. Typically, auction sites use a reputation based framework for aiding users to assess the trustworthiness of each other. However, it is not difficult for a fraudster to manipulate such reputation systems. As a result, the problem of auction fraud has continued to worsen over the past few years, causing serious concern to auction site users and owners alike.

2.2 Internal Fraud

Internal fraud detection is concerned with determining fraudulent financial reporting by management [25] and abnormal retail transactions by employees [24].

2.3 Insurance Fraud

Insurance fraud is divided into four types: Home insurance fraud [26], Crop insurance fraud [27], Medical insurance fraud [28] and Automobile insurance fraud [3].

2.4 Customs Fraud

Customs is a main body of a country to administrate the action of importing or exporting commodity and to collect the duties. At the same time, the duties collected are one of the primary sources of the income of a country. To avoid administrate regulation or the duties, some lawless persons take the cheating measures while their commodities pass the customs, such as hiding, declaring less or making false reports. To assure the truthfulness of the import or export commodity declaring report data, many countries spend a grate deal of manpower and material resources in checking the import and export commodities to avoid fraud behavior in commodity reports [46]. However, the situation is not satisfactory. On the hand, because of the huge commodity volume and the time limit of commerce activities, customs only have the ability of inspecting 10% commodities. On the other hand, only 1% commodities are detected as fraud in all inspected commodities. The method to improve the inspection rate of the customs and to reduce the loss of the country is a new application field for data mining.

2.5 Credit Fraud

Credit fraud is divided into two types: credit application fraud and credit card transaction fraud.

- **Credit Application Fraud.** Credit bureaus accumulate millions of enquiries involving credit applications. Each credit application contains sparse identity attributes such as personal names, addresses, telephone numbers, social security numbers, dates of birth, other personal identifiers, and these are potentially accessible to the credit bureau (if local privacy laws permit it). Application fraud, a manifestation of identity crime, is present when application form(s) contain plausible and synthetic identity information (identity fraud), or real but stolen identity information (identity theft) [16].
- **Credit Card Transaction Fraud.** Credit card transaction fraud is divided into two types [20]: offline fraud and online fraud. Offline fraud is committed by using a stolen physical card at storefront or call center. In the most cases, the institution issuing the card can lock it before it is used in a fraudulent manner [9]. Online fraud is committed via web, phone shopping or cardholder not-present. Only the card's details are needed, and a manual signature and card imprint are not required at the time of purchase [4]. In today's increasingly electronic commerce on the Internet, the use of credit cards for purchases has become convenient and necessary. There arc millions of credit card transactions processed each day. Mining such enormous

amounts of data requires highly efficient techniques that scale. The data are highly skewed-many more transactions are legitimate than fraudulent. Typical accuracy-based mining techniques can generate highly accurate fraud detectors by simply predicting that all [40, 11].

2.6 Computer Intrusion

An intrusion can be defined as any set of actions that attempt to compromise the integrity, confidentiality or availability of a resource. The elements central to intrusion detection are: resources to be protected in a target system, user accounts, file systems, system kernels, etc; models that characterize the normal or legitimate behavior of these resources; techniques that compare the actual system activities with the established models, and identify those that are abnormal or intrusive. Intrusion detection techniques can be categorized into misuse detection, which uses patterns of well known attacks or weak spots of the system to identify intrusions; and anomaly detection, which tries to determine whether deviation from the established normal usage patterns can be flagged as intrusions. Anomaly detection is about finding the normal usage patterns from the audit data, whereas misuse detection is about encoding and matching the intrusion patterns using the audit data [32]. In the context of intrusion detection, supervised methods are sometimes called misuse detection, while the unsupervised methods used are generally methods of anomaly detection, based on profiles of usage patterns for each legitimate user.

In anomaly detection systems, in the first, the normal profiles is defined (training data), then the learned profiles is applied to new data (test data). In anomaly detection intrusive activity is a subset of anomalous activity [41]. There are two advantages for these systems. First, they have the capability to detect insider attacks. For instance, if a user or someone using a stolen account starts performing actions that are outside the normal user-profile, an anomaly detection system generates an alarm. Second, an anomaly detection system has the ability to detect previously unknown attacks. This is due to the fact that a profile of intrusive activity is not based on specific signatures representing known intrusive activity. An intrusive activity generates an alarm because it deviates from normal activity, not because someone configured the system to look for a specific attack signature [17]. Also there are some disadvantages in these systems such as: the system must go through a training period in which appropriate user profiles are created by defining normal traffic profiles. Moreover, creating a normal traffic profile is a challenging task. The creation of an inappropriate normal traffic profile can lead to poor performance. Maintenance of the profiles can also be time-consuming. Since, anomaly detection systems are looking for anomalous events rather than attacks, they are prone to be affected by time consuming false alarms [1].

2.7 Telecommunication Fraud

The telecommunications industry suffers major losses due to Fraud. There are many different types of telecommunications fraud and these can occur at various levels. The two most types of fraud are subscription fraud and superimposed fraud.

- **Subscription Fraud.** In subscription fraud, fraudsters obtain an account without intention to pay the bill. This is thus at the level of a phone number, all transactions from this number will be fraudulent. In such cases, abnormal usage occurs throughout the active period of the account. The account is usually used for call selling or intensive self usage. Cases of bad debt, where customers who do not necessarily have fraudulent intentions never pay a single bill also fall into this category [4].

- **Superimposed Fraud.** In Superimposed fraud, fraudsters take over a legitimate account. In such cases, the abnormal usage is superimposed upon the normal usage of the legitimate customers. There are several ways to carry out superimposed fraud, including mobile phone cloning and obtaining calling card authorization details. Examples of such cases include cellular cloning, calling card theft and cellular handset theft. Superimposed fraud will generally occur at the level of individual calls; the fraudulent calls will be mixed in with the legitimate ones. Subscription fraud will generally be detected at some point through the billing process, though one would aim to detect it well before that, since large costs can quickly be run up. Other types of telecommunications fraud include ghosting (technology that tricks the network in order to obtain free calls) and insider fraud where telecommunication company employees sell information to criminals that can be exploited for fraudulent gain [33, 11, and 4].

3 Methods and Techniques

There are some methods based on data mining for fraud detection that include: Batch time techniques and Real time techniques [10].

3.1 Supervised Methods

In supervised methods, records of both fraudulent and non-fraudulent are used to construct models which yield a suspicion score for new cases and allow one to assign new observations into one of the two classes [11]. Of course, this requires one to be confident about the true classes of the original data used to build the models. Furthermore, it can only be used to detect frauds of a type which have previously occurred [22]. Neural networks [21, 25], Decision trees, rule induction, case-based reasoning are popular and support vector machines (SVMs) have been applied. Rule-based methods are also supervised learning algorithms that produce classifiers using rules such as BAYES [37], RIPPER [38].

Tree-based algorithms like C4.5 and CART also produce classifiers. Also, Link analysis using records linkage relates known fraudsters to other individuals. For example, in telecommunication industry, after an account has been disconnected for fraud, the fraudster will often call the same numbers from another account [36]. Some of articles in fraud detection such as:

1. The neural network and Bayesian network [9] comparison study [33] uses the STAGE algorithm for Bayesian networks and back propagation algorithm for neural networks in credit transactional fraud detection.
2. Rough sets are an emergent technique of soft computing that have been used in many applications. The Rough sets algorithm proves that it is a powerful technique with application in fraud detection. The concepts of rough sets (mainly

lower approximation, reduct and MDA [34]) were used to reach a reduced information system, which is a classification rule system. In the cases that two or more examples have the same attribute values and different class label, generation the same classification rule for these examples are impossible. Rough set theory proposes a solution for this problem. After data preprocessing and data transformation, use rough set algorithm for data mining that include these stage: in the first stage, applying reduct algorithm for deleting additional attributes and generating columns reduction of the information system. In second stage, the main objective is to detect fraudulent profiles and putting them to x set. In the third stage, select subset from x. The four stages generate rules for these examples that represent all rules of fraudulent examples. In the five stages, applying MDA (Minimal Decision Algorithm) [34] to comparing each rule, one-to-one, and reducing the number of Conditional rule attributes. In the final, the rest rules is applying for the test set [5].

3. Using least squares regression and stepwise selection of predictors that is better than other algorithm such as C4.5. Their regression model obtained significantly lesser misclassification costs for telecommunications bankruptcy prediction [35].

4. Multi-algorithmic and adaptive CBR(case-based reasoning) techniques to be used for fraud classification and filtering and comparison a suite of algorithms such as Probabilistic curve Algorithm, Best Match Algorithm, Density Selection Algorithm, Negative Selection Algorithm and Multiple algorithms for better performance is reported [6].

3.2 Semi-supervised Methods

In semi-supervised approach, the input contains both unlabeled and labeled data. Semi supervised is a small amount of labeled data with a large pool of unlabeled data. In many situations assigning classes is expensive because it requires human insight. In these cases, it would be enormously attractive to be able to leverage a large pool of unlabeled data to obtain excellent performance from just a few labeled examples. The unlabeled data can help you learn the classes. For example, a simple idea to improve classification by unlabeled data is Using Naïve Bayes to learn classes from a small labeled dataset, and then extend it to a large unlabeled dataset using the EM (expectation–maximization) iterative clustering algorithm. This approach is this: First, train a classifier using the labeled data. Second, apply it to the unlabeled data to label it with class probabilities (the expectation step). Third, train a new classifier using the labels for all the data (the maximization step). Fourth, iterate until convergence.

A novel fraud detection semi-supervised method propose in five steps: First, generate rules randomly using association rules algorithm Apriori and increase diversity by a calendar schema; second, apply rules on known legitimate transaction database, discard any rule which matches this data; third, use remaining rules to monitor actual system, discard any rule which detects no anomalies; fourth, replicate any rule which detects anomalies by adding tiny random mutations; and fifth, retain the successful rules. This system has been and currently being tested for internal fraud by employees within the retail transaction processing system [24].

3.3 Combining Multiple Algorithm

An obvious approach to making decisions more reliable is to combine the output of different models. These approaches include bagging, Stacking and stacking-bagging. Using this approach increases predictive performance than a single model.

- **Bagging.** Bagging is combining the decisions of different algorithm. Several training datasets of the same size are choosing randomly. Then take a vote (perhaps a weighted vote) from various outputs of multiple models for a single prediction. In this approach, the models receive equal weight [39].

- **Stacking.** Stacking approach is a different way of combining multiple models. Stacking is not normally used to combine models of the same type. For example, set of decision trees. Instead it is applied to models built by different learning algorithms. Suppose you have a decision tree inducer, a Naïve Bayes learner, and an instance-based learning method and you want to form a classifier for a given dataset. The usual procedure would be to estimate the expected error of each algorithm by cross-validation and to choose the best one to form a model for prediction on future data. If two of the three classifiers make predictions that are grossly incorrect, we will be in trouble! Instead, stacking introduces the concept of a meta-learner, which replaces the voting procedure. The problem with voting is that it's not clear which classifier to trust. Stacking tries to learn which classifiers are the reliable ones, using another learning algorithm (the meta-learner) to discover how best to combine the output of the base learners. To classify an instance, the base classifiers from the three algorithms present their predictions to the meta-classifier (meta-learner) which then makes the final prediction [39]. For example one approach that using stacking method is utilizes naive Bayes, C4.5, CART, and RIPPER as base classifiers and stacking to combine them. The results indicate high cost savings and better efficiency in fraud.

- **Stacking-Bagging.** Stacking-bagging is a hybrid technique. This approach is to train the simplest learning algorithm first, followed by the complex ones. For example [3] uses this method. In this way, naive Bayes (NB) base classifiers are computed, followed by the C4.5 and then the back propagation (BP) base classifiers. The NB predictions can be quickly obtained and analyzed while the other predictions, which take longer training and scoring times, are being processed. As most of the classification work has been done by the base classifiers, the NB algorithm, which is simple and fast, is used as the meta-classifier [40]. In order to select the most reliable base classifiers, stacking-bagging uses stacking to learn the relationship between classifier predictions and the correct class. For a data instance, these chosen base classifiers' predictions then contribute their individual votes and the class with the most votes is the final prediction.

3.4 Unsupervised Methods

In unsupervised methods, there are no prior class labels of legitimate or fraudulent observation. Techniques employed here are usually a combination of profiling and outlier detection methods. We model a baseline distribution that represents normal behavior and then attempt to detect observations that show greatest departure from this norm. Unsupervised neural networks have been applied. Also, unsupervised approaches such as outlier detection, spike detection, and other forms of scoring have been applied [2].

3.5 Real Time Approaches

There are many fraud detection approaches that have capability of detecting frauds in real-time. Data stream mining [18] involves detecting real-time patterns to detecting records which are indicative of anomalies. At the same time, the detection system has to handle continuous and rapid examples (also known as records and instances) where the recent examples have no class-labels. In this techniques every current records which arrives in the fraud detection System, detect as fraud or normal in real-time.

Some of these approaches are unsupervised. There is a unifying method for outlier and change detection from data streams that is closely related to fraud detection, abnormality discovery and intrusion detection. The forward and backward prediction errors over a sliding window are used to represent the deviation extent of an outlier and the change degree of a change point [14, 19, and 28].

In addition, link analysis and graph mining are hot research topics in fraud detection in real time. For instance, there is a method that has used large dynamic graphs for telecommunication fraud detection [47]. Multi-attribute pair-wise matching is an approach applies an algorithm to find interesting patterns in corrected students' answers for multiple-choice questions, which can be indicative of teachers cheating by changing their students' answers in the exams [48].

Data stream mining also is the hot research and one article propose a new concept of active stream data mining that has three processes to avoid refreshing models periodically. A novel approach (active mining of data stream) proposed by [18]. These three simple steps contain: 1-Detect potential changes of data streams on the fly when the existing model classifies continuous data streams. The detection process does not use or know any true labels of the stream. One of the change detection methods is a guess of the actual loss or error rate of the model on the new data stream. 2-If the guessed loss or error rate of the model in step 1 is much higher than an application-specific tolerable maximum; they choose a small number of data records in the new data stream to investigate their true class labels. With these true class labels, they statistically estimate the true loss of the model. 3-If the statistically estimated loss in step 2 is verified to be higher than the tolerable maximum; they reconstruct the old model by using the same true class labels sampled in the previous step. To detect events of interest in streams, stream mining systems should be highly automated [49]. Also, communal analyses suspicion scoring (CASS) [13, 43] algorithm is a technique that uses pair-wise matching [15] and numeric suspicion scoring [50] in streaming credit applications. CASS is a low false alarm, rapid credit application fraud detection tool and technique which are complementary to those already in existence. [42]. CASS ïïïï to one-scan of data, real-time responsiveness, and incremental maintenance of knowledge, reduces false positives by accounting for real-world, genuine communal links between people (family, friend, housemate, colleague, or neighbor) using continually updated white lists and improves data quality. Their approach tested on credit application data set that generated by [23].

A survey on techniques for finding anomalies in time for disease outbreaks is presented in [53]. Also, surveys on stream mining techniques used for topic detection and tracking of frauds are presented in [52, 51]. Spike detection is inspired by Stanford Stream Data Manager (STREAM) [54] and AURORA [55], which are two significant stream processing systems. STREAM details the Continuous Query Language (CQL) and uses it in the extraction of example-based or time-based sliding windows, optional elimination of exact duplicates, and enforcement of increasing time-stamp order in queues. AURORA has been applied to financial services, highway toll billing, battalion and environmental monitoring. In addition, spike analysis for monitoring sparse personal identity streams is presented in [56]. Analyzing sparse attributes exists in document and bio-surveillance streams [12]. In [57], time series analysis is used for a simulated bio-terrorism attack to track early symptoms of synthetic anthrax outbreaks from daily sales of retail medication (throat, cough, and nasal) and some grocery items (facial tissues, orange juice, and soup).

4 Comparing Methods

Many types of fraud tend to be very creative and adaptive and fraudsters adapt to new prevention and detection measures. Fraudsters find new ways of committing their crimes so fraud detection needs to be adaptive and evolve over time and we need a dynamic detection model that continually learn from examples seen and keep automatically adjusting their parameters and their values to match the latest patterns of these criminals. However, legitimate users may gradually change their behavior over a longer period of time, and it is important to avoid false alarms.

In the most previously proposed mining methods on data sets the assumption is that the data has labeled and the labeled data is ready for classification at anytime. In credit card fraud detection, we usually do not know if a particular transaction is a fraud until at least one month later. Due to this fact, most current applications obtain class labels and update existing models in preset frequency, usually synchronized with data refresh. Also, in some case, it may have unnecessary model refresh is a waste of resources.

Supervised methods, using samples from the fraudulent or legitimate classes as the basis to construct classification rules detecting future cases of fraud, suffer from the problem of unbalanced class sizes: the legitimate transactions generally much more of the fraudulent ones. Most works focus on how to detect the change in patterns and how to update the model. ï ïïï ïïmost of these systems unable to detecting frauds that the same as legal profiles or transactions. Besides, only some of fraud detection and anomaly detection studies claim to be implemented and few fraud detection studies which explicitly utilize temporal information and virtually none use spatial information. In addition, novel approaches such as real time techniques that using graph theory unable to detect standalone fraud transactions. Because, there will be unlinked transactions which seem to be standalone and not link to other transactions that these techniques will not be able to detect them.

5 Conclusions

This paper defines the adversary, all kinds of fraud, and the methods and techniques in detecting fraud, and review novel methods and techniques in fraud detection. After that, we identify the limitations in methods of fraud detection. Finally, we were comparing these methods.

References

1. Patch, A., Park, J.M.: An overview of anomaly detection techniques: Existing solutions and latest technological trends. Elsevier B.V (2007)
2. Phua, C., Gayler, R., Lee, V., Smith-Miles, K.: A Comprehensive Survey of Data Mining-based Fraud Detection Research (2005)
3. Phua, C., Lee, V., Smith-Miles, K., Damminda, A.: Minority Report in Fraud Detection: Classification of Skewed Data. SIGKDD Explorations 6(1), 50–59 (2004)
4. Kou, Y., Lu, C., Sinvongwattana, S., Huang, Y.P.: Survey of Fraud Detection Techniques. In: Proc. of IEEE Networking, Taiwan, March 21-23 (2004)
5. Cabral, J.E., And Gontijo, E.M.: Fraud Detection in Electrical Energy Consumers Using Rough Sets. In: Proc. of IEEE Conf. on Systems, Man and Cybernetics (2004)
6. Wheeler, R., Aitken, S.: Multiple algorithms for fraud detection. Knowledge-Based Systems 13, 93–99 (2000)
7. Abbott, D.W., Matkovsky, I.P., Elder IV, J.F.: An Evaluation of High-End Data Mining Tools for Fraud Detection. In: Proc. of IEEE SMC 1998 (1998)
8. Metwally, A., Agrawal, D., Abbadi, A.E.: Using Association Rules for Fraud Detection in Web Advertising Networks. In: Proc. of VLDB, Trondheim, Norway (2005)
9. Maes, S., Tuyls, K., Vanschoenwinkel, B., Manderick, B.: Credit Card Fraud Detection using Bayesian and Neural Networks. In: Proc. of NAISO Congress on Neuro Fuzzy Technologies (2002)
10. Mena, J.: Investigative Data Mining for Security and Criminal Detection. Butterworth Heinemann (2003)
11. Bolton, R.J., Hand, D.J.: Statistical Fraud Detection: A Review (January 2002)
12. Phua, C., Gayler, R., Lee, V., Smith-Miles, K.: Adaptive Spike Detection for Resilient Data Stream Mining. In: Proc. of 6th Australasian Data Mining Conf. (2007)
13. Phua, C., Gayler, R., Lee, V., Smith-Miles, K.: Adaptive Communal Detection in Search of Adversarial Identity Crime. ACM SIGKDD (2007)
14. Zhi, L., Hong, M., Yongdao, Z.: A Unifying Method for Outlier and Change Detection from Data Streams. In: Proc. Computational Intelligence and Security, pp. 580–585 (2006)
15. Phua, C., Lee, V., Smith-Miles, K.: The Personal Name Problem and a Recommended Data Mining Solution. In: Encyclopedia of Data Warehousing and Mining, Australian (2007)
16. Phua, C., Gayler, R., Lee, V., Smith-Miles, K.: On the Approximate Communal Fraud Scoring of Credit Applications. In: Proc. of Credit Scoring and Credit Control IX (2005)
17. Maloof, M.A.: Machine Learning and Data Mining for Computer Security (Methods and Application). Springer, Heidelberg (2006)
18. Fan, W., Huang, Y., Wang, H., Yu, P.S.: Active Mining of Data Streams. IBM T. J. Watson Research 10532 (2006)
19. Abe, N., Zadrozny, B., Langford, J.: Outlier Detection by Active Learning. In: Proc. of the 12th ACM SIGKDD, pp. 504–509 (2006)
20. Stolfo, S.J., Fan, D.W., Lee, W., Prodromidis, A.L.: Credit Card Fraud Detection Using Meta-Learning: Issues and Initial Results (1999)
21. Dorronsoro, J., Ginel, F., Sanchez, C., Cruz, C.: Neural Fraud Detection in Credit Card Operations. Proc. of IEEE Transactions on Neural Networks 8(4), 827–834 (1997)
22. Bonchi, F., Giannotti, F., Mainetto, G., Pedreschi, D.: A Classification-based Methodology for Planning Auditing Strategies in Fraud Detection. In: Proc. of SIGKDD 1999, pp. 175–184 (1999)

23. Christen, P.: Probabilistic Data Generation for Deduplication and Data Linkage. Department of Computer Science, Australian National University (2008), http://datamining.anu.edu.au/linkage.html
24. Kim, J., Ong, A., Overill, R.: Design of an Artificial Immune System as a Novel Anomaly Detector for Combating Financial Fraud in Retail Sector. Congress on Evolutionary Computation (2003)
25. Lin, J., Hwang, M., Becker, J.A.: Fuzzy Neural Network for Assessing the Risk of Fraudulent inancial Reporting. Managerial Auditing Journal 18(8), 657–665 (2003)
26. Bentley, P.: Evolutionary, my dear Watson: Investigating Committee-based Evolution of Fuzzy Rules for the Detection of Suspicious Insurance Claims. In: Proc. of GECCO (2000)
27. Little, B., Johnston, W., Lovell, A., Rejesus, R., Steed, S.: Collusion in the US Crop Insurance ogram: Applied Data Mining. In: Proc. of SIGKDD 2002, pp. 594–598 (2002)
28. Yamanishi, K., Takeuchi, J., Williams, G., Milne, P.: On-Line Unsupervised Outlier Detection Using Finite Mixtures with Discounting Learning Algorithms. Data Mining and Knowledge Discovery 8, 275–300 (2004)
29. Reiter, M., Anupam, V., Mayer, A.: Detecting Hit-Shaving in Click-Through Payment Schemes. In: Proc. of USENIX Workshop on Electronic Commerce, pp. 155–166 (1998)
30. Metwally, A., Agrawal, D., Abbadi, A.E.: Duplicate Detection in Click Streams. In: Proc. of Int'l. World Wide Web Conf. (1999)
31. Anupam, V., Mayer, A., Nissim, K., Pinkas, B., Reiter, M.: On the Security of Pay-Per-Click and Other Web Advertising Schemes. In: Proc. of the 8th WWW Int'l. World Wide Web Conf., pp. 1091–1100 (1999)
32. Lee, W., Stolfo, S.J.: Data Mining Approaches for Intrusion Detection (1996)
33. Rosset, S., Murad, U., Neumann, E., Idan, Y., Pinkas, G.: Discovery of Fraud Rules for Telecommunications Challenges and Solutions. In: Proc. of KDD 1990 (1999)
34. Polkowski, L., Kacprzyk, J., Skowron, A.: Rough Sets in Knowledge Discovery. Applications, Case Studies and Sofmare Systems, vol. 2. Physica-Verlag (1998)
35. Foster, D., Stine, R.: Variable Selection in Data Mining: Building a Predictive Model for Bankruptcy. Journal of American Statistical Association 99, 303–313 (2004)
36. Cortes, C., Pregibon, D., Volinsky, C.: Communities of interest. In: Hoffmann, F., Adams, N., Fisher, D., Guimarães, G., Hand, D.J. (eds.) IDA 2001. LNCS, vol. 2189, p. 105. Springer, Heidelberg (2001)
37. Clark, P., Niblett, T.: The CN2 induction algorithm. Machine Learning, 261–285 (1989)
38. Cohen, W.: Fast effective rule induction. In: Proc. 12th Int'l. Conf. of Machine Learning, pp. 115–123 (1995)
39. Witten, I., Frank, E.: Data Mining: Practical Attribute Ranking Machine Learning Tools and Techniques. Morgan Kaufmann, San Francisco (2005)
40. Chan, P., Fan, W., Prodromidis, A., Stolfo, S.: Distributed Data Mining in Credit Card Fraud Detection. In: Proc. of IEEE Intelligent Systems, pp. 67–74 (1999)
41. Kumar, S., Spafford, E.H.: An application of pattern matching in intrusion detection. The COAST Project, Department of Computer Sciences, Purdue University, West Lafayette, IN, USA, Technical Report CSD-TR-94-013, June 17 (1994)
42. Phua, C., Gayler, R., Lee, V., Smith-Miles, K.: On the Communal Analysis Suspicion Scoring for Identity Crime in Streaming Credit Applications. European J. of Operational Research (2006)
43. Phua, C., Gayler, R., Lee, V., Smith-Miles, K.: Communal detection of implicit personal identity streams. In: Proc. of IEEE ICDM 2006 on Mining Streaming Data (2006)
44. Fawcett, T., Provost, F.: Adaptive Fraud Detection. Data Mining and Knowledge Discovery 1(I), 291–316 (1997)

45. Pandit, S., Chau, D.H., Wang, S., Faloutsos, C.: A Fast and Scalable System for Fraud Detection in Online Auction Networks. ACM, New York (2007)
46. Shao, H., Zhao, H., Chang, G.R.: Applying Data Mining to Detect Fraud Behavior in Customs Declaration. In: Proc. of machine learning and Cybernetic, Beijing (2002)
47. Cortes, C., Pregibon, D., Volinsky, C.: Computational Methods for Dynamic Graphs. J. of Computational and Graphical Statistics 12, 950–970 (2003)
48. Levitt, S., Dubner, S., Freakonomics: A Rogue Economist Explores the Hidden Side of Everything. Penguin Books, Great Britain (2005)
49. Fawcett, T., Provost, F.: Activity Monitoring: Noticing Interesting Changes in Behaviour. In: Proc. of SIGKDD 1999, pp. 53–62 (1999)
50. Macskassy, S., Provost, F.: Suspicion Scoring based on Guilt-by-Association, Collective Inference. In: Proc. of Conf. on Intelligence Analysis (2005)
51. Kleinberg, J.: Bursty and Hierarchical Structure in Streams. In: Proc. of SIGKDD 2002, pp. 91–101 (2002)
52. Kleinberg, J.: Temporal Dynamics of On-Line Information Streams. In: Garofalakis, M., Gehrke, J., Rastogi, R. (eds.) Data Stream Management: Processing High-Speed Data Streams. Springer, Heidelberg (2005)
53. Wong, W.: Data Mining for Early Disease Outbreak Detection. Ph.D. Thesis, Carnegie Mellon University (2004)
54. Arasu, A., et al.: STREAM: The Stanford Stream Data Manager Demonstration Description-Short Overview of System Status and Plans. In: Proc. of SIGMOD 2003 (2003)
55. Balakrishnan, H., et al.: Retrospective on Aurora. VLDB Journal 13(4), 370–383 (2004)
56. Phua, C., Gayler, R., Lee, V., Smith, K.: Temporal representation in spike detection of sparse personal identity streams. In: Chen, H., Wang, F.-Y., Yang, C.C., Zeng, D., Chau, M., Chang, K. (eds.) WISI 2006. LNCS, vol. 3917, pp. 115–126. Springer, Heidelberg (2006)
57. Goldenberg, A., Shmueli, G., Caruana, R.: Using Grocery Sales Data for the Detection of Bio-Terrorist Attacks. Statistical Medicine (2002)

A Hybrid Approach for Knowledge-Based Product Recommendation

Manish Godse, Rajendra Sonar, and Anil Jadhav

SJM School of Management,
Indian Institute of Technology Bombay, Mumbai, India
{manishgodse,rm_sonar,aniljadhav}@iitb.ac.in

Abstract. Knowledge-based recommendation has proven to be useful approach for product recommendation where individual products are described in terms of well defined set of features. However existing knowledge-based recommendation systems lack on some issues: it does not allow customers to set importance to the feature with matching options and, it is not possible to set cutoff at individual feature as well as case level at runtime during product recommendation process. In this paper, we have presented a hybrid approach, which integrates rule based reasoning (RBR) and case based reasoning (CBR) techniques to address these issues. We have also described how case based reasoning can be used for clustering i.e. identifying products, which are similar to each other in the product catalog. This is useful when user likes a particular product and wants to see other similar products.

Keywords: knowledge-based recommendation, product recommendation.

1 Introduction

The recommendation systems are becoming an essential part of the E-commerce business to increase sale by assisting customer to choose right product. These systems help customers to select best suitable choice among multiple choices without having previous experience of all those alternatives. The earlier recommender system's definition was limited to "people provide recommendations as inputs, which the system then aggregates and directs to appropriate recipients" [5]. The definition indicates that center of attention is group behaviour. With time, personal taste is added to recommendation that made recommendation more specialized rather than generalized. The new age recommendation systems are focused on personalization techniques such as user needs, requirements, profile, past purchase patterns, demographic data etc [6,7,10,13].

The four types of recommender techniques have been proposed so far are: collaborative, content-based, knowledge-based, and demographic. Each of these techniques has strengths and weaknesses. These techniques can be used individually or combined together depending upon the context [6]. *Collaborative filtering* technique simulates word-of-mouth promotion where buying decisions are influenced by the opinion of friends and benchmarking reports [11]. It relies on interest or opinion of the customer

S.K. Prasad et al. (Eds.): ICISTM 2009, CCIS 31, pp. 268–279, 2009.

having similar profile. In this approach, shared interests are analyzed and products of the same interest are recommended to the group members [22]. *Content-based* recommendation use product features and user rating [6]. *Demographic-based* recommendation systems use demographic data such as age, gender, or profession [15]. Though these techniques are dependent of user profile, the collaborative technique is group dependent recommendation while other two techniques are user specific.

The main weaknesses of these techniques are: sparsity (deficiency) [4], cold-start (late-start) [18,21] or ramp-up (increase) [9]. These terms refer to distinct but related problems. These problems lead to poor quality of recommendation. A *sparsity* problem is [4]: (I) *Deficiency of users*: when number of user are less then it becomes more difficult to find other users with whom to correlate, since the overlap between users is small on average. (II) *Deficiency of rated items*: Until all items are rated, many users will experience the first-rater problem of finding articles with no prediction whatsoever. A *cold-start* problem is related to [18,21]: (I) recommendation to *new user*, and (II) recommendation of *new* or *updated item*. In this problem, recommendation of item or recommendation to new user starts very slowly. It is hard to recommend because there is no correlation of it with other users or items [18,21]. Accuracy of the system depends on the number of active users with long know history and number of rated items. These factors contribute to a *ramp-up* problem: (I) system cannot be useful for most users until there are a large number of users whose habits are known, and (II) until a sufficient number of rated items has been collected, the system cannot be useful for a particular user [9].

A *knowledge-based* recommender systems avoids these drawbacks as, recommendation is independent of user rating and individual taste [9]. A knowledge-based recommender suggests product based on inference about user needs and knowledge of the product domain. Domain knowledge engineering plays a key role to make recommendations more effective. A main advantage of knowledge-based recommender system is the ease in using the system with a general knowledge. Being systems are domain dependent it helps users to understand domain during interaction with a system [23]. Compared to other recommendation techniques, knowledge-based approach is used mainly for recommending products and services such as consumer goods, technical equipment, or (financial) services [11].

In product recommendation, customer has to submit requirements using predefined interfaces. This information is then compared with the descriptions of the available products from catalogue to identify potential product candidates. The product suitable to customer are displayed in ranked order or grouped on some logic. This comparison may use similarity-based approach, which does comparison based on specific similarity measures. Some of the problems with the present recommender system are: (I) it doesn't allow flexible weight systems. Usually predefined feature weights are used or user can set importance (weight) to the feature, but both methods are allowed. (II) Same system can't have different approach depending knowledge level of user, that is, basic and easier approach for general user, and advanced level for knowledgeable user, (III) It does not allow user to set matching cut off at individual feature or case level, (IV) It does not allow clustering, which is useful feature in case user likes particular product and interested to see other similar products. In this paper, we have proposed hybrid approach for product recommendation to address these problems.

2 Related Work

A common techniques used for recommendation are collaborative filtering [3], content based recommendation [14] and knowledge-based systems [9,11]. In all these techniques, demographic is poorly reported because of difficulty in getting demographic data. These techniques have weaknesses if used individually for recommendation methods. Hence, common practice is to use hybrid recommendation systems. Hybrid systems combine techniques of different types with the expectation that the strengths of one will compensate for the weaknesses of another [17]. Robin has covered a good survey on hybrid approach. He has reported 53 different possible hybrid recommender systems. The maximum used combination of techniques is collaborative and content-based combination [9,18].

In knowledge-based recommendation, RBR [20] and CBR-based [13] approaches are used. A hybrid approach of CBR is used more compared to the hybrid RBR. CBR can be combined with collaborative filtering [19] or similarity-based recommendation [12] etc. Robin has suggested 20 different hybrids for CBR [16].

3 Hybrid Approach for Product Recommendation

Rule-based system is a deductive reasoning approach, which mimics the problem-solving behaviour of human experts. The knowledge-base of a rule-based system comprise the knowledge that is specific to the domain of application. It is appropriate for problem solving tasks that are well constructed. But if there are much content then for every content there will be one rule, which is not feasible to maintain. Also for every addition or omission of content, rules need to be modified. CBR is an inductive approach, which solves problem by adapting solution of more similar case that has been solved in the past. A new problem is matched against cases in the case-base and similar cases are retrieved [1]. Integration of rule based and case based reasoning is intelligent way of eliminating drawbacks of pure inductive and deductive reasoning.

In this paper, we have implemented a hybrid approach of RBR and CBR for product recommendation. RBR is used to: (I) capture user requirements, (II) set importance (weight) of feature, (III) set cut-off at feature and case level, (IV) control CBR at runtime. The user requirements are captured along with feature importance and cut off at feature and case level. Then, these are submitted to the CBR system. CBR technique is used to: (I) compare requirements of the desired product with the description of the products stored as cases in the case-base of the system, (II) retrieve ranked set of most similar products, (III) clustering i.e. to find out products which are most similar to each product in the case base. This feature is useful when user likes a particular product and interested to see other similar products. Figure-1 describes an integrated approach of RBR and CBR used for product recommendation. If you have more than one surname, please make sure that the Volume Editor knows how you are to be listed in the author index.

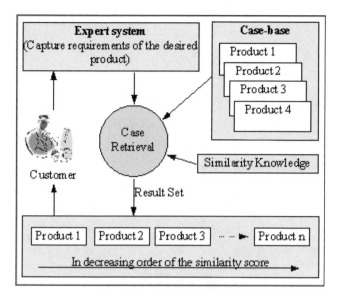

Fig. 1. Hybrid approach flow

4   An Illustrative Example for Hybrid Approach

We have developed a system for recommendation of mobile phones using an enter-prise intelligent system development and solution framework described by Sonar [8]. The solution framework supports to develop, deploy and run web based applications and decision support systems backed by hybrid approach of rule-based reasoning, case-based reasoning. This framework allows direct connection with databases hence existing cases stored in database can be directly used. Also, there is no necessity of restructuring database. This framework also allows controlling CBR using RBR. This framework supports many custom as well as user defined functions to reduce number of rules.

Rule-based reasoning is used to specify requirements using well-defined set of questions or features. We have used two levels for requirement collection depend-ing upon the user knowledge level. At the start of recommendation, user is asked to select a search method: advanced or expert question & answer (Q&A) as shown in figure-2.

- User having less domain knowledge is supposed to select expert advice based Q&A. (Figure-3). These Q&A series is controlled by RBR. The inputs of user during consultation session are converted into variable values and submitted to CBR for further processing.
- Advanced user consultation is in two stages. In first stage user has to select the features of interest (Figure-4). Depending upon features selected, next form is shown dynamically to collect feature specific user requirements (Figure-5). This form has features and list of possible values for every feature.

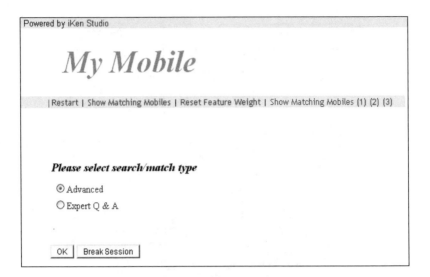

Fig. 2. Select user level

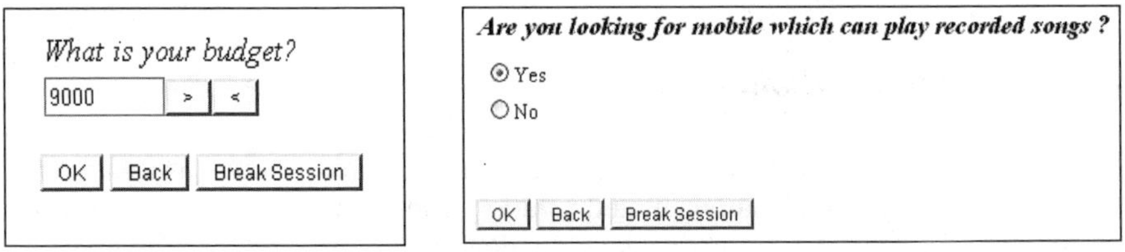

Fig. 3. Representative interfaces for general level user

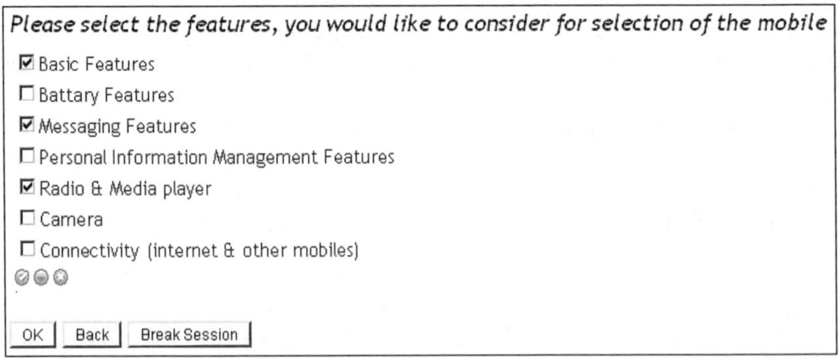

Fig. 4. Interface for advanced user to select features

User is supposed to select the feature value. In this form user can use predefined feature weights or can change weights depending on her requirements. User defined weights are necessary because most of the users doesn't have all their preferences at the start. They usually form during decision making, so there shall be provision to revise them during the process, using a range of interaction styles [2]. This flexible weight helps in getting better product recommendation with maximum satisfying criteria.

User is also provided with a criteria-matching facility like Exact, Almost, Mostly, At least 50%, Any etc. This facility helps in narrowing or broadening a search space. User can set cutoff for features and case levels. These cutoff drops the products for recommendation, which are not meeting the minimum percentage matching.

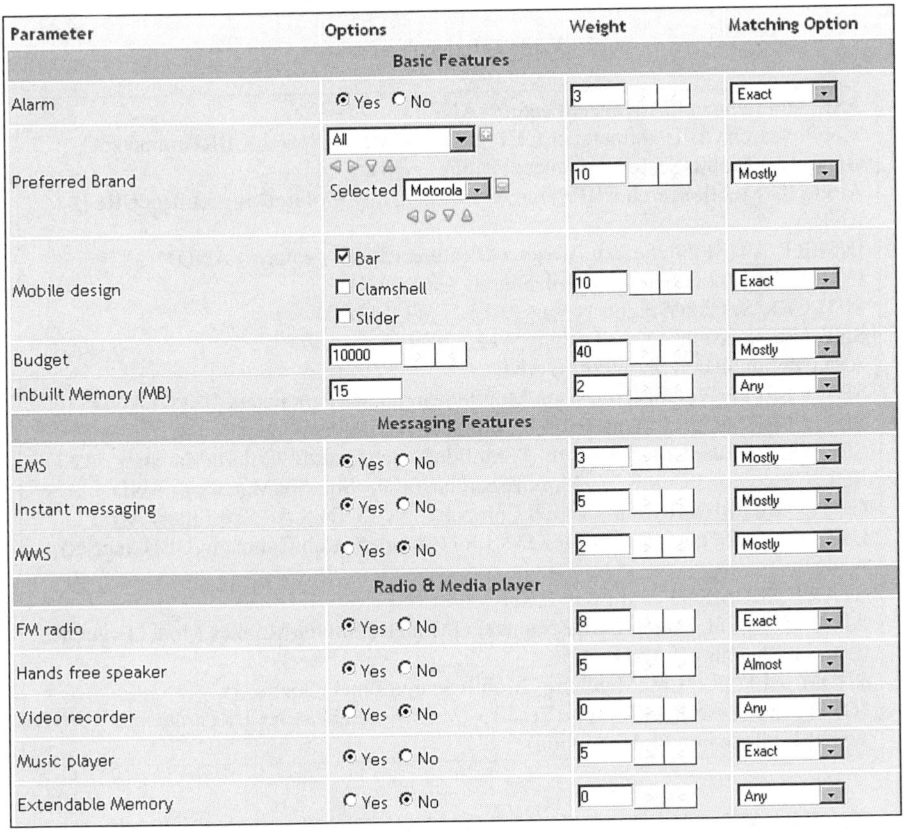

Parameter	Options	Weight	Matching Option
Basic Features			
Alarm	⦿ Yes ○ No	3	Exact
Preferred Brand	All / Selected Motorola	10	Mostly
Mobile design	☑ Bar □ Clamshell □ Slider	10	Exact
Budget	10000	40	Mostly
Inbuilt Memory (MB)	15	2	Any
Messaging Features			
EMS	⦿ Yes ○ No	3	Mostly
Instant messaging	⦿ Yes ○ No	5	Mostly
MMS	○ Yes ⦿ No	2	Mostly
Radio & Media player			
FM radio	⦿ Yes ○ No	8	Exact
Hands free speaker	⦿ Yes ○ No	5	Almost
Video recorder	○ Yes ⦿ No	0	Any
Music player	⦿ Yes ○ No	5	Exact
Extendable Memory	○ Yes ⦿ No	0	Any

Fig. 5. Interface for advanced user (Partial view)

Rules in the system are written in the simple IF-THEN-ELSE format as per user guidelines of shell. These rules have complete control on calling required input forms, passing user filled values to CBR, calling CBR, appending the query values, displaying output etc. The role of experts system in this application is best depicted as host to control full application and CBR act as an assistant.

An example of one of the rules written to perform these tasks is given in table-1. Various functions used in example rule have various purposes. For example:

- The *ASK* function is used to call an input form, while *SHOW_REPORT* is used to show report to a user.

- *APPEND(MobileSearch.CBRParameters,[Company,MobileDesign,ImageURL])*:
  appends the parameters value (Company,MobileDesign,ImageURL) to the variable
  (CBRParameters).
- *SET_CBR(VarWeightsToSession,MobileSearch.CBRParameters,"Wt_")*:
  It sets session weights of key variables. *VarWeightsToSession* is a predefined process to set session weights. *Wt_* is predefined expression for weights, while *MobileSearch.CBRParameters* is a key variable used.

Table 1. Example of rule

ASK(MobileSearch.AdvancedFeature) AND MobileSearch.CBRParameters:=GET_MENU("MobileSearch.CBRParameters", OptionList,MobileSearch.AdvancedFeature) AND APPEND(MobileSearch.CBRParameters,[Company,MobileDesign,ImageURL]) AND INSERT_AT(MobileSearch.AdvancedFeature,1,Basic Features) AND INIT_CBR(MobileSearch.MobileSearch_Case) AND SET_CBR(SessionWeightsToVars,"ALL","Wt_") AND SET_CBR(SessionCutoffsToVars,"ALL","Cutoff_") AND ASK(MobileSearch.ShowForm) AND SET_CBR(VarWeightsToSession,MobileSearch.CBRParameters,"Wt_") AND SET_CBR(VarCutoffsToSession,MobileSearch.CBRParameters,"CutOff_") AND SET_CBR(GlobalCriteriaCutoff," ", MobileSearch.CutOff_CriteriaMatching) AND SET_CBR(GlobalCutoff," ", MobileSearch.CutOff_SolutionMatching) AND GOTO_SEGMENT(MobileSearch.Company INCLUDES All,SkipFilter) AND CBR_FILTER("company in "+CONVERT(MobileSearch.Company,DB_CharSet)) AND START_SEGMENT SkipFilter AND APPEND(MobileSearch.CBRParameters,[ModelId,SimilarMobiles,Model,PageURL, Criteria_Matching]) AND SHOW_REPORT(CaseMatching,"MobileSearch.Final_Goal", "OKButton=Yes&BackButton=Yes&ListVar=MobileSearch.CBRParameters") AND ASK(MobileSearch.FinishSession)

Case base of the system consist of detailed description of each product in the catalog. Each product is described using well-defined set of features with feature values. We have used MS-Access database to store the cases. Advantages of using database management system to store the cases are: easy integration with the existing product database, no separate indexing mechanism is required, and it is easy to add, update, and remove the cases. In CBR, case schema is collection of case features such as key, index, solution, and output. Assessing similarity at the case level involves combining the individual feature level similarities for the relevant features. Formula used to calculate case similarity is given below in equation-1.

Similarity between target query and candidate case is the weighted sum of the individual similarities between the corresponding features of the target query and candidate case. Each weight (*Wi*) encodes the relative importance of a particular feature in the similarity assessment process, and each individual feature similarity is calculated according to a similarity function defined for that feature, *Sim(fs,ft)*.

$$\frac{\sum_{i=1}^{n} Wi \times Sim(fs, ft)}{\sum_{i=1}^{n} Wi} \quad (1)$$

4.1 Results of Recommendation

Result of the recommendation consists of top three most similar products as shown in the figure-6. If user needs more results then there is provision to see top 25 results. The result of recommendation shows product features and their values with similarity score. A *"view"* link provided helps in finding similar mobiles that is clustering the products.

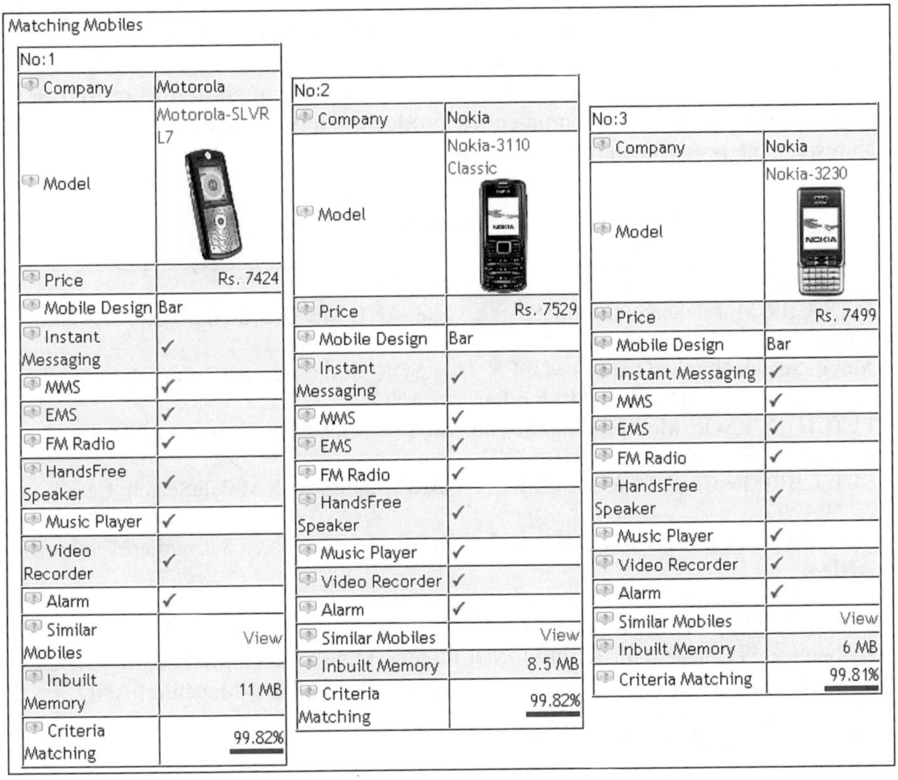

Matching Mobiles							
No:1							
Company	Motorola	**No:2**					
Model	Motorola-SLVR L7	Company	Nokia	**No:3**			
		Model	Nokia-3110 Classic	Company	Nokia		
Price	Rs. 7424			Model	Nokia-3230		
Mobile Design	Bar	Price	Rs. 7529				
Instant Messaging	✓	Mobile Design	Bar	Price	Rs. 7499		
MMS	✓	Instant Messaging	✓	Mobile Design	Bar		
EMS	✓	MMS	✓	Instant Messaging	✓		
FM Radio	✓	EMS	✓	MMS	✓		
HandsFree Speaker	✓	FM Radio	✓	EMS	✓		
Music Player	✓	HandsFree Speaker	✓	FM Radio	✓		
Video Recorder	✓	Music Player	✓	HandsFree Speaker	✓		
Alarm	✓	Video Recorder	✓	Music Player	✓		
Similar Mobiles	View	Alarm	✓	Video Recorder	✓		
Inbuilt Memory	11 MB	Similar Mobiles	View	Alarm	✓		
Criteria Matching	99.82%	Inbuilt Memory	8.5 MB	Similar Mobiles	View		
		Criteria Matching	99.82%	Inbuilt Memory	6 MB		
				Criteria Matching	99.81%		

Fig. 6. Result of consultation

4.2 Product Clustering Based on CBR

Clustering is a grouping of products into subsets, so that the data in each subset share some common traits. This feature is useful when user likes particular product and wants to see other similar products. We have used CBR technique for product clustering i.e. to identify products that are similar to product selected by customer. CBR approach used for recommendation is based on finding products matching to input query of user, while CBR based clustering uses product-to-product comparison approach. These two approaches are different. The CBR based approach helps to do contextual and conceptual clustering.

In clustering, description of the product for which similar products are to be identified is submitted as input case to the CBR. Then, CBR returns ranked list of products, which are most similar to the input case. Our approach of clustering is static and not dynamic. In our approach, we do clustering initially when any new product is added into the catalogue. We store product and its respective cluster list into the database. Thus, every product has a one-cluster list. Depending upon product selected for clustering respective list is shown. This static way of clustering saves computation time during consultation. User can also set cutoff for identifying similar products, for example if user wants to make clustering of products which are 90% similar then cutoff can be set to 90. An example of the rule for CBR based clustering is shown in table-2. User can click on the link provided as *"view"* (Figure-6) for clustering. If user clicks *"view"* link of first recommended product then result of clustering are shown in Figure-7, while for second recommended product results are shown in Figure-8. In both results, the products shown are different.

Table 2. Example of Rule for Clustering

IF MobileSearch.MainTask IS Set next push AND INIT_CBR(MobileSearch.MobileSearch_Case) AND MobileSearch.SeqNo:=1 AND START_SEGMENT Myloop AND MobileSearch.NumberOfRows:=OPEN_CURSOR("MobileSearch.TempCursor", "MobileSearch.GetData#GetAllMobiles"," ") AND FETCH_CURSOR(MobileSearch.SeqNo,"MobileSearch.TempCursor") AND SET_CBR(GlobalCutOff," ",90.0) AND SET_CBR(MaxCases," ",25) AND RUN_CBR(MobileSearch.MobileSearch_Case," "," ") AND MobileSearch.UpdateStatus:=UPDATE_TO_DB("MobileSearch.SetSimilarMobiles") AND MobileSearch.SeqNo:=MobileSearch.SeqNo+1 AND GOTO_SEGMENT(MobileSearch.SeqNo<2,Myloop) AND MobileSearch.UpdateStatus:=SPLIT_SOLUTION(MobileSearch,[Mobile,ModelID, SimilarMobiles,Matching],[SimilarProducts,ModelID,SimilarID,Matching])AND MobileSearch.TempCursor:="" AND ASK(MobileSearch.FinishSession) THEN MobileSearch.Final_Goal IS complete

4.3 Data Capture and Analysis

A user session is tracked by RBR hence different types of data are generated. This data is useful for designers, marketing people, manufactures, and other decision makers. Experts decide the initial predefined weights, but subsequently those are calculated from the data analysis. The data is analyzed automatically at particular interval to modify the feature weights. Thus, weights are dynamic and not static. The data analysis is also helpful to understand customer requirements and expectation thus recommendation can be improved. This analysis is also useful to the mobile manufactures to understand the customer's requirements.

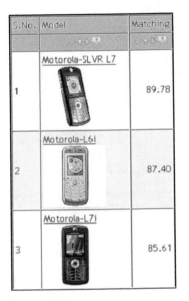

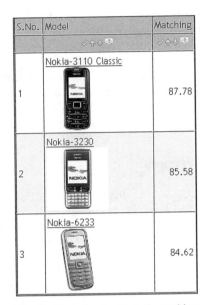

Fig. 7. Clustering for Rank-1 matching

Fig. 8. Clustering for Rank-2 matching

5 Conclusion

Compared to collaborative filtering and content-based recommendation, a knowledge-based recommendation is not dependent of user profile and rating. Thus it is independent of sparsity, cold-start and ramp-up problems.

Domain knowledge of user is crucial in selection of product features, their weights, matching options etc. Domain knowledge varies from customer to customer hence the same system is not suitable for all types of customers. Our approach treats customer at two different levels depending upon their domain knowledge: general customer with basic domain knowledge and advanced user with detailed domain knowledge. For a general user, Q&A session is carried to understand user requirements, while for advanced user a form is shown so as to fill in her requirements. In our hybrid approach of RBR and CBR, a RBR assists customers to specify their requirements, while CBR does recommendation. RBR also controls CBR and, helps CBR to filter products. The CBR technique is useful to retrieve most similar products matching with customer requirements. Being based on similarity approach user gets ranked list of products recommended.

Usually during every consultation user never has predefined feature weights, hence she can use predefined parameter weights. If user is not satisfied with recommendation results then she can modify the weights This flexible weight approach helps in getting better product recommendation with maximum satisfying criteria. The user-level cutoff setting adds more flexibility in reducing list of recommendations. It omits the results which are not satisfying minimum cutoff requirements.

A user session is tracked by RBR hence different types of data are generated. This data is analyzed periodically to modify the predefined weights. Thus, weights are dynamic and not static. The data analysis is also helpful to understand customer requirements and expectation thus recommendation can be improved. This analysis is also useful to the mobile manufactures to understand the customer's requirements.

It is quite likely that user may focus initially on few features of product to get results. After initial consultation, user may explore more products satisfying required features of her choice. For exploring products, a clustering approach helps user. An illustrative example of this approach shows that it is an effective form of recommendations, which is well suited to for product recommendations.

References

1. Aamodt, A., Plaza, E.: Case-Based Reasoning: Foundational Issues, Methodological Variations, and System Approaches. AI Communications 7(1), 39–52 (1994)
2. Ricci, F., Nguyen, Q.N.: Acquiring and Revising Preferences in a Critique-Based Mobile Recommender System. IEEE Intelligent systems 22(3), 22–29 (2007)
3. Linden, G., Smith, B., York, J.: Amazon.com Recommendations Item-to-Item Collaborative Filtering. IEEE Internet Computing, 76–80 (January-February 2003)
4. Konstan, J.A., Miller, B.N., Maltz, D., Herlocker, J.L., Gordon, L.R., Riedl, J.: GroupLens: applying collaborative filtering to Usenet news. Communications of the ACM 40(3), 77–87 (1997)
5. Resnick, P., Varian, H.R.: Recommander Systems. Communications of the ACM 40(3), 56–58 (1997)
6. Burke, R.: Hybrid recommender systems: Survey and experiments. User Modeling and User-Adapted Interaction 12(4), 331–370 (2002)
7. Lawrence, R.D., Almasi, G.S., Kotlyar, V., Viveros, M.S., Duri, S.S.: Personalization of supermarket product recommendations. Data Mining and Knowledge Discovery 5(1-2), 11–32 (2001)
8. Sonar, R.M.: An Enterprise Intelligent System Development and Solution Framework. International journal of applied science, engineering and technology 4(1), 1307–4318 (2007)
9. Burke, R.: Knowledge-based Recommender Systems. In: Kent, A. (ed.) Encyclopedia of Library and Information Systems, New York, vol. 69(32) (2000)
10. Micarelli, A., Gasparetti, F., Sciarrone, F., Gauch, S.: Personalized search on the world wide web. In: Brusilovsky, P., Kobsa, A., Nejdl, W. (eds.) Adaptive Web 2007. LNCS, vol. 4321, pp. 195–230. Springer, Heidelberg (2007)
11. Felfernig, A., Gula, B., Leitner, G., Maier, M., Melcher, R., Teppan, E.: Persuasion in knowledge-based recommendation. In: Oinas-Kukkonen, H., Hasle, P., Harjumaa, M., Segerståhl, K., Øhrstrøm, P. (eds.) PERSUASIVE 2008. LNCS, vol. 5033, pp. 71–82. Springer, Heidelberg (2008)
12. Stahl, A.: Combining case-based and similarity-based product recommendation. In: Roth-Berghofer, T.R., Göker, M.H., Güvenir, H.A. (eds.) ECCBR 2006. LNCS (LNAI), vol. 4106, pp. 355–369. Springer, Heidelberg (2006)

13. Smyth, B.: Case-based recommendation. In: Brusilovsky, P., Kobsa, A., Nejdl, W. (eds.) Adaptive Web 2007. LNCS, vol. 4321, pp. 342–376. Springer, Heidelberg (2007)
14. Pazzani, M.J., Billsus, D.: Content-based recommendation systems. In: Brusilovsky, P., Kobsa, A., Nejdl, W. (eds.) Adaptive Web 2007. LNCS, vol. 4321, pp. 325–341. Springer, Heidelberg (2007)
15. Lim, M., Kim, J.: An adaptive recommendation system with a coordinator agent. In: Zhong, N., Yao, Y., Ohsuga, S., Liu, J. (eds.) WI 2001. LNCS, vol. 2198, pp. 438–442. Springer, Heidelberg (2001)
16. Burke, R.: Hybrid recommender systems with case-based components. In: Funk, P., González Calero, P.A. (eds.) ECCBR 2004. LNCS (LNAI), vol. 3155, pp. 91–105. Springer, Heidelberg (2004)
17. Burke, R.: Hybrid systems for personalized recommendations. In: Mobasher, B., Anand, S.S. (eds.) ITWP 2003. LNCS (LNAI), vol. 3169, pp. 133–152. Springer, Heidelberg (2005)
18. Burke, R.: Hybrid web recommender systems. In: Brusilovsky, P., Kobsa, A., Nejdl, W. (eds.) Adaptive Web 2007. LNCS, vol. 4321, pp. 377–408. Springer, Heidelberg (2007)
19. Aguzzoli, S., Avesani, P., Massa, P.: Collaborative case-based recommender systems. In: Craw, S., Preece, A.D. (eds.) ECCBR 2002. LNCS (LNAI), vol. 2416, p. 460. Springer, Heidelberg (2002)
20. Nguyen, A., Denos, N., Berrut, C.: Improving New User Recommendations with Rule-based Induction on Cold User Data. In: ACM conference on Recommender systems, pp. 121–128 (2007)
21. Ishikawa, M., Geczy, P., Izumi, N., Morita, T., Yamaguchi, T.: Information Diffusion Approach to Cold-Start Problem. In: 2007 IEEE/WIC/ACM International Conference on Web Intelligence and Intelligent Agent Technology - Workshops (wi-iatw), pp. 129–132 (2007)
22. Rubens, N., Sugiyama, M.: Influence-based collaborative active learning. In: ACM Conference On Recommender Systems, pp. 145–148 (2007)
23. Burke, R.: Integrating Knowledge-Based and Collaborative-Filtering Recommender Systems. In: AAAI Workshop on AI in Electronic Commerce, pp. 69–72. AAAI, Menlo Park (1999)

An Integrated Rule-Based and Case-Based Reasoning Approach for Selection of the Software Packages

Anil Jadhav and Rajendra Sonar

Indian Institute of Technology Bombay, Powai, Mumbai, India
aniljadhav@iitb.ac.in, rm_sonar@iitb.ac.in

Abstract. This paper presents an integrated rule-based and case-based reasoning approach for evaluation and selection of the software packages. Rule-based reasoning is used to (i) store knowledge about the software evaluation criteria (ii) guide user to capture user needs of the software package. Case-based reasoning is used to (i) determine the fit between candidate software packages and user needs of the software package (ii) rank the candidate software packages according to their score. We have implemented this approach and performed usability test to verify functionality, efficiency and convenience of this approach.

Keywords: Software selection, Case-based reasoning, Rule-based reasoning.

1 Introduction

Decision making in the field of software selection has become more complex due to availability of large number of software products in the market, ongoing improvements in information technology, incompatibilities between various hardware and software systems, and lack of technical knowledge and experience for software selection decision making [13]. The task of software selection is often assigned under schedule pressure and evaluators are often first timers at the task therefore they may not use the most appropriate method for software selection [11].

Software selection can be formulated as multiple criteria decision making (MCDM) problem. MCDM refers to making preference decisions over the available alternatives that are characterized by multiple, usually conflicting, attributes [30]. The goal of the MCDM is (i) to help decision makers to choose the best alternative of those studied, (ii) to help sort out alternatives that seems good among the set of the alternatives studied, and (iii) to help rank the alternatives in decreasing order of their performance [17].

Knowledge based system (KBS) are computer based information systems that represents the knowledge of experts, and manipulates the expertise to solve problems at an expert's level of performance. Case-based reasoning (CBR) and rule-based reasoning (RBR) are the two fundamental and complementary reasoning methods in the KBS. KBS has potential to play significant role in the process of evaluation and selection of the software packages [8, 28].

S.K. Prasad et al. (Eds.): ICISTM 2009, CCIS 31, pp. 280–291, 2009.
© Springer-Verlag Berlin Heidelberg 2009

2 Related Work

In this section we review the major literature for identifying evaluation technique and criteria for evaluation and selection of the software packages. Table 1 gives a summary of the major literature reviewed in the field of evaluation and selection of the software packages.

Table 1. A summary of the major literature reviewed for selection of the software packages

Sr. No	Author(s)	Type of the software system	Evaluation technique
1	Blanc and Jelassi, 1989	DSS software	Weighted scoring method
2	Bozdag et al., 2003	Computer integrated manufacturing system	Fuzzy multi-attribute group decision-making method
3	Colombo and Francalanci, 2004	CRM Packages	AHP
4	Cochran and Chen, 2005	Object oriented simulation software	Fuzzy-based approach
5	Collier et al., 1999	Data mining software	Weighted scoring method
6	Davis and Williams, 1994	Manufacturing simulation software	AHP
7	Kim and Moon, 1997	Workflow management system	AHP
8	Kim and Yoon, 1992	Expert system shell	AHP
9	Kontio et al., 1996	Reusable component	AHP
10	Lai et al., 1999	Multimedia authoring system	AHP
11	Lin et al., 2006	Data warehouse system	Fuzzy-based approach
12	Mohanty and Venkataraman, 1993	Automated manufacturing system	AHP
13	Morera, 2002	COTS	DESMET and AHP
14	Ngai and Chan, 2005	Knowledge management tools	AHP
15	Ossadnik and Lange, 1999	AHP software	AHP
16	Perez and Rojas, 2000	Workflow type software	Weighted scoring method
17	Phillips-Wren et al., 2004	Decision support system	AHP
18	Sarkis and Talluri, 2004	e-commerce software and communication systems for a supply chain	AHP and Goal programming model
19	Shtub et al., 1998	Operations Management Software	AHP
20	Teltumbde, 2000	ERP	AHP and nominal group technique
21	Toshtzar, 1988	Computer software	AHP
22	Wei et al., 2005	ERP system	AHP
23	Zahedi, 1998	Database management system	AHP

Evaluation Techniques: Analytic hierarchy process (AHP) has been widely used for evaluation of the software packages. Other common techniques discussed in the literature for evaluation of the software packages are weighted scoring method, and fuzzy based approach. Each technique has its own advantages and disadvantages.

AHP technique enables decision makers to structure a decision making problem into a hierarchy, helping them to understand and simplify the problem. But it is time consuming process because of mathematical calculations and number of pair wise comparisons which increases as number of alternatives and criteria increases. Weighted scoring method is easy to understand and use, but weights to the attribute are assigned arbitrary and it is very difficult to assign weight when number of criteria are high. Another problem with this method is that common numerical scaling is required to obtain the score. In fuzzy based approach decision makers can use linguistic terms to evaluate alternatives which improves decision making procedure by accommodating the vagueness and ambiguity in human decision making. But in this method it is difficult to compute fuzzy appropriateness index and ranking values for all the alternatives.

The evaluation techniques discussed above help to rank the candidate software packages and choose the best one with the highest score, but lacks in indicating how well each candidate software package meets user needs of the software package. CBR system has the capability to indicate how well each candidate package meets user needs of the software package in terms of similarity score.

Evaluation criteria: Many literature concerning evaluation and selection of the software packages provides criteria to evaluate specific software package such as ERP, CRM but does not provides a generic list of criteria which can be used for evaluation of any software package. On the basis of review of literature we have identified criteria which can be used for evaluation of any software package and those are broadly categorized into seven different groups: functional, quality, vendor, cost & benefits, hardware and software, opinion, and output. These criteria are further decomposed into sub-criteria and these one into measurable attributes.

3   An Integrated Approach for Selection of the Software Packages

In this section we describe an integrated rule-based and case-based reasoning approach for evaluation and selection of the software packages. So far this kind of approach for evaluation and selection of the software packages has not been discussed in the reviewed literature.

Rule-based system, a deductive reasoning approach, is reasoning system which mimics the problem solving behavior of the human experts. The knowledge base of rule-based system comprises the knowledge that is specific to domain of the application. It is appropriate for problem solving tasks that are well constructed. However, it does not make substantial use of the knowledge embedded in previous cases. Case-based reasoning, an inductive reasoning approach, is problem solving approach that solves problem by adapting solution of more similar cases that has been solved in the past. A new problem is matched against cases in the case-base and one or more similar cases are retrieved. Solution suggested by matching cases is then reused and tested for success [1]. Integration of rule-based and case-based reasoning approach eliminates the drawbacks of pure inductive reasoning and pure deductive reasoning and solves problem in more intelligent way.

We have developed a system for evaluation and selection of the software packages using an integrated rule-based and case-based reasoning approach. Figure 1 describes how an integrated RBR and CBR approach is used for evaluation of the software packages. Rule-base of the system stores knowledge about the software evaluation criteria and it is used to (i) guide decision makers to select the evaluation criteria which he/she want to consider for evaluation of the software packages, and (ii) capture user needs of the software package. Requirements of the desired software package are captured using well defined set of evaluation criteria (features), with each criterion (feature) associated with well defined space of criterion (feature) values. Once user needs of the software package are captured using RBR these requirements are then submitted to the CBR system.

Case-based reasoning is used to (1) compare user needs of the software package with description of the candidate software packages stored as cases in case-base of the system (2) retrieve software packages which most closely meet user needs of the software package and rank them according to the similarity score.

Software packages to be evaluated are stored as cases in case-base of the system. Case-base is made up of detailed description of the candidate software packages to be evaluated. Each software package is described using well defined set of features and feature values. Features used to describe software packages have the same vocabulary as that of criteria used for evaluation of the software packages. When user submits user needs of the software package through expert system, CBR returns ranked set of most similar software packages in descending order of the similarity score. Similarity score indicates how well each candidate software package meets user needs of the software package.

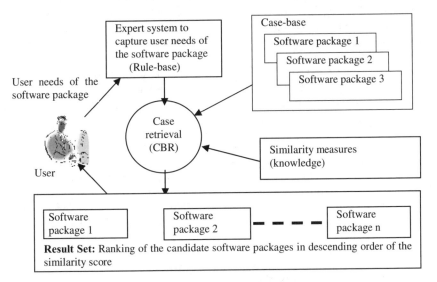

Fig. 1. An integrated RBR and CBR approach for software selection

4   Case Study: Selection of the Campus Management System

An integrated rule-based and case-based reasoning approach has been tested on number of selection problems. In this paper, selection of the campus management system for educational institute will be discussed as verification for this approach. One of the management institutes in India, in order to strengthen its position and to provide unparalleled value to its students, is undertaking a series of initiatives. One of such initiatives is to deploy campus management system to streamline processes with respect to the institute and provide value through information to all stakeholders of the institute i.e. students, faculty, management, administrative staff, parents of the student, alumni, and corporate bodies associated with the institute for placements. In order to select the best suitable campus management software package which caters needs of all stakeholders of the system, five experts with strong academic and IT background, including director of the institute, formed a committee to make the selection decision.

After having series of discussion with all stakeholders of the system, the experts committee decided that package to be purchased should give comprehensive support to each of the following module: Admin, Institute, Admission, Principle/Directors, Faculty/HOD, Student, Parent, Alumni, Accounts management, HR and Payroll management, Inventory management, Hostel management, Examination, Library, Placement and corporate, Transport module, Knowledge management, and Quiz module.

Detail proposals were invited from three different vendors, after eliminating two vendors that doesn't meet mandatory requirements of the system. Proposals of the three vendors were then evaluated against functional requirements, and other common criteria grouped into quality, vendor, hardware and software requirements, cost and benefits, opinion, and output related characteristics of the software package. A system for selection of the campus management software package has been developed using an enterprise intelligent system development and solution framework described by Sonar [25]. The solution framework supports to develop, deploy and run web based applications and decision support systems backed by an integrated architecture of rule-based reasoning, case-based reasoning, neural network and genetic algorithms. We have used only rule-based and case-based reasoning component of the solution framework for development of the system for selection of the campus management system.

Rule-based reasoning: Rule-based reasoning is used to guide user to select evaluation criteria, and capture user needs of the software package. An example of how system assists user to select technical evaluation criteria and capture technical requirements of the desired software package is illustrated in Figure 2 and Figure 3 respectively. Rules of the system are written in simple IF...THEN...ELSE format. An example of one of the rules written to select technical evaluation criteria and capture technical requirements of the software package is given below.

Rule No=6[94] ID=SoftwareSelection.Hardware And Software
IF SoftwareSelection.MainTask IS Software Evaluation
AND SoftwareSelection.MainCriteriaCategory INCLUDES ANY OF [Other]
AND SoftwareSelection.OtherCommonCriteria INCLUDES ANY OF [Hardware and Software]
AND ASK(SoftwareSelection.TechnicalCriteria)
AND ASK(SoftwareSelection.ShowForm_TechnicalCriteria)
THEN SoftwareSelection.Final_Goal IS Hardware and Software Criteria

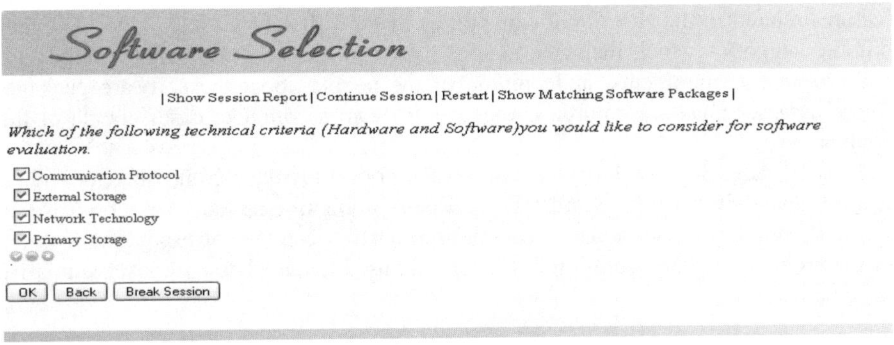

Software Selection

| Show Session Report | Continue Session | Restart | Show Matching Software Packages |

Which of the following technical criteria (Hardware and Software) you would like to consider for software evaluation.

☑ Communication Protocol
☑ External Storage
☑ Network Technology
☑ Primary Storage

[OK] [Back] [Break Session]

Fig. 2. Interface to guide user to select technical criteria

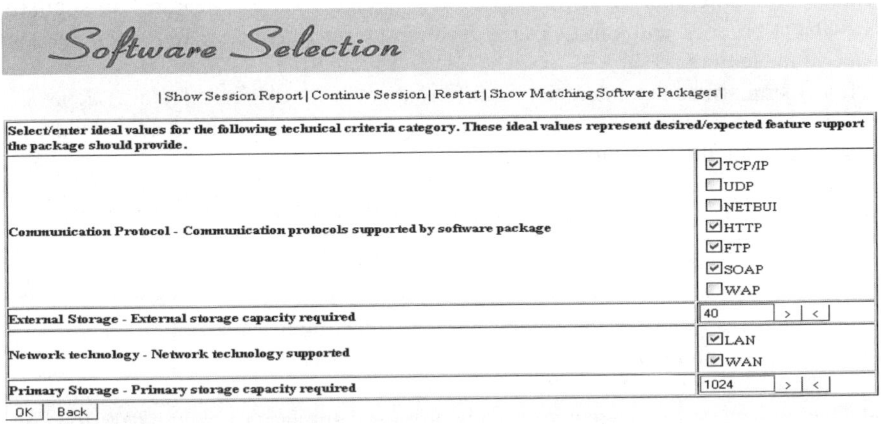

Software Selection

| Show Session Report | Continue Session | Restart | Show Matching Software Packages |

Select/enter ideal values for the following technical criteria category. These ideal values represent desired/expected feature support the package should provide.	
Communication Protocol - Communication protocols supported by software package	☑TCP/IP ☐UDP ☐NETBUI ☑HTTP ☑FTP ☑SOAP ☐WAP
External Storage - External storage capacity required	40 > <
Network technology - Network technology supported	☑LAN ☑WAN
Primary Storage - Primary storage capacity required	1024 > <

OK   Back

Fig. 3. Interface to capture technical requirements of the software package

Case-based reasoning: Case-based reasoning is used to find similarity between user needs of the software package and capabilities of the candidate software packages, and rank them according to similarity score. Candidate software packages to be evaluated are stored as cases in case base of the system. Each case in the case base is described using well defined set of features and feature values. We have used MS-access database to store the cases. Advantages of using database management system for storing cases are: (1) It is very easy to add new cases and manage the cases, (2) Data security and integrity is assured, (3) Same database can also be used for other applications.

In CBR system case schema is a collection of case features. Each case feature is linked to similarity measure, a function that assesses the similarity of a problem case to the cases in case base of the system. In our application scenario problem case is nothing but user needs of the software package and solution cases are the candidate software packages to be evaluated. We have defined case schema for each criteria category: (1) Quality criteria category - which includes quality features of the software package, (2) Vendor criteria category - which includes features related to the vendor of the software package, (3) Functional criteria category - which includes features related to the functional capabilities of the software package, and (4) Other criteria category - which includes features related to the cost and benefits, hardware and software requirements, and opinion of the people about the software package. These case schemas are combined together to form a complete case schema of the application.

In CBR assessing similarity at case level (global) involves combining individual feature level similarities. Similarity between problem case and solution case is weighted sum of the individual (local) similarities between the corresponding features of the problem case and solution case. Formula used to calculate case level similarity is given below

$$\frac{\sum_{i=1}^{n} Wi * sim \ (qv \ , cv \ )}{\sum_{i=1}^{n} Wi}$$

Where, Wi represents relative importance (weight) of the feature in similarity assessment process, and sim(qv,cv) is local similarity between query value and case value of the feature.

Local similarities are calculated in different ways for different features depending on type of the feature. For example, in case of numeric feature, local similarity is calculated using the following formula.

$$sim(qv,cv) = 1 - \frac{d(qv,cv)}{range(v)}$$

Where $d(qv,cv) = | qv - cv|$, range(v) = max_v - min_v, qv - query value of the feature, cv - case value of the feature, max_v - maximum value which feature v can takes, min_v - minimum value which feature v can takes.

In our application scenario all case features are not numeric types. It also contains boolean or symbolic features and in such case local similarities can be described by a table that defines the similarities for all possible pairs of attribute values.

Thus local similarity for each feature is calculated using similarity function linked to that feature. After calculating local similarity of each feature, system calculates similarity score for case schema defined for each criteria category and then for complete case. An example of how system calculates similarity score for vendor 1 (name changed) is illustrated in Table 2.

Table 2. Similarity calculations

Criteria Category	Similarity Score	Criteria Category Weight	Weighted Score
Functional	96.00	0.25	24.00
Quality	94.53	0.25	23.63
Vendor	75.00	0.25	18.75
Other	84.76	0.25	21.19
Total Score (Case Matching Score)			87.57

Results of the system: Result set of the system consists of ranked set of software packages in descending order of the similarity score as shown in Figure 4. Case matching column represents similarity score of each candidate software package. The

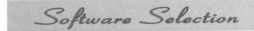

Software Selection

| Show Session Report | Continue Session | Restart | Show Matching Software Packages |

Vendor List- Click on Vendor ID to see more detail
Results from:1 to 3 (Total:3)

VendorID	VendorName	Vendor Criteria	Quality Criteria	Other Common Criteria	Functional Criteria	Case Matching
1	Vendor 1	75.00	94.53	84.76	96.00	87.57%
3	Vendor 3	65.00	88.27	77.93	68.00	74.80%
2	Vendor 2	40.00	55.13	65.71	51.00	52.96%

Fig. 4. System results

similarity score indicates how well each candidate software package meets user needs of the software package.

Advantages of using an integrated RBR and CBR approach for software evaluation

- An integrated approach works well with both qualitative as well as quantitative parameters.
- It assists user not only in selecting criteria for evaluation but also specifying user needs of the software package as it stores knowledge about the software evaluation criteria in rule-base of the system.
- Addition or deletion of the candidate software packages is easy as it uses case-base for storing details of the candidate software packages.
- This approach is easy to use even when number of evaluation criteria or a number of alternatives to be evaluated are large in number. For example, Assume that number of criterions considered for evaluation are 100 and number of alternatives to be evaluated are 10. If we use AHP for evaluation, we need to do 45 pair wise comparisons to obtain score of each alternative with respect to any one evaluation criterion. Since we have considered 100 evaluation criterions, total number of pair wise comparisons required to be done are 45*100=4500, which is quite complicated and time consuming task. If we use weighted sum method first we need to convert all qualitative values to numeric type as weighted sum method works only with numeric scales. If we use an integrated RBR and CBR approach for software evaluation, system calculates similarity between each candidate software package and user needs of that package using similarity knowledge stored in the system. No need of pair wise comparison or no human interventions is required once user needs of the software package are captured and submitted to the CBR system.

Table 3 shows the comparison of proposed evaluation technique with the other widely used existing evaluation techniques: analytical hierarchy process (AHP) and weighted average sum (WAS).

Table 3. Comparison of evaluation techniques

	AHP	WAS	An Integrated RBR & CBR approach
Support for qualitative parameters	Yes	No	Yes
Support for quantitative parameters	Yes	Yes	Yes
If number of alternatives changes	pair wise comparison needs to be done again	Calculation Wi*Ri needs to be done again. Where Wi is weight and Ri is rating	Any number of alternatives can be added or removed with no extra efforts required to calculate similarity score
If number of evaluation criteria changes	pair wise comparison needs to be done again	Calculation Wi*Ri needs to be done again	Any number of criteria can be added or removed with no extra efforts required to calculate similarity score
If user needs changes	pair wise comparison needs to be done again	Calculation Wi*Ri needs to be done again	Provides flexibility to change requirements and calculate similarity score accordingly

5 Evaluation of an Integrated RBR and CBR Approach

We developed knowledge based system for evaluation and selection of the software packages and performed usability test to verify functionality, efficiency, and convenience of the system. To conduct this experiment we followed the approach used by Ricci and Nguyen [22] to test usability of the mobile phone recommender system.

In our experiment 3 experts (testers) having knowledge of evaluating and selecting the software packages were involved so that they can give us proper feedback on the system. The experiment comprised three phases: Training, testing, and evaluation

During training, we introduced knowledge-based (rule-based and case-based reasoning) system to the testers. We explained testers how system assists to (1) select criteria for evaluation of the software packages, (2) specify user needs of the software packages. Training phase typically lasted for 15 to 20 minutes.

During testing, we asked testers to (1) select only those criteria/features which he/she wants to consider for evaluation of the software packages, and (2) specify user needs of the software package for selected criteria/features. We also asked testers to check results (ranking of the candidate software packages) produced by the system by changing system requirements and importance (weight) of the evaluation criteria.

During evaluation, testers evaluated systems performance by completing usability questionnaire. The questionnaire contained predefined list of 10 questions, few of them are taken from post-study system usability questionnaire [14]. We also provided free-text space for comments. Testers answered questions using a seven point likert scale where 1 is "strongly agree" and 7 is "strongly disagree".

Table 4 shows testers average rating of the questionnaire statements which expressed testers subjective evaluation of the system's performance. In particular, testers found that system effectively helped them in (1) selecting criteria for evaluation of the software packages, (2) specifying user needs of the software package (3) determining fit between software package and user needs of that package. All testers explicitly mentioned that it is easy to specify user needs of the software package and determine the fit between software package and user needs of that package using this system. They also mentioned that results (ranking of candidate software packages) produced by the system in the form of percentage case matching (i.e. percentage similarity between user needs of the software package and capabilities of each candidate software package) are quite impressive. Overall, testers were satisfied with the knowledge-based system for evaluation and selection of the software packages.

Table 4. Results of evaluation of knowledge based system

Statement	Average rating
1. This system effectively helped me in selecting criteria for evaluation of the software packages.	2
2. Evaluation criteria used in this system are enough to evaluate and rank the candidate software packages.	1
3. This system has used proper hierarchy of evaluation criteria.	2
4. This system effectively helped me to specify user needs of the software package.	2
5. I can effectively complete task of evaluation and selection of the software package using this system.	1
6. I can efficiently complete task of evaluation and selection of the software package using this system.	1
7. Results i.e. ranking of the candidate software packages, produced by the system in the form of case matching is easy to understand.	2
8. Results produced by the system effectively helped me in determining the fit between software package and user needs of the package.	1.5
9. This system has all the functions and capabilities I expected it to have.	3.0
10. Overall, I am satisfied with this system.	2.5
11. It was simple to use this system.	2.5

6 Conclusions

In this paper we have proposed an integrated rule-based and case-based reasoning approach for evaluation and selection of the software packages. The results of evaluation of the system shows that the system effectively helped decision makers to specify user needs of the software package and determine the fit between software package and user needs of the software package. The proposed approach works well with both qualitative as well as quantitative software evaluation parameters. It helps decision makers not only in evaluation but also reduce the time and efforts required for evaluation and selection of the software packages.

References

1. Aamodt, A., Plaza, E.: Case based reasoning foundational issues, methodical variations and system approaches. AI communications 7(1), 39–59 (1994)
2. Bozdag, C., Kahraman, C., Ruan, D.: Fuzzy group decision making for selection among computer integrated manufacturing systems. Computers in Industry 51, 13–29 (2003)
3. Blanc, L., Jelassi, M.: DSS software selection: A multiple criteria decision methodology. Information and management 17, 49–65 (1989)
4. Cochran, J.K., Chen, H.: Fuzzy multi-criteria selection of object-oriented simulation software for production system analysis. Computers and operations research 32, 153–168 (2005)
5. Collier, K., Carey, B., Sautter, D., Marjanierni, C.: A methodology for evaluating and selecting data mining software. In: Proceedings of 32nd Hawaii International conference on system sciences, pp. 1–11 (1999)
6. Colombo, E., Francalanci, C.: Selecting CRM packages based on architectural, functional, and cost requirements: empirical validation of a hierarchical ranking model. Requirements Engineering 9, 186–203 (2004)
7. Davis, L., Williams, G.: Evaluation and selecting simulation software using the analytic hierarchy process. Integrated manufacturing systems 5(1), 23–32 (1994)
8. Kathuria, R., Anandarajan, M., Igbaria, M.: Selecting IT applications in manufacturing: A KBS approach. Omega 27, 605–616 (1997)
9. Kim, J., Moon, J.Y.: An AHP and survey for selecting workflow management systems. International journal of intelligent systems in accounting, finance, and management 6, 141–161 (1997)
10. Kim, C.S., Yoon, Y.: Selection of good expert system shell for instructional purposes in business. Information and management 23(5), 249–262 (1992)
11. Kontio, J., Caldiera, G., Basili, V.R.: Defining factors, goals and criteria for reusable component evaluation. In: CASCON 1996 Conference, Toronto, Canada, November 12-14 (1996)
12. Lai, V.S., Trueblood, R.P., Wong, B.K.: Software selection: a case study of the application of the analytical hierarchical process to the selection of a multimedia authoring system. Information and Management 36, 221–232 (1999)
13. Lin, H.-Y., Hsu, P.-Y., Sheen, G.-J.: A fuzzy-based decision making procedure for data warehouse system selection. Expert systems with applications (2006)
14. Lewis, J.R.: IBM Computer Usability Satisfaction Questionnaire: Psychometric Evaluation and instructions for use. International journal of human computer interaction 7(1), 57–78 (1995)
15. Mohanty, R.P., Venkataraman, S.: Use of Analytic hierarchy process for selecting automated manufacturing systems. International journal of operations and production management 13(8), 45–57 (1993)
16. Morera, D.: COTS Evaluation Using Desmet Methodology & Analytic Hierarchy Process (AHP). In: Oivo, M., Komi-Sirviö, S. (eds.) PROFES 2002. LNCS, vol. 2559, pp. 485–493. Springer, Heidelberg (2002)
17. Mollaghasemi, M., Pet-Edwards, J.: Technical briefing: making multiple objective decisions. IEEE computer society press, Los Alamitos (1997)
18. Ngai, E.W.T., Chan, E.W.C.: Evaluation of knowledge management tools using AHP. Expert system with applications, 1–11 (2005)
19. Ossadnik, W., Lange, O.: AHP-based evaluation of AHP-Software. European journal of operational research 118, 578–588 (1999)

20. Phillips-Wren, G.E., Hahn, E.D., Forgionne, G.A.: A multiple criteria framework for evaluation of decision support systems. Omega 32(4), 323–332 (2004)
21. Perez, M., Rojas, T.: Evaluation of workflow type software products: a case study. Information and software technology 42, 489–503 (2000)
22. Ricci, F., Nguyen, Q.N.: Acquiring and Revising Preferences in a Critique-Based Mobile Recommender System. IEEE Intelligent Systems, 22–29 (2007)
23. Sarkis, J., Talluri, S.: Evaluating and selecting e-commerce software and communication systems for supply chain. European journal of operational research 159, 318–329 (2004)
24. Shtub, A., Spiegler, I., Kapaliuk, A.: Using DSS methods in selecting operations management software. Computer-integrated manufacturing systems 1(4) (1998)
25. Sonar, R.M.: An Enterprise Intelligent System Development and Solution Framework. International journal of applied science, engineering and technology 4(1) (2007)
26. Teltumbde, A.: A framework for evaluating ERP projects. International journal of production research 38(17), 4507–4520 (2000)
27. Toshtzar, M.: Multi-criteria decision making approach to computer software evaluation: application of the Analytical Hierarchical Process. Mathl. and comput. modeling 11, 276–281 (1988)
28. Vlahavas, I., Stamelos, I., Refanidis, I., Tsoukias, A.: ESSE: an expert system for software evaluation. Knowledge-based systems 12, 183–197 (1999)
29. Wei, C., Chien, C., Wang, M.J.: An AHP based approach to ERP system selection. International journal of production economics 96(1), 47–62 (2005)
30. Yoon, K., Hwang, C.: Multiple Attribute Decision-Making: an Introduction. Sage Publisher, Thousand Oaks (1995)
31. Zahedi, F.: Database management system evaluation and selection decisions. Decision sciences 16(1), 91–116 (1998)

Learning Agents in Automated Negotiations

Hemalatha Chandrashekhar and Bharat Bhasker

Indian Institute of Management, Lucknow, Prabandh Nagar, Off Sitapur road,
Lucknow-226013, India
{fpm5001,bhasker}@iiml.ac.in

Abstract. In bilateral multi-issue negotiations involving two-sided information uncertainty, selfish agents participating in a distributed search of the solution space need to learn the opponent's preferences from the on-going negotiation interactions and utilize such knowledge to construct future proposals in order to hope to arrive at efficient outcomes. Besides, negotiation support systems that inhibit strategic misrepresentation of information need to be in place in order to assist the protagonists to obtain truly efficient solutions. To this end, this work suggests an automated negotiation procedure that while protecting the information privacy of the participating agents encourages truthful revelation of information through successive proposals. Further we present an algorithm for proposal construction in the case of two continuous issues. When both the negotiating agents implement the algorithm the negotiation trace shall be confined to the Pareto frontier. The Pareto-optimal deal close to the Nash solution shall be located whenever such a deal exists.

Keywords: Automated Negotiation, bilateral multi-issue negotiations, Negotiating agents, distributed search, Learning agents.

1 Introduction

Bilateral multi issue negotiation is the process by which two parties iteratively resolve a conflict over how to settle the various issues under negotiation and the negotiation problem as such is one of searching the joint solution space of the negotiating parties and arriving at a mutually agreeable deal. According to economists individual bargaining abilities influence the negotiation process and the strategic interaction of the bargainers entirely determines the final outcome [1]. Owing to this strategic aspect of the negotiation process, several Game-theoretic models of bargaining have emerged. These approaches treat bilateral negotiations as two person non-zero sum games and proceed to find values for the game as in [2] or solutions under perfect equilibrium as in [3].

Game theoretic approaches in general assume complete rationality of the individuals participating in the bargaining and equal bargaining skill. As mentioned in [2] they also assume that each player is capable of accurately comparing ones own desires for various things and also possess complete knowledge of the opponent's tastes and preferences. Approaches that assume such idealized conditions hence become less appropriate for real world situations that run contrary to these assumptions.

S.K. Prasad et al. (Eds.): ICISTM 2009, CCIS 31, pp. 292–302, 2009.

Besides, most of these models focus on the solution properties while the actual procedure to arrive at a good solution remains an open problem especially when there is two sided information uncertainty. When uncertainties loom large, the negotiating parties become too skeptic, and tend to misrepresent information to the opponents for the fear of being exploited by the opponents if they reveal their private information truthfully. However such strategic misrepresentation becomes detrimental to the quality of the final outcome. If both parties strategically misrepresent their value tradeoffs, inefficient contracts will result [5 (page 144)]. Hence there arises the need for the right atmosphere and some truthful communication of values to locate efficient contracts that are mutually agreeable.

In pursuit of efficient outcomes under information uncertainty two broad streams may be identified in the Automated Negotiations literature; one that embraces an integrative search of the solution space and the other that employs a distributed search of the solution space. The integrative search methods make use of the services of a mediating agent which shall interact with the negotiating parties and demonstrate its neutrality by proceeding with the search of the solution space only in the direction which improves the joint gains of the negotiating parties, till a solution which can no further be improved in terms of joint gains is reached [10, 11].

In contrast to the integrative search methods, the distributed search methods allow the protagonists to haggle over the negotiation objects to find a mutually agreeable deal. The agents in this kind of set up are self-centered in the sense that they try to maximize their own utility. Despite the selfish attitudes of the participating agents, the distributed search methods are also capable of converging to a mutually agreeable deal mainly due to the threat of an aborted negotiation. Striking a deal anywhere in the agreeable zone is always considered better than breaking off a negotiation; this forces an agent to try to appease the opponent through an offer that shall be more agreeable to the opponent while not compromising too much on one's own utility. This force over several iterations drives the agents finally to a mutually agreeable solution point. The negotiation models in [6, 7, 8, 9, 12, 13, 14, 15, 16] adopt such distributed search techniques.

What should be the choice between the two search methods – integrative and distributed, could be a debatable issue. Generally an integrative search method would seem reasonable from an outsider's perspective. But when one becomes a negotiator himself, he would most probably prefer a distributed search since he would not trust a third party mediator fully, would never like to disclose any private information not even partial information to the mediator and would like to take full control of the situation even if it were at the cost of foregoing a possibility to gain more through the services of a mediator. Supporting arguments for the need for decentralized approaches may be found in [6, 7, 8].

When negotiators prefer distributed search, there arises an absolute need to develop approaches that would give the negotiators full confidence that the derived solution would be nothing but the best to themselves under the circumstances. This research attempts to address this need through a new kind of learning method that can be adopted by the selfish negotiating agents to arrive at a Pareto optimal deal that is also close to the Nash solution. Besides, the negotiation procedure suggested here seems to set the stage for truthful revelation of information through successive proposals by the negotiating agents thereby assisting them to locate efficient solutions.

2 The Proposed Approach

2.1 Motivation for This Work

Quite a few of the distributed search methods [9, 14, 15, 16] adopt some form of learning to judge the preferences of the opponent and modify the successive proposals such that they become progressively more agreeable to the opponent. It is this learning aspect that becomes key to the success of these approaches and this realization opens the door of opportunities to try out new types of learning agents that can pick up their leads and clues and also utilize them in novel ways.

The possibility of a new learning technique combined with the need for good distributed search methods for bilateral multi-issue negotiations under two sided information uncertainty and the need for the right kind of atmosphere that inhibits strategic misrepresentation of information, has motivated the development of the negotiation approach and the proposal construction algorithm suggested in this paper. Many of the published Automated Negotiation approaches have also influenced us towards evaluating the solution derived by our approach based on the following three parameters:

(1) Pareto efficiency of the derived solution: "An economic allocation or decision is efficient if and only if there is no other feasible allocation that makes some individuals better off without making other individuals worse off" [4].
(2) Closeness to the Nash Solution: Nash solution would be the point where the product of the corresponding individual utilities is maximized. The existence and uniqueness of such a point has been demonstrated in [2].
(3) Search time: The number of rounds of negotiation required to locate the solution measures Search time.

2.2 Context of the Proposed Work

The negotiation procedure suggested here might be applicable to bilateral multi issue negotiations under two-sided information uncertainty. The proposal construction algorithm in particular shall suit the requirements of selfish agents haggling over two continuous issues. Other assumptions are:

1. The negotiations are one off negotiations (No negotiation history is assumed. Hence all the learning has to come from the interactions of the on-going negotiation.)
2. The negotiation decisions are not constrained by time. (This could be a reasonable assumption given the fact that automated negotiations of the type suggested here are definite to conclude (agreement or no agreement) within finite time. They are unlike the manual negotiations that can stretch to very long time periods (days or months) and sometimes to indefinitely long periods thereby causing the utility of the various proposals put forward to depend on the time dimension too.)
3. The negotiations are of the alternating offers type. The negotiation ends when there is an agreement or when one or both of the parties prefer to break off the negotiation.

This model shall particularly suit negotiations involving (1) unsophisticated bargainers who need a quick deal and the best under the given circumstances (2) parties who would like to avoid a deadlock situation likely to arise due to comparable bargaining skill/power or information uncertainty or both.

2.3 The Proposed Negotiation Procedure

The procedure suggested here advocates a distributed search of the solution space by the negotiating agents in the presence of an intervener. Among the four different types of interveners identified in [5, pages22-23], the current procedure advocates the services of an intervener assuming the roles of a facilitator as well as a rules manipulator. As a facilitator he arranges for the negotiating agents to come to the negotiation table and as a rules manipulator he constrains the process of negotiation by imposing a procedure to resolve the conflict.

The intervener in the capacity of a facilitator makes the agents to (1) agree on the negotiation objects (issues under negotiation) (2) announce their initial offer (3) commonly agree upon a scale over which they shall relatively evaluate the various issues under negotiation and (4) unanimously fix on the least concession step size for each issue under negotiation.

Since the initial offers of both the agents are already on the table, the negotiation typically begins with one agent (does not matter which of the two) making the first counter offer. It is at this point that the intervener in the capacity of a rules manipulator mandates that *the successive proposals from each agent should be non-decreasing in utility to the opponent.* But the utility of a proposal to any agent is private information and neither the opponent nor the intervener has access to such information. Still the intervener judges an agent's compliance with the above rule at any stage by checking if the agent's previous proposal has been modified in the current round to suit the opposing agent's preferences as exhibited by the opposing agent through its own previous proposal(s).

The intervener's rule in effect necessitates each agent to learn the opponent's relative flexibility over the negotiation objects from the opponent's past proposals, utilize this knowledge to construct the current proposal and thereby ensure that the current proposal is not of a lesser utility to the opponent in comparison to the previous proposal. Additionally this rule indirectly necessitates each agent to truthfully put forward its own proposals in successive rounds of negotiation because it is only through an honest revelation of information through successive proposals that any agent can hope to receive successive offers from the opponent that keep progressively increasing in utility to oneself. In essence, this rule while putting a constraint on the agent making an offer, subtly indicates to the agent receiving the offer that 'what you show is what you get'. This rule gains further importance due to two other important phenomena. (1) It ensures that the negotiation trace remains confined to the Pareto frontier despite the fact that nobody knows the frontier in advance. (2) It ensures the convergence of the negotiation process within finite time.

The intervener in this model also ensures that as the negotiation progresses, the agents become not only committed to their current proposals but also to all their previous proposals. This condition however would be advantageous to both the negotiating agents because the negotiating agents in this set up being coerced to truthfully reveal information would put forth their successive proposals in decreasing order of their own utility, which causes one's earlier proposal to be of greater utility to oneself and therefore more welcome to become the deal. It shall also be the duty of the rules manipulator/intervener to ensure that if and when deals are struck, they shall be enforceable and inviolable.

Each agent in the negotiation would have its own domain of acceptable values for the various issues under negotiation. If there is an overlap between the acceptable regions of both the agents, a deal (the best one under the circumstances) in the overlapping region will be located by the said procedure. If there is no overlap, the negotiation will get aborted when one or both agents reach their reservation values in all the issues and hence desire to withdraw from the negotiation.

2.4   The Proposal Construction Algorithm

Since the negotiation procedure suggested in section 2.3 necessitates the individual agents to learn the opponent's preferences from their past proposals and accordingly construct one's own future proposals, we develop a proposal construction algorithm that may be implemented by both the agents so as to be confident of complying with the rules of the suggested negotiation procedure.

Let us assume that there are two continuous issues (Issue A and Issue B) over which two agents (Agent 1 and Agent 2) are haggling. For the sake of simplicity we assume linear utility functions for both agents over both the issues. Let us also assume that both agents have commonly agreed upon a 5-point scale (Table 1) to relatively evaluate the negotiation issues with the help of which they determine their own flexibility vectors and normalized weight vectors for the negotiation issues. Flexibility and weight as conceived here are opposite concepts - the more the weight of an issue, lesser the flexibility in conceding on the issue.

Table 1. 5-point scale to compare issues

Equally important	1
Slightly more important	2
Moderately more important	3
Highly important	4
Extremely important	5

For example if Agent 1 feels that Issue B is highly important (equivalent to 4 on the 5-point scale) as compared to Issue A, then its flexibility vector would be [4, 1] and its normalized weight vector would be [0.2, 0.8] and if Agent 2 feels that Issue A is slightly more important (equivalent to 2 on the 5-point scale) than Issue B then its flexibility vector would be [1, 2] and its normalized weight vector would be [0.667, 0.333]. One could as well use values in between the scale such as 1.8 or 3.3 to express one's finer comparisons if at all they can be meaningfully done.

The agents have to be very careful in choosing one's own flexibility vector because it is this vector that approximates one's tradeoff preferences over the negotiation issues. As an example, a flexibility vector [1, 1.5] implicitly means that the loss in utility to an agent by making 2 concessions in the first issue is roughly equivalent to the loss in utility by making 3 concessions in the second issue. In other words it means that the agent can tradeoff 3 concessions of the second issue for 2 concessions in the first issue. (In terms of ratios it may be stated that the tradeoff ratio of Issue A to Issue B is 2/3.) Thus if the agents have to make a reasonably accurate estimate of their own flexibility vector, they have to take into consideration the overall deviation between the initial offers of the two agents and the commonly agreed least concession step size for each issue and the commonly agreed upon scale for comparing the various issues.

The above observation is only meant to serve as a guideline for the agents in determining their own flexibility vector. In practical situations, the determination of the flexibility vector may not be as complex as it may sound because there would be enough room for errors in judgment. At the beginning of the negotiation, each agent knows its own flexibility vector and initializes the opponent's flexibility vector as [1, 1] meaning it assumes that the opponent weighs both issues equally and therefore is equally flexible in both the issues. The agent then updates the opponent's flexibility vector as exhibited by the opponent with every proposal it receives from the opponent. To update the opponent's flexibility vector it does the following.

1. The agent compares the opponent's most recent proposal (proposal N) with the opponent's previous proposal (proposal N-1). For each of the issues under negotiation, if there has been a change in the value of the issue in the latest offer as compared to the value of the issue in the previous offer, it checks if the change is welcome. (A change will be welcome if it is towards closing the gap between the value of the issue in the opponent's previous offer and value of the issue in its own initial offer)
2. If the change is indeed welcome, the agent subsequently checks how many concessions have been given. (The number of concessions will depend upon the commonly agreed least concession step size for the issue.)
3. The agent then increments the vector component (in the opponent's flexibility vector) corresponding to the issue by the number of concessions received in that issue.

At any stage in the negotiation, when an agent receives a new offer from the opponent and finds it unacceptable (an offer shall be unacceptable if the utility of the opponent's new offer to oneself is far less than the utility of one's own new counter offer prepared to be put forth in one's own current round) it rejects the opponent's offer tentatively and plans to put forth a new counter offer. The agent follows the following steps to construct the new proposal:

(1) The agent updates the opponent's flexibility vector as exhibited by the opponent.
(2) It compares the opponent's flexibility vector at that stage with that of its own. If they have complementary preferences (one of them is more flexible in Issue A and the other is more flexible in Issue B) then it gives one (or one more) concession in its own most flexible issue while maintaining status quo in the other issue.
(3) If they have similar preferences (both of them are more flexible in the same issue) then it compares its own flexibility in its most flexible issue with that of the opponent. If the opponent's flexibility in the most flexible issue is less than that of its own, it gives one (or one more) concession in its most flexible issue while maintaining status quo in the other issue.
(4) If the opponent's flexibility in the most flexible issue is greater than or equal to that of its own, it explores the possibility of a tradeoff. If there is no possibility of

a tradeoff at that stage, it offers one (or one more) concession in its most flexible issue and maintains status quo in the other issue.

(5) If there is a possibility of a tradeoff then it replaces some concessions in the most flexible issue with some in the other issue (this depends on the agent's tradeoff ratio between the two issues) in addition to giving one (or one more) concession in its most flexible issue.

3 Experiments

The proposed negotiation procedure and the proposal construction algorithm was coded and tested in MATLAB 7.3. The negotiations between 2 agents over two continuous issues were simulated. In the simulation, each agent is totally ignorant of the other agent's reservation values, utility functions and relative preferences over the negotiation issues. Thus they proceed with the negotiation using the proposed algorithm and have no idea of the overall solution space, or its bounds.

After the negotiation concludes, we plot the negotiation trace by obtaining the corresponding utilities of the negotiating agents for each of the proposals put forth by each of the agents in the course of the negotiation. Then we gather the private information of both agents to determine the entire solution space after which we superimpose the plot of the entire solution space on the plot of the negotiation trace. This superimposed plot clearly demonstrates the confinement of the negotiation trace to the Pareto frontier and thus the Pareto optimality of the derived solution. The plot also shows the number of rounds of negotiation and the closeness of the derived solution to the Nash solution.

For all the experiments, we assume that both the agents have linear utility functions for each of the negotiation issues and the domains of values for each of the issues are the same for both agents. These assumptions are similar to those in [9] and are common in many other past approaches. Specifically in all our experiments, we assume that each of the issues (Issue A and Issue B) can take any continuous value in the interval [1, 10]. The initial offer of Agent 1 is (1, 1) and that of Agent 2 is (10, 10). The utility values are mapped to the interval [0, 1] such that each agent gets a utility value of 1 for its own initial offer. For Agent 1, the utility of a particular offer is calculated using equation 1 while for Agent 2 the same is calculated using equation 2.

$$1 - \left[\frac{MinValue_A - Currentvalue_A}{MinValue_A - MaxValue_A}(Weight_A) + \frac{MinValue_B - Currentvalue_B}{MinValue_B - MaxValue_B}(Weight_B) \right] \qquad (1)$$

$$1 - \left[\frac{MaxValue_A - Currentvalue_A}{MaxValue_A - MinValue_A}(Weight_A) + \frac{MaxValue_B - Currentvalue_B}{MaxValue_B - MinValue_B}(Weight_B) \right] \qquad (2)$$

For each of the agents, the *MaxValue* and *MinValue* correspond to the maximum and minimum values respectively of the corresponding issue (denoted by the subscript) from the domain of acceptable values for the issue for that agent while *Weight* corresponds to the vector component of the corresponding issue from the normalized weight vector of that agent.

Table 2. Negotiation simulation experiments

Exp. No.	Flexibility vector		Total No of feasible solutions	Search time	Gain		Product of gains	
	Agent 1	Agent 2			Agent1	Agent 2	Derived solution	Nash Solution
1	[1, 3]	[1, 3]	100	24	0.5	0.5	0.25	0.25
2	[1, 3]	[1, 5]	100	24	0.5	0.5185	0.25926	0.2778
3	[1, 3]	[1, 2]	100	19	0.5	0.5556	0.27778	0.27778
4	[1, 3]	[3, 1]	100	10	0.75	0.75	0.5625	0.5625
5	[1, 3]	[4, 1]	100	10	0.75	0.8	0.6	0.6
6	[1, 3]	[2, 1]	100	10	0.75	0.6667	0.5	0.5000025
7	[1, 3]	[1, 3]	361	47	0.4861	0.5139	0.24981	0.25
8	[1, 3]	[1, 5]	361	46	0.5278	0.5185	0.27366	0.2778
9	[1, 3]	[1, 2]	361	37	0.5	0.5556	0.27778	0.27778
10	[1, 3]	[3, 1]	361	19	0.75	0.75	0.5625	0.5625
11	[1, 3]	[4, 1]	361	19	0.75	0.8	0.6	0.6
12	[1, 3]	[2, 1]	361	19	0.75	0.6667	0.5	0.5000025

The experiments were designed to test the procedure when the agents had various combinations of preferences and when the least concession step sizes for the issues were varied. (As the concession step sizes are reduced, the search space becomes larger). The results are presented in Table 2.

For Exp 1-6 the least concession step size for each issue was set to 1. Hence the solution space contains 100 points (10X10=100). For Exp 7-12 the least concession step size for each issue was set to 0.5. Hence the solution space contains 361 points (19X19=361). Experiment 1 and 7 has exactly similar preferences from both agents. Experiment 2, 3, 8, 9 have more or less similar preferences from both agents. Experiment 4 and 10 depict the scenario where the agents have exactly complementary preferences while Experiment 5, 6, 11, 12 depict the scenario where the agents' preferences are more or less complementary.

Due to space restriction, the negotiation trace plots (superimposed on the solution space plot) for Exp No 3 (almost similar preferences; search space containing 100 feasible solutions) (Figure 1) and Exp No 11 (nearly complementary preferences; search space containing 361 feasible solutions) (Figure 2) alone are included.

In Figures 1 and 2, the dark square on the frontier in the middle is the derived solution; the trace (points marked by a circle) below the deal point is made by the successive proposals of Agent 1 while the trace (points marked by a diamond) above the deal point is made by the successive proposals of Agent 2.

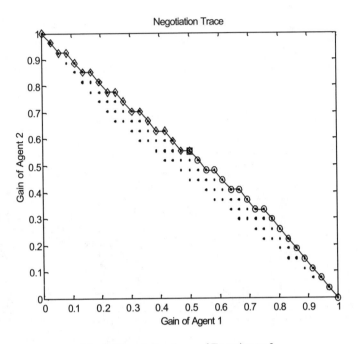

Fig. 1. Negotiation trace of Experiment 3

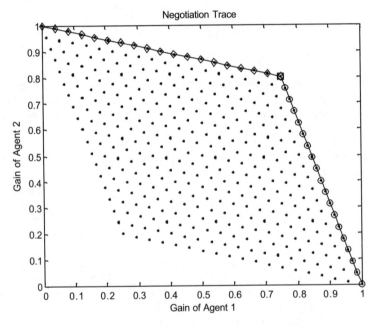

Fig. 2. Negotiation trace of Experiment 11

3.1 Discussion of Results

(1) The negotiation trace invariably remains confined to the Pareto frontier, thereby rendering the derived solution to be Pareto efficient.

(2) The derived solution most of the times coincides with the Nash solution (Point where the product of the individual gains of the agents is maximized). The deviation is marginal wherever it does not coincide.

(3) When both agents have similar preferences, the approach takes longer to converge to a solution because more number of points lies on the Pareto frontier. When the preferences are exactly similar all the points lie on the frontier line.

(4) When the agents have complementary preferences, the approach converges to a solution quicker because the number of points on the frontier becomes lesser and lesser as the agents' preferences become more and more complementary.

(5) The negotiation simulations with different flexibility vectors for each of the agents reveals the sensitivity of the derived solution to only the relative preferences of the agents. This leads us to the inference that small errors in judgment (in determining one's own flexibility vector and weight vector) will not alter the derived deal drastically as long as they do not alter the relative preferences of the agents.

(6) The approach fares excellently on scalability. As the search space becomes larger and larger, the number of points searched as a percentage of the total number of points in the search space becomes lesser and lesser.

4 Limitations and Future Work

The proposed algorithm was only tested for two issues. But the approach seems to have enough potential to be extended to more than two issues. When there are only two issues, assessing/learning the opponent's relative flexibility on the two issues is easier. Hence the Pareto frontier can be traced exactly through successive proposals. However with more number of issues, there may be some errors in the assessments/learning. Still the agents should be able to approximately follow the frontier and derive a solution very close to the frontier if not on the frontier itself. Our future work would be towards extending this approach to more than two issues.

Two other limitations, which we have identified, are that the approach has been tested on continuous issues only and the agents have used linear utility functions. It would be interesting to extend the approach to negotiation over issues that can take discrete or categorical values too. As part of our future work, we would also like to test the approach when agents use non-linear scoring functions to define the utility of the various issues in various ranges in their own domain of acceptable values.

5 Conclusions

The two broad approaches to deal with the negotiation problem under two-sided information uncertainty are the integrative search and the distributed search approaches. This paper first addressed the need for an efficient decentralized procedure for automated negotiations and then the need for a congenial atmosphere to encourage honest revelation of information from the negotiating parties in order to identify efficient solutions. A negotiation procedure that more or less satisfies these needs was presented. Further, a proposal construction algorithm that can be implemented by selfish negotiating agents to comply with the needs of the suggested negotiation procedure was presented.

Simulation experiments of bilateral negotiations over two continuous issues demonstrated the confinement of the negotiation trace to the Pareto frontier and thereby the identification of Pareto optimal deals. The other points that add further promise to the proposed approach are the proximity of the derived solution to the Nash solution, the fast convergence to a solution and the scalability of the algorithm.

References

1. Roth, A.E.: Game-theoretic models of bargaining. Cambridge University Press, New York (2005)
2. Nash, J.F.: The bargaining problem. Econometrica 18, 155–162 (1950)
3. Rubinstein, A.: Perfect equilibrium in a bargaining model. Econometrica 50, 97–109 (1982)
4. Holmstrom, B., Myerson, R.B.: Efficient and durable decision rules with Incomplete information. Econometrica 51, 1799–1819 (1983)
5. Raiffa, H.: The Art and Science of Negotiation. Harvard University Press, Cambridge (1982)
6. Heiskanen, P.: Decentralized method for computing Pareto solutions in multiparty negotiations. European Journal of Operational Research 117, 578–590 (1999)
7. Heiskanen, P., Ehtamo, H., Hämäläinen, R.P.: Constraint proposal method for computing Pareto solutions in multi-party negotiations. European J. Oper. Res. 133(1), 44–61 (2001)
8. Luo, X., Jennings, N.R., Shadbolt, N., Leung, H., Lee, J.H.: A fuzzy constraint based model for bilateral multi-issue negotiations in semi-competitive environments. Artificial Intelligence 148, 53–102 (2003)
9. Faratin, P., Sierra, C., Jennings, N.R.: Using similarity criteria to make issue trade-offs in automated negotiations. Artificial Intelligence 142, 205–237 (2002)
10. Lin, R.J., Chou, S.T.: Mediating a bilateral multi-issue negotiation. Electronic Commerce Research and Applications 3, 126–138 (2004)
11. Ehtamo, H., Verkama, M., Hamalainen, R.P.: How to Select Fair Improving Directions in a Negotiation Model over Continuous Issues. IEEE Transactions on Systems, Man, and Cybernetics – Part C: Applications and Reviews 29(1) (February 1999)
12. Barbuceanu, M., Lo, W.: A Multi-Attribute Utility Theoretic Negotiation Architecture for Electronic Commerce. In: Proceedings of 4th Int. Conf. on Autonomous Agents, Barcelona, Spain, pp. 239–247 (2000)
13. Shakun, M.F.: Multi-bilateral Multi-issue E-negotiation in E-commerce with a Tit-for-Tat Computer Agent. Group Decision and Negotiation 14, 383–392 (2005)
14. Coehoorn, R.M., Jennings, N.R.: Learning an Opponent's Preferences to Make Effective Multi-Issue Negotiation Trade-Offs. In: Proceedings of Sixth International Conference on Electronic Commerce, ICEC 2004 (2004)
15. Zeng, Z., Meng, B., Zeng, Y.: An Adaptive Learning Method in Automated Negotiation based on Artificial Neural Network. In: Proceedings of the Fourth International Conference on Machine Learning and Cybernetics (August 2005)
16. Lau, R.Y.K., Tang, M., Wong, O.: Towards Genetically Optimized Responsive Negotiation Agents. In: Proceedings of the IEEE/WIC/ACM Int. Conf. on Intelligent Agent Technology, IAT 2004 (2004)

On Performance of Linear Multiuser Detectors for Wireless Multimedia Applications

Rekha Agarwal[1], B.V.R. Reddy[1], E. Bindu[1], and Pinki Nayak[2]

[1] Amity School of Engineering and Technology, New Delhi
rarun96@yahoo.com, bindusugathan@gmail.com, pinki_dua@yahoo.com
[2] Professor, USIT, GGSIP University, Delhi
bvrreddy64@rediffmail.com

Abstract. In this paper, performance of different multi-rate schemes in DS-CDMA system is evaluated. The analysis of multirate linear multiuser detectors with multiprocessing gain is analyzed for synchronous Code Division Multiple Access (CDMA) systems. Variable data rate is achieved by varying the processing gain. Our conclusion is that bit error rate for multirate and single rate systems can be made same with a tradeoff with number of users in linear multiuser detectors.

Keywords: Additive white Gaussian noise, bit error rate, minimum mean square error, multiple access interference.

1 Introduction

In current and near future digital mobile communication systems, multiple user access is supported by Direct Sequence (DS) Code-Division Multiple Access (CDMA) technology. One of the most important performance degrading factors in such a system is the Multiple Access Interference (MAI). One solution to reduce the effect of MAI is multiuser detection [1]. Although various multiuser detection schemes promise a huge capacity increase, the computational complexities are a burden to existing hardware technologies. This is why reduced complexity, sub-optimum algorithms are preferred to optimal detection algorithms. These sub-optimal detectors are further classified into linear and nonlinear type detector. The linear detectors include decorrelating detector and Minimum Mean Squared Error (MMSE) detectors [7]-[9]. The detectors need to compute the inverse of cross-correlation matrix [10], the complexity of which is linear. The decorrelating detector applies inverse of correlation matrix to the conventional detector output [10] so that the output of this detector completely removes the MAI term. Thus, we see that decorrelating detector completely eliminates MAI. The decorrelating detector was exhaustively analyzed by Lupas and Verdu in [20] and was shown to have many good features, which include substantial capacity improvement over conventional detector, and less complexity. The second linear detector (MMSE) [7]-[9] applies linear mapping, which minimizes mean squared error between actual data and soft output of the conventional detector. An extensive literature survey about linear and nonlinear multiuser detection algorithms can be found in [10]-[12]. These receivers use single rate system.

S.K. Prasad et al. (Eds.): ICISTM 2009, CCIS 31, pp. 303–312, 2009.

Recently, the advent of third generation wireless communication for multimedia applications initiated investigations of multiuser detection for multirate systems [2]. The third generation systems are expected to support various kinds of communication services, such as voice, video and data [2]. To handle mixture of different applications, the system must support variable data rates and their performance requirements will vary from user to user. It is desirable to develop a system that operates on multiple data rates without much degradation in the performance [2]. A typical network should be able to support voice transmission having relatively low data rates (e.g. 9.6 kbps) and relatively high bit error rate requirement (e.g. 10^{-4} ... 10^{-2}). Also the network should serve data transmission with higher data rate (up to 5 Mbps) and low bit error rate (BER) requirement (e.g. 10^{-7} ... 10^{-5}) [3].   There are several multiplexing strategies to design a multiple data rate, henceforth, called as multirate multiuser communication system.

Motivated by the requirements of 3G systems, an attempt is made in this paper to combine multiuser detection and multirate handling capacity. To take advantage of reduced complexity, mostly used sub-optimal multiuser detection algorithms i.e. decorrelator and MMSE are considered in this paper. These receivers apply linear transformation to the output of matched filter detector. They are less computationally complex (linear complexity with number of users) compared to optimum receiver, have optimum near-far resistance [20] and can also be adapted blindly. Moreover, the BER of these receivers is also independent of the power of interfering users. In this paper, the effect of increase in number of data rates with total number of users is analyzed and verified by simulation for constant bit error rate. A simplified general expression is also given with which effective number of users in multirate multiuser environment communication system can be calculated. In this paper the results are given only for sub-optimal detectors i.e. for decorrelator and MMSE detector. In section 2, a basic model is given. Section 3 discusses all multirate schemes and their performance is compared. In section 4, decorrelator and MMSE detector for multirate systems are analyzed. The simulation results, tables and conclusion are explained in section 5.

2   Signal Model

Consider a baseband direct sequence (DS) CDMA network of N users. The received signal can be modeled as

$$r(t) = S(t) + \sigma n(t), \tag{1}$$

where n(t) is white guassian noise with unit power spectral density, and S(t) is the superposition of the data signals ok N users, given by

$$S(t) = \sum_{k=1}^{N} A_k \sum_{i=-P}^{P} b_k(i) s_k(t - iT - \tau_k), \tag{2}$$

Where $2P+1$ is the number of data symbols per user per frame, T is the symbol interval, where A_k, τ_k, $\{b_k(i) ; i = 0, \pm 1, ..., \pm P\}$ and $\{ s_k(t); 0 \le t \le T\}$ denote, respectively, the received amplitude, delay, symbol stream, and normalized signaling waveform of the kth user. It is assumed that $s_k(t)$ is supported only on the interval [0,T] and has unit energy, and that $\{ b_k(i) \}$ is a collection of independent equiprobable ±1 random variables. For a synchronous system model, $\tau_1 = \tau_2 =... = \tau_k =0$. It is then sufficient to consider the received signal during one symbol interval, and the received signal model becomes

$$r(t) = \sum_{k=1}^{N} A_k b_k s_k(t) + n(t).$$ (3)

The system under consideration has N users transmitting with equal power and that the pulses are strictly band limited. The transmitted symbols have duration T and spreading factor of the system is G. The effective single sided noise spectral density encountered by a user in a bandwidth of W Hz is the sum of background noise and multiple access interference, i.e. the received interfering power from the remaining (N-1) users [9]. For equal power users with known power, the effective noise spectral density is given by [9]

$$I_0 = N_0 + \frac{(N-1)R_b E_b}{W}.$$ (4)

where the first term N_o represents the thermal noise, and the second term represents the multiple access interference, i.e. the received interfering power from remaining (N-1) users for total bandwidth and for fixed bit rate per user. It follows that the signal to interference ratio can be written as [16]

$$\frac{E_b}{I_o} = \frac{\dfrac{2E_b}{N_o}}{1 + 2(N-1)\left(\dfrac{R_b}{W}\right)\left(\dfrac{E_b}{N_o}\right)},$$ (5)

or

$$\frac{E_b}{I_o} = \frac{1}{\dfrac{1}{2\dfrac{E_b}{N_o}} + \dfrac{(N-1)}{G}},$$ (6)

where E_b/N_o is signal to noise ratio. It is assumed that multiple access is Gaussian distributed with zero mean and variance proportional to the number of users, Thus, bit error probability for matched filter detector can be written as

$$P_b = Q(\sqrt{(E_b / I_o)}),$$ (7)

where Q (.) is the complementary gaussian error function.

3 Multirate Schemes

3.1 Multi-modulation Systems

In this scheme, the constellation size for M-ary QAM modulation is chosen according to the required data rate. We take BPSK for the lowest rate r_n, and at other rates users use M-ary square lattice QAM where the constellation size is given by

$$M = 2^{(r/r_{in})}$$

for the bit rate r_i. The processing gain is W/r_n for the system bandwidth W and same for all users. The average signal to interference ratio in this method for asynchronous users (j^{th} user) is given by [15]

$$\frac{E_b}{I_o} = \frac{1}{2} \left[\frac{M-1}{3} \left(\frac{No}{\log_2(M) E_b} + \frac{2}{3G} \left(\sum_{i=1}^{n} \frac{r_i}{r_j} N_i - 1 \right) \right) \right]^{-1} . \tag{8}$$

3.2 Multi Processing Gain System

In this scheme, processing gain for user i with rate r_i is given by W/r_i . All users use BPSK modulation and processing gain is not constant for all users. The average signal to interference ratio in this method for asynchronous users (j^{th} user) is given by [15]

$$\frac{E_b}{I_o} = \frac{1}{2} \left[\left(\frac{No}{2E_b} + \frac{1}{3G_j} \left(\sum_{i=1}^{n} \frac{r_i}{r_j} N_i - 1 \right) \right) \right]^{-1} . \tag{9}$$

3.3 Multi Channel Systems

In this scheme processing gain is constant for all users and equal to W/r_n. The users with bit rate r_i is transmitted on r_i/r_n parallel channels. The average signal to interference ratio in this method for asynchronous users (j^{th} user) is given by [15]

$$\frac{E_b}{I_o} = \frac{1}{2} \left[\left(\frac{No}{2E_b} + \frac{2}{3G} \left(\sum_{i=1}^{n} \frac{r_i}{r_j} N_i - 1 \right) \right) \right]^{-1} . \tag{10}$$

We have investigated different multirate schemes and found that multi channel and multi processing gain scheme perform almost in the same way. The disadvantage of multi channel scheme is the need of linear amplifiers for mobile transmitting at high rate. It is also possible to use multi modulation scheme, which only degrades the performance for users with high bit rates, that is, users that use higher level of modulation than QPSK. The multi processing gain scheme has better performance among all the schemes with only disadvantage of being sensitive for external interferences. In this paper, we have considered multi processing gain scheme for further discussions.

4 Multi-processing Gain Schemes in Linear Multiuser Detection

In the same bandwidth W, different data rates to different users can be allocated by varying the processing gain. The rate is given by W/G, hence high rate means low processing gain and vice versa. Without loss of generality, we assume that for a Multi Processing Gain (MPG) system with N users and rate $r_1 > r_2 > \ldots r_N$, all users have same energy per bit to noise ratio E_b/N_o. We divide total number of users in two groups with bit rates r_1 and r_2 where $r_1>r_2$ in dual rate system. Similarly for three different bit rates, all users are divided in three groups with each group of users occupying one data rate and so on. The system is defined in a way such that total users of all groups is equal to total number of users i.e.

$$\sum_{i=1}^{n} N_i = N ,$$

where n is number of groups with distinct bit rates [16]. The performance of user k with rate r_k in a synchronous CDMA system with matched filter can be written as [16]

$$P_{b_{ij}} = Q\left(\sqrt{\dfrac{1}{\dfrac{1}{2\dfrac{E_b}{N_o}} + \dfrac{1}{G_j}\left(\displaystyle\sum_{i=1}^{n}\dfrac{r_i}{r_j}N_i - 1\right)}} \right). \tag{11}$$

It can be seen from (11) that the processing gain has an impact on the value of BER. The equivalent number of users in a multirate can be expressed as

$$N = \sum_{i=1}^{n} \dfrac{r_i}{r_j} N_i . \tag{12}$$

The result can also be applied to sub-optimal multiuser detectors i.e. decorrelating detector and MMSE detector. The expression for bit error probability in terms of processing gain and number of users can be derived. We already know that bit error probability in any linear multiuser detector can be expressed as

$$P_b = Q\left(\sqrt{\dfrac{2E_b}{N_o}\bar{\eta}} \right), \tag{13}$$

Here, $\bar{\eta}$ is near-far resistance. In case of the decorrelating detector $\bar{\eta}$ can be expressed as

$$\bar{\eta} = E\left[\dfrac{1}{R_{k,k}} \right], \tag{14}$$

where $R_{k,k}$ is the cross-correlation NxN matrix for spreading sequences and E[.] is the average operator. The near-far resistance in simplified form from equation (14) can be written as [9]

$$\bar{\eta} = 1 - \dfrac{N-1}{G}. \tag{15}$$

Hence the bit error probability P_b for the decorrelating detector with perfect power control in a synchronous system on AWGN channel, can be given by

$$P_b = Q\left(\sqrt{\frac{2E_b}{N_o}\left(1-\left(\frac{N-1}{G}\right)\right)}\right).\tag{16}$$

The bit error probability for MMSE detector in term of number of users and processing gain can also be calculated as [9]

$$P_b = Q\left(\sqrt{\frac{2E_b}{N_o}-\frac{1}{4}F\left(\frac{2E_b}{N_o},\frac{N-1}{G}\right)}\right),\tag{17}$$

where,

$$F(x,z)\underset{=}{def}\left(\sqrt{x(1+\sqrt{z})^2+1}-\sqrt{x(1-\sqrt{z})^2+1}\right)^2.\tag{18}$$

As we have seen from (12) that in a multi-rate environment, the total number of users can be replaced in terms of rates of different groups with their equivalent number of users, the performance of decorrelating and MMSE detector can also be found. For decorrelating detector, replacing the number of users from (12) into (16), we get

$$P_{bij}=Q\left(\sqrt{\frac{2E_b}{N_o}\frac{(1-(\sum\limits_{i=1}^{n}\frac{r_i}{r_j}N_i-1))}{G}}\right).\tag{19}$$

For the MMSE detector it can be shown that P_{bij} can be evaluated as

$$P_{bij}=Q\left(\sqrt{\frac{2E_b}{N_o}-\frac{1}{4}F(\frac{2E_b}{N_o},\frac{(\sum\limits_{i=1}^{n}\frac{r_i}{r_j}N_i-1)}{G}}\right).\tag{20}$$

Equations (19) and (20) give bit error probabilities for decorrelator and MMSE detector in multirate environment.

5 Results and Conclusion

In the following simulation results, we first discuss various multirate schemes as shown in Fig. 1. We have assumed random spreading sequences and AWGN channel. We see that multi processing gain and multi channel scheme have almost the same performance and these schemes are better than multi modulation (16 QAM) scheme. The multi channel scheme has highest processing gain and therefore suppression level. The multi-modulation scheme has much worse performance for the higher level of modulation (here 16 QAM). Therefore, we conclude that multi modulation scheme has low multi-rate support. As stated, the performance of multi-processing gain and multi-channel scheme is about the same. In a more realistic situation, multi-processing gain scheme is

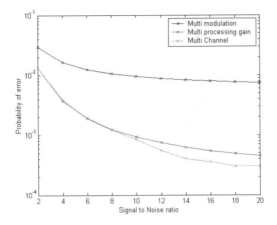

Fig. 1. Comparison of multirate schemes

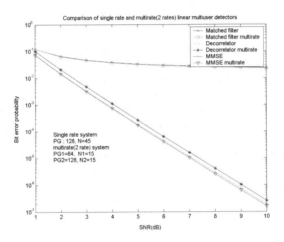

Fig. 2. Analysis of single rate system

easy to implement and therefore we consider it for the analysis in this paper. For the simulation of multiprocessing gain scheme with linear multiuser detectors, we have considered dual rate case to 5 different rates and the results can be extended easily to n different rates. It is shown that for a fixed value of bit rates in multirate system, the bit error probability is same in a single rate system. This helps to find equivalent number of users in multirate system as that of single rate system without degradation in the performance as the bit error probability is kept constant. Fig. 2 shows that bit error rate of all linear multiuser detectors is constant for a fixed values of processing gains in multirate system with a constant rate in single rate system. Similarly Fig. 3, Fig. 4, and Fig. 5 give the performance of multirate linear multi-user detectors for three rates, four rates and five rates system respectively. Hence by choosing the processing gains carefully, the performance is not degraded in

310   R. Agarwal et al.

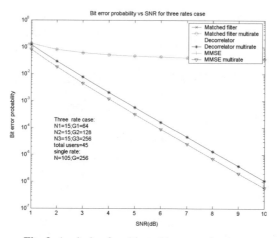

Fig. 3. Analysis of multirate (three rates) system

multirate multiuser communication system when compromise is made with the number of users.

In all the cases, we have taken processing gains in ascending order (or rates in descending order), which are in multiples of 2^n where n is number of groups with different rates. One very important observation in all the figures given is that, the BER of single rate system and in multirate system irrespective of number of different rates is exactly same. This is clear since two lines of each linear multiuser detector overlap with each other. A conclusion can be drawn from Fig. 6 that a tradeoff can be analyzed in number of rates with the number of users which means that increase in number of rates reduces the percentage of effective number of users allowed in multirate multiuser communication system. Therefore the system with single rate system can be used for multirate system with less number of users without change in BER. Considering various types of noise sources and channel model can further extend this work.

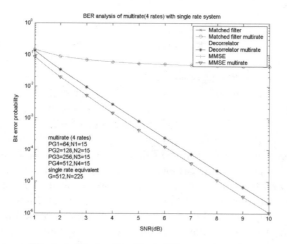

Fig. 4. Analysis of multirate (four rates) system

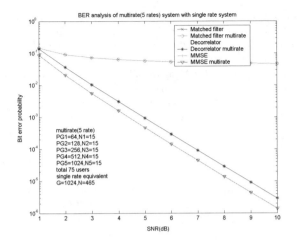

Fig. 5. Analysis of multirate (five rates) system

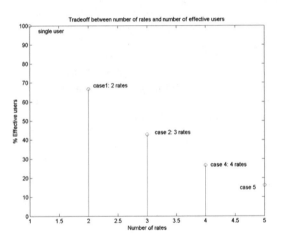

Fig. 6. Tradeoff in a multirate system between number of rates and effective number of users

References

1. Thit, M., Kai-Yeung, S.: Dynamic Assignment of Orthogonal Variable Spreading Factor Codes in W-CDMA. IEEE Journal on Selected Areas in Comm. 18, 1429–1440 (2000)
2. Sabharwal, Mitra, U., Moses, R.: MMSE Receivers for Multirate DS-CDMA Systems. IEEE Trans. on Comm. 49, 2184–2196 (2001)
3. Juntti, M.J.: System Concept Comparisons for Multirate CDMA with Multiuser Detection. In: Proc. IEEE VTC 1998, Canada, pp. 36–40 (1998)
4. Peter, S., Muller, R.R.: Spectral Efficiency of CDMA Systems with Linear MMSE Interference Suppression. IEEE Trans. on Comm. 47, 722–731 (1999)
5. Proakis, J.G.: Digital Communications. Mc Graw Hill, New York (1989)

6. Verdu, S.: Minimum Probability of Error for Asynchronous Gaussian Multiple Access Channels. IEEE Trans. on Information Theory 32, 85–96 (1986)
7. Lupas, R., Verdu, S.: Linear Multiuser Detectors for Synchronous Code Division Multiple Access. IEEE Trans. on Information Theory 35, 123–136 (1989)
8. Xie, Z., Short, R., Rushforth, C.: A Family of Sub-Optimum Detectors for Coherent Multiuser Communications. IEEE Journal on Selected Areas in Comm. 8, 683–690 (1990)
9. Madhow, U., Honif, M.: MMSE Interference Suppression for Direct Sequence Spread Spectrum CDMA. IEEE Trans. on Comm. 42, 3178–3188 (1994)
10. Moshavi, S.: Multi-User Detection for DS-CDMA Communications. IEEE Comm. Magazine 11, 137–150 (1996)
11. Hallen, D., Holtzman, J., Zvonar, Z.: Multiuser Detection for CDMA Systems. IEEE Personal Comm., 151–163 (1995)
12. Kuan, E.-L., Hanzo, L.: Burst-by-Burst Adaptive Multiuser Detection CDMA: A Frame Work for Existing and Future Wireless Standards. Proc. IEEE 91, 278–301 (2003)
13. Honig, M., Kim, J.B.: Allocation of DS-CDMA Parameters to Achieve Multiple Rates and Qualities of Service. Proc. IEEE 3, 1974–1979 (1996)
14. Chen, J., Mitra, U.: Analysis of Decorrelator Based Receivers for Multirate CDMA Communications. IEEE Trans. on Vehic. Tech. 48, 1966–1983 (1999)
15. Johansson, A.L., Svensson, A.: On Multirate DS / CDMA Schemes with Interference Cancellation. Journal of Wireless Personal Comm. 9, 1–29 (1999)
16. Verdu, S., Shamai, S.: Spectral Efficiency of CDMA with Random Spreading. IEEE Trans. on Inform. Theory 45, 622–640 (1999)
17. Mitra, U.: Comparison of Maximum Likelihood Based Detection for Two Multi-Rate Access Schemes for CDMA Signals. IEEE Trans. on Comm. 47, 64–77 (1999)
18. Lupas, R., Verdu, S.: Near-Far Resistance of Multiuser Detectors in Asynchronous Channels. IEEE Trans. on Comm. 38, 496–508 (1990)

A Trust Based Clustering Framework for Securing Ad Hoc Networks

Pushpita Chatterjee, Indranil Sengupta, and S.K. Ghosh

School of Information Technology
Indian Institute of Technology, Kharagpur
{pushpita.chatterjee,isg,skg}@sit.iitkgp.ernet.in

Abstract. In this paper we present a distributed self-organizing trust based clustering framework for securing ad hoc networks. The mobile nodes are vulnerable to security attacks, so ensuring the security of the network is essential. To enhance security, it is important to evaluate the trustworthiness of nodes without depending on central authorities. In our proposal the evidence of trustworthiness is captured in an efficient manner and from broader perspectives including direct interactions with neighbors, observing interactions of neighbors and through recommendations. Our prediction scheme uses a trust evaluation algorithm at each node to calculate the direct trust rating normalized as a fuzzy value between zero and one. The evidence theory of Dempster-Shafer [9], [10] is used in order to combine the evidences collected by a clusterhead itself and the recommendations from other neighbor nodes. Moreover, in our scheme we do not restrict to a single gateway node for inter cluster routing.

Keywords: ad hoc networks, trust, cluster, security.

1 Introduction

Mobile ad hoc networks can be deployed in the situation where no infrastructure is present such as disaster recovery or battle fields. It is a group of mobile nodes which do not require a centralized administration or a fixed network infrastructure. In contrast to wired networks, each node in an ad-hoc networks acts like a router. Also wireless links are susceptible to link attacks ranging from passive eavesdropping to active interfering. Unlike fixed hardwired networks with physical defense at firewalls and gateways, attacks on ad hoc networks can come from all directions and may target any node. Autonomous nodes have inadequate physical protection,and can be captured, compromised, and hijacked easily. Attacks from a compromised node are more dangerous and much harder to detect. Damage includes leaking secret information, interfering message and impersonating nodes, thus violating the basic security requirements. All these mean that every node must be prepared for encounter with an adversary directly or indirectly. Due to dynamic topology of the networks any security solution with static configuration would not be sufficient. Moreover, a authority responsible for distribution of keys for the whole network is vulnerable to single point failure.

S.K. Prasad et al. (Eds.): ICISTM 2009, CCIS 31, pp. 313–324, 2009.

So we require a distributed architecture for this kind of network for its proper functionality. Any node must be prepared to operate in a mode that should not immediately trust on any peer. If the trust relationship among the network nodes is available for every node, it will be much easier to select proper security measure to establish the required protection. Moreover, it will be more sensible to reject or ignore hostile service requests [8]. As the overall environment in ad hoc network is cooperative by default, these trust relationships are extremely susceptible to attacks. Also, the absence of fixed infrastructure, limited resources, ephemeral connectivity and availability, shared wireless medium and physical vulnerability, make trust establishment very complex.

In this paper we introduce a trust based distributed clustering approach where the ad hoc network can be viewed as a set of self organizing clusters, which are formed in the basis of trust relationship between the neighbor nodes. To formalize the trust of a particular node, nodes monitor the behavior of other nodes and collect information from its neighbors and then take the decision about the node. We have used a trust evaluation algorithm at each node to evaluate the direct trust of its neighbor nodes. To get the direct trust values and recommendation from other nodes we use Dempster-Shafer Theory of combining evidences [9], [10] and calculate the trust value of a particular node. Each node is periodically monitored by other nodes to check whether any malicious or selfish activity can be traced.

We have proposed a mechanism for energy efficient distributed clusterhead (CH) election, new node join, and a group based key distribution protocol. If some nodes are compromised, our protocol ensures that the non-compromised nodes can still communicate with full secrecy. Our scheme is always unconditionally secure against node capture attacks and node unjoin due to node mobility. We clearly specify the functionality of clusterheads, intra and inter-cluster routing in our proposal and also give importance to the security aspects and possibility of malicious attacks of the network. The rest of this paper is organized as follows.

In Section 2, we discuss related work. In Section 3 we present the philosophy of our design, trust evaluation algorithm and overall protocol details. Our conclusion and future work are discussed in Section 4.

2 Related Works

Several works have been proposed in the literature to deal with the security problems in ad hoc networks. Several algorithms like WCA [12], are proposed for clustering the ad hoc networks but security is not taken care of. Some clustering mechanism and routing mechanism are proposed in [13],[7],[5]. But none of them is able to completely handle the secure clustering . Moreover they did not specify any well defined clustering mechanism, new cluster head selection, etc. Among several secure solutions based on clustering ad hoc networks, Varadharajan et al [1], uses NDTR architecture but does not deal with partitioning and merging of clusters. A cluster based security architecture is proposed by Becheler et al. [3], which uses threshold cryptography scheme to distribute CA(Certification Authority). This approach is not realistic, because the warrantors do not have any information about the new node to be guaranteed. The network traffic generated by each new node is very high thereby causing wasting of both bandwidth and energy. Also, to renew the network key, the intervention of a trusted third party is needed so that it can subdivide the new key and distribute the fragment of the key over Cluster heads. Rachedi et al [2] proposes a clustering algorithm based on trust and a DDMZ(Dynamic Demilitarized Zone) for protecting CAs for overcoming the drawbacks of [3] by hierarchical monitoring of nodes. But in that paper it is not clearly described how such a firewall like RA can be implemented in a self organized pure ad hoc network and protect against different kind of DoS attacks. Moreover, intra and inter cluster routing is not properly formulated. The distributed trust model adopted by Abdul-Rahman and Hailes [8] is a decentralized approach for trust management. It uses a recommendation protocol to exchange trust-related information. It is applicable to any distributed system and not specifically targeted for ad hoc networks. Pretty Good Privacy [6] is an example of system proposed by using a web-of-trust authentication model; it uses the public key certificate. Hubaux et al. [4] proposed self-organized public key management system for fully self-organized ad-hoc network; the idea is that each user maintains a local certificate repository. This approach has two drawbacks. First, each user is required to build his local certificate repository before being able to use the system. Second, this approach assumes that trust is transitive, which is not always true. Virendra et al [14] proposes a technique for quantifying trust; in our model we are adapting a similar process to measure the trustworthiness of a neighbor node depending upon some metrics based on system requirements.There are several algorithms proposed for distributed leader election[16], [15], but the malicious nature of the nodes is not considered.

3   Protocol Details

Our primary goal is to develop a distributed trust based framework for securing ad hoc networks and to devise a prediction scheme to evaluate degree of trust of each mobile node in the network. All nodes communicate via a shared bi-directional channel and operate in promiscuous mode. In other words, after each forwarding the node can hear if the intermediate node has forwarded the message to the destination or not. All nodes are identical in their physical characteristics, that is, if a node A is within the transmission range of B then B is also within the transmission range of A. It is also assumed that all nodes are equipped with a residual energy detection device and some energy consumption model is offered for each node to estimate how much time the node will be in active mode (can relay traffic and perform other basic functionalities) with its remaining battery capacity, this metric we refer as Battery power. After deployment of the network the nodes use an efficient secure distributed leader election algorithm [1] that can adapt arbitrary topological changes. Using the pair-wise key pre-distribution scheme, keys are distributed over the nodes of the network. After election of

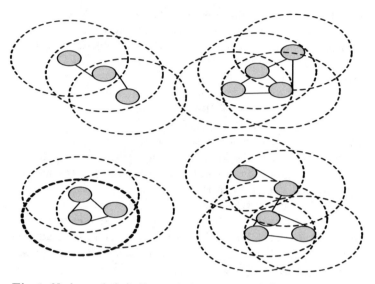

Fig. 1. Nodes and their Transmission range and Cluster Formation

CHs a network key is generated by the CHs. Any node wants to become a CH has to get access to the network key which is only sharable by the CHs. There are other keys also for secure communication, CH-group-key, the pairwise secret key generated by pair of neighboring CHs to communicate to each other. Initial deployment of mobile nodes(15 nodes) and their transmission range and probable cluster formation shown in Fig 1.Small dark circles are nodes and concentric large circles are their transmission range. We also consider that any node having status other than "Trusted" will not be able to communicate outside the cluster. Thus the communication is secured. Each mobile node maintains a TRUST table of its one hop neighbors along with trusted pair-wise key for peer to peer communication without intervention of CH. Maximum distance between two nodes will be 2-hop and maximum distance for any mobile node and CH will be one.

3.1 Notation

We will use the following notations to describe the methods of initialization, key generation, trust evaluation, node registration, and intra and inter cluster routing and node unjoin due to mobility for secure end-to end communication between mobile nodes.

K_N	Network key
M_{id}	mobile node identity
C_{id}	Cluster Identity
CH_{id}	Cluster Head Identity
$CG\text{-}K$	Cluster Group Key (CHs Public key)
$CH\text{-}k$	Private key for CH
K_{X-Y}	Pair-Wise Encryption Key
$List$	Member List generated by CH
K_{CH}	Cluster Head Group Key or Network Key
$K_{CH_X-CH_Y}$	Shared secret key between two clusterheads
ST_X	Status of Node
$CERT$	Trust Certificate
K_{SYM-M}	Symmetric key of mobile node

3.2   Secure Distributed Leader Election Algorithm

After deployment mobile nodes executes a secure distributed leader election mechanism(Algorithm 1) to find the cluster head of corresponding clusters.To reduce the computation overhead the CH selection mechanism only resumes if the existing CH runs off its battery, that is, if the residual battery capacity and remaining time to be active with this battery capacity reaches a threshold or the CH has to move from its previous position. We are using some metrics for

---

Algorithm 1. Secure Distrubuted Leader Election

Step 1 :   A node (say M) wants to be CH, broadcasts START-ELECTION message with its mobility, battery value to all its one hop neighbors.

Step 2 :   Getting this message each node calculates the global weight of that candidate node using a global function. $G_w = w_1*trust\text{-}val + w_2*mobi\text{-}val + w_3 * battery$

Step 3 :   If G_w is greater than a predefined threshold, the node will vote for M by signing a Leader Certificate. Sends it to M.

Step 4 :   After a certain time interval (say T_{Elect}), the candidate node will count how many certificate it has received.

Step 5 :   If this is greater than $n/2$ (where n is the number of neighbor nodes), it advertises itself as leader and broadcasts the leader message with the set of $node_{id}$s who has voted for it.

Step 6 :   If any node finds that its id is falsely included, it will generate a warning message to all its neighbors.

Step 7 :   After certain time say T_{CH}, neighbor nodes will sign a $Trust_{Cert}$ for Leader, sends it to M.(as self-organized Public Key Infrastructure)[see reference[4]]

Step 8 :   Thus M becomes a Leader and the elector nodes who has signed the certificate become its member.

---

making decision whether to elect a node as CH using a global function in each mobile node. The metrics are $Trust_{val}$, Mobility and Battery Power.

3.3   Node Join or Registration

After deployment, CHs communicate between each other and find out their neighbor CH and generates shared key $K_{CH_X - CH_Y}$ between them. When a new node wants to join in the network , the registration procedure of a new node is described in Algorithm 2.

Node under review shown in dark circle and the node marked as 1 is the corresponding CH and collection of recommendation trust from the one-hop neighbor nodes of the node and generation of trust certificate(CERT) are shown in Fig. 3.

For secure distribution of Trust Certificate, the certificate is encrypted by the public key of the particular node. Thus, if any malicious node may able to sniff the certificate, it will not be able to get the CERT until it knows the symmetric key of M. So security is assured here.

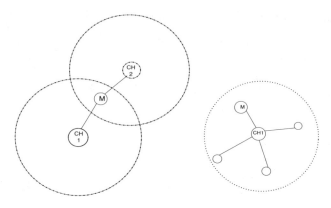

Fig. 2. New Node Join

Algorithm 2. Node Registration

Step 1 :   Each CH starts to broadcast CH beacon and attracts some nodes to join its cluster.

Step 2 :   As the node M gets the CH beacon, it sends join beacon to join the network with its public key.

Step 3 :   CH checks whether it is a duplicate message or not. If it not duplicate CH stores the public key of M as its id and generates a pair wise shared key to communicate between CH and M. Also send *CG-K* for secure intra cluster communication.

Step 4 :   CH gives the node as *"Suspicious"* status and allows it to register subject to periodic review.

Step 5 :   Then CH sends message along with the status of the node to those member nodes to review the status of the newly joined node.

Step 6 :   CH executes the Algorithm 5 and calculates its direct trust about M.CH asks one hop members of M to send their recommendation for M.

Step 7 :   CH executes Dempster-Shafer theory of combining evidences (described in Section 3.8) to find the probable belief (T_{f_M}) of M.

Step 8 :   if T_{f_M} is higher than a threshold CH sends a trust certificate CERT. Thus M becomes a Trusted Member of the cluster.

If any node with *Suspicious* status does not cooperate to the network, that is having a lower trust value, the CH sends a *Warning* message. In the next review if the CH finds the warned node is still misbehaving, CH isolates it from the cluster and informs others that the node has been isolated.

3.4   Intra Cluster Routing

Assume that mobile nodes M_1, M_2 are in same cluster and under same CH. M_1 wants to communicate with M_2. For this intra cluster communication we propose an algorithm described in Algorithm 3.

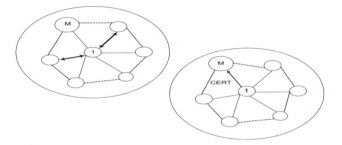

Fig. 3. (a) Trust Information Collection (b) Trust Certificate Generation

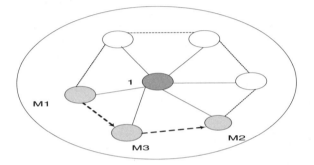

Fig. 4. Intra Cluster Communication

The pictorial representation of this procedure is given in Fig.4, where (M_1 communicates to M_2 via M_3) in a same cluster and the CH represents as 1. As the message is encrypted with the public key of M_2, only M_2 can have the access of it. So passive eavesdropping is not possible.

Algorithm 3. Intra Cluster Routing

Step 1 : M_1 will check its neighbor list. If M_2 is in the one hop neighbor list then it just encrypts the message by the pair-wise key $K_{M_1-M_2}$ generated by M_1 and M_2 and sends it to M_2.

Step 2 : If it is not, M_1 will ask the CH_1 for the public key of M_2. And asks its one hop neighbors that any node has any path to M_2. Say, any node M_3 (other than CH) responds that it has a path to M_2.

Step 3 : If no node responds to M_1, M_1 asks for the public key of M_2 to CH_1.

Step 3 : M_1 generates a key (K_s)and encrypts K_s) with the public key of M_2 and encrypts the message with K_s,and sends it via M_3 or CH_1 .

Step 4 : Getting the message M_2 sends an acknowledgement via the reverse path.

Algorithm 4. Inter Cluster Routing

Step 1 : M_1 sends the route request to CH_1.

Step 2 : CH_1 checks the status of the M_1; if M_1 is not *"Trusted"*, CH_1 just drops the request and generates a message. If M_1 is trusted, CH_1 generates a OK message.

Step 3 : M_1 starts Route discovery to get M_2. Do Step 4 to Step 6 , if M_3 under CH_1 responds in positive. If no member node replies do Step 7 to Step 9 .

Step 4 : M_1 sends the route to CH_1 . CH_1 checks the status of M_3; if it is trusted, the CH_1 generates a session key, K_s for inter cluster communication for M_3.

Step 5 : M_3 gives reply to M_1 with the public key of M_2. M_1 encrypts the K_s with the public key of M_2 and encrypts the message with K_s and sends via M_3.

Step 6 : M_3 gets the message and generates a session key with M_2 and encrypts the total message with that session key and sends the message to M_2.

Step 7 : M_1 sends the the the message to CH_1. CH_1 multicasts *"Who can sense query"* to all it neighbor CHs for having a communication to CH_2.

Step 8 : If CH_2 replies or any other CH replies that it can sense CH_2, CH_1 initiates a route discovery request and asks for the public key of M_2.

Step 9 : M_3 getting the public key CH_1 encrypts the K_s with the public key of M_2 and encrypts the message with K_s and sends over the discovered route.

Step 10 : After a successful receipt M_2 sends an acknowledgement via the same path.

3.5   Inter Cluster Routing

Assume that mobile nodes M_1, M_2 are in different clusters C_1 and C_2 under Clusterheads CH_1 and CH_2. M_1 wants to communicate with M_2. For this inter cluster communication we devise the Algorithm 4 through trusted member node or through Clusterheads. Inter cluster communication using CHs as intermediate nodes is shown in Fig.5. The proposed inter cluster routing can prevent Man-in-the-middle attack and passive eavesdropping.

3.6   Node Unjoin

Each node has to send an *Alive* beacon to CH at a certain time interval. If the CH cannot hear from a node at a certain time out, there will be two possible reasons: One is due to mobility the silent node may move to such place, from where it cannot sense the CH or the node is damaged. CH broadcasts a *"Who can sense"* message and tries to sense the *Silent* node. If any node gives any reply to this message, the CH will try to establish a path to the silent node through that answering node and ask for its location and detail information about new CH. If the old CH gets the information about the new CH, the old CH sends the $Trust_{cert}$ for that node to the new CH. Or if the CH is not getting any reply from any node about the *Silent* node the CH thinks either the node is damaged (that is no node can sense the particular node). Or it goes beyond the communication range of CH, CH just removes the information of the *Silent* node from its list.

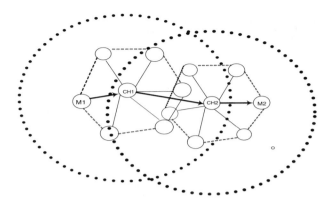

Fig. 5. Inter Cluster Communication

3.7 Trust Evalutaion

Initially, CH sets the status of a newly joined node as Suspicious and tries to evaluate the trustworthiness of the node. There are a lot of parameters for determining the trust level of a node. As we want to devise a mechanism for evaluating the trust of a node according to its contribution towards proper functioning of the network and minimizing the number of bad nodes from the network, we are dealing with the metrics given below:

$$
\begin{aligned}
f &= \text{No. of packets forwarded} \\
d &= \text{No. of packets dropped} \\
m &= \text{No. of packets misrouted} \\
i &= \text{No. of packets falsely injected} \\
R_p &= \text{Total no. of packets received by B sent from A} \\
S_p &= \text{Total no. of packets sent by B to A}
\end{aligned}
$$

After collecting the information about B, A will run the Algorithm 5 and calculates its Direct trust about B. Whenever CH asks A's opinion about B, it will send the trust value.

And w_1, w_2, w_3, w_4 are predefined weights for cooperative and non-cooperative behaviors. Using the similar algorithm CH will calculate its direct trust about B.

3.8 Dempster Shafer Theory of Combining Evidences and Its Application to Trust Prediction

The Dempster-Shafer (DS) theory for uncertainty developed by Arthur Dempster [9] and extended by Glenn Shafer [10]is used to combine information from different sources i.e evidence and gives a more accurate prediction to final decision. The theory provides necessary tools to combine various evidences and gives them various weightings, according to their importance in the final decision making, their quality and relevance. The DS theory was used in several fields of ad hoc networks[11].We justify the use of the DS theory by the uncertain nature of the trust prediction problem and the need to combine the different criteria (evidences). We suppose that we are concerned with the value of some quantity u,

Algorithm 5. Trust Evaluation

Step 1 : Collect data for R_p, S_p, f, d, m, i.

Step 2 : Find the threshold values associated to each behavior f_n, d_n, m_n.

Step 3 : Calculate ratio f_s, d_s, m_s, i_s of each behavior and R_p, S_p total sent or received packet accordingly.

Step 4 : Calculate the deviation f_d, d_d, m_d, i_d from the corresponding threshold.

$$f_s = f/R_p \text{ and } f_d = f_n/f_s$$
$$d_s = d/R_p \text{ and } d_d = d_n/d_s$$
$$m_s = m/R_p \text{ and } m_d = m_n/m_s$$
$$i_s = i/S_p \text{ and } i_d = i_n/i_s$$

Step 5 : Calculate the corresponding direct trust value using the formula

$$Trust(t) = (w_1 * f_d) - (w_2 * d_d) + (w_3 * m_d) + (w_4 * i_d)$$

and the set of its possible values is **U**. The set **U** is called frame of discernment. In our prediction scheme, the frame of discernment **U** is a set of mobile nodes which are able to become the trusted nodes in future. The frame of discernment is $\mathbf{U}\{T, \neg T\}$, m(A) represents the exact belief committed to A, according to the evidence associated with each node's opinion about the *Suspicious* node. To each subset of **U** is assigned a score that represents the belief affected by the evidence. This confidence value is usually computed based on a density function $m : 2^{\mathbf{U}} \rightarrow [0, 1]$ called a basic probability assignment (*bpa*) function. If $m(A) > 0$ then A is called a *focal element*. The focal elements and the associated *bpa* define a body of evidence.

$$m(\Phi) = 0, \sum_{A \subseteq U} m(A) = 1$$

From any neighbor node CH has got the information and the following probability assignments are given. If received trust value $t > 0.5$, the node is treated as *trusted*. If received trust value $t < 0.5$, node is treated as *untrusted* and the probability is assigned accordingly.

$$m_1(T) = 0.8$$
$$m_1(\neg T) = 0$$
$$m_1(T, \neg T) = 0.2 (Suspicious)$$

And the CH has the probability assignments on the same node.

$$m_2 T) = 0.6$$
$$m_2(\neg T) = 0$$
$$m_2(T, \neg T) = 0.4 (Suspicious)$$

The Dempster Combination Rule. Let m_1 and m_2 be the *bpa* associated with two independent bodies of evidence defined in a frame of discernment U. The new body of evidence is defined by a *bpa*, m on the same frame U.

$$m(A) = m_1 \bigotimes m_2 = \frac{\sum_{B \cap C = A} m_1(B)m_2(C)}{\sum_{B \cap C = \phi} m_1(B)m_2(C)}$$

The rule focuses only on those propositions that both bodies of evidence support. The new *bpa* regards for the *bpa* associated with the propositions in both bodies that yield the propositions of the combined body. The K of the above equation is a normalization factor that ensures that m is a *bpa*. The combination rule is commutative and associative. In our approach, the clusterhead computes the trust of each node according to each criterion (evidence) and combines them two by two.Therefore,

m 1	m 2		
	$\{T\} : 0.6$	$\{\neg T\} : 0$	$\{T, \neg T\} : 0.4$
$\{T\} : 0.8$	0.24	0	0.32
$\{\neg T\} : 0$	0	0	0
$\{T, \neg T\} : 0.2$	0.12	0	0.08

$$m_1 \otimes m_2(T) = (1)(.24 + .32 + .12) = 0.68$$
$$m_1 \otimes m_2(\neg T) = (1)(0) = 0$$
$$m_1 \otimes m_2(T, \neg T) = (1)(0.08) = 0.08$$

So the given evidence presented here by m_1 and m_2 , the most probable belief for this universe of discourse is T with probability 0.68.

4   Conclusion and Future Work

In this paper we have proposed a new approach based on trust based self-organizing clustering algorithm. Only few works have been done in this field. The majority of security solutions were based on traditional cryptography which may not be well-suited with dynamic nature of ad hoc networks. We have used the trust evaluation mechanism depending on the behavior of a node towards proper functionality of the network. Initially a node given the status *"Suspicious"* node should be restricted to intra cluster communication until it gets trust certificate. This certificate is also subject to review. As the trust value of a particular node depends on its participation towards proper functionality of the network each node must cooperate and the network can be prevented from inside malicious attacks. Our trust evaluation model gives a secure solution as well as stimulates the cooperation between the nodes of the network. We are not only restricting to direct observation for predicting trust but also recommendation from one hop neighbors of any node under review. We use mathematical model of Dempster-Shafer theory of combining evidences. The advantage of this theory is its capability to represent uncertainty which is the main problem of trust prediction. The originality of our work consists of combining different metrics for quantifying trust and the use of DS theory in order to predict the trust of mobile node more accurately. We have not completely implemented our proposal yet . We are working on simulation studies using NS-2 [17]. In future we plan to compare our proposal with other existing proposals and to consider other issues like secure movement and location management of individual node to provide a better robust and secured solution.

References

1. Vardhanrajan, et al.: Security for cluster based ad hoc networks. In: Proc. of Computer Communications, vol. 27, pp. 488–501 (2004)
2. Rachedi, A., Benslimane, A.: A secure architecture for mobile ad hoc networks. In: Cao, J., Stojmenovic, I., Jia, X., Das, S.K. (eds.) MSN 2006. LNCS, vol. 4325, pp. 424–435. Springer, Heidelberg (2006)
3. Bechler, M., Hof, H.-J., Kraft, D., Pahlke, F., Wolf, L.: A Cluster-Based Security Architecture for Ad Hoc Networks. In: INFOCOM 2004 (2004)
4. Hubaux, J.P., Buttyan, L., Capkun, S.: The Quest for Security in Mobile Ad Hoc Networks. In: Proc. of ACM Symposium on Mobile Ad Hoc Networking and Computing, pp. 146–155 (2001)
5. Zhou, L., Haas, Z.J.: Securing Ad Hoc Networks. IEEE Network Magazine 13(6) (1999)
6. Garfinkel, S.: PGP: Pretty Good Privacy. O'Reilly Associates, Inc., Sebastopol (1995)
7. Pirzada, A.A., McDonald, C.: Establishing Trust In Pure Ad-hoc Networks. In: Estivill-Castro, V. (ed.) Proc. Twenty-Seventh Australasian Computer Science Conference (ACSC 2004), Dunedin, New Zealand. CRPIT, vol. 26, pp. 47–54. ACS (2004)
8. Rahman, A.A., Hailes, S.: A Distributed Trust Model. In: Proc. of the ACM New Security Paradigms Workshop, pp. 48–60 (1997)
9. Dempster, A.P.: A generalization of Bayesian interface. Journal of Royal Statistical Society 30, 205–447 (1968)
10. Shafer, G.: A Mathematical theory of Evidence. Princeton University Press, Princeton (1976)
11. Dekar, L., Kheddouci, H.: A cluster based mobility prediction scheme for ad hoc networks. Ad Hoc Networks 6(2), 168–194 (2008)
12. Chatterjee, M., Das, S.K., Turgut, D.: An on-demand weighted clustering algorithm (WCA) for ad hoc networks. In: Proc. of IEEE GLOBECOM 2000, San Francisco, pp. 1697–1701 (November 2000)
13. Basagni, S.: Distributed clustering for ad hoc networks. In: Proc. of International Symposium on Parallel Architectures, Algorithms and Networks, pp. 310–315 (June 1999)
14. Virendra, et al.: Quantifying trust in Mobile ad hoc networks. In: Proc. of KIMAS 2005, Waltham USA, April 18-21 (2005)
15. Malpani, N., Welch, J., Vaidya, N.: Leader Election Algorithms for Mobile Ad Hoc Networks. In: Fourth International Workshop on Discrete Algorithms and Methods for Mobile Computing and Communications, Boston, MA (August 2000)
16. Vasudevan, S., Decleene, B., Immerman, N., Kurose, J., Towsley, D.: Leader Election Algorithms for Wireless Ad Hoc Networks. In: Proceedings of DARPA Information Survivability Conference and Exposition (2003)
17. NS-2: The Network Simulator, UC Berkeley, http://www.isi.edu/nsnam/ns/

Security Enhancement in Data Encryption Standard

Jatin Verma and Sunita Prasad

Centre for Development of Advanced Computing
NOIDA, India
er.jatin83@gmail.com, sunitaprasad@cdacnoida.in

Abstract. Data Encryption Standard (DES) is the most widely used cryptosystem. Faster computers raised the need of high security cryptosystems. Due to its small key length and simple feistel network, many cryptanalysts developed various methods, like parallel and exhaustive attack, to break DES. DES is most vulnerable to differential cryptanalysis attack. This paper proposes modifications in DES to ensure security enhancement by improving the key length and the weak round function against cryptographic attacks. We show analytically that the modified DES is stronger against cryptographic attacks. The proposed modifications divide the expanded 48 bits of right block into two equal blocks of 24 bits. Then two different functions are performed on these two blocks. The effective key length is increased to 112 bits by using 2 keys. These two keys are used in special sequence with round numbers. The analysis shows that for the proposed approach, the probability of characteristic for differential cryptanalysis is reduced and the unicity distance is better as compared to standard DES.

Keywords: Data Encryption Standard (DES), Differential Cryptanalysis, Unicity Distance.

1 Introduction

Cryptosystem is one of the most efficient techniques for secure information transmission. The DES, published by NBS [3], contained many problems. Trapdoors and hidden weaknesses in the Feistel structure were not disclosed at the time of publishing, but known to the IBM team that designed DES. The published report did not disclose that DES's design was based upon differential cryptanalysis technique. Because of faster hardware, process time for cryptanalysis is shorter as compared to the process time required a decade ago. In current scenario, even an enormous amount plaintext (say about 10^{15} chosen plaintext) messages required for differential cryptanalysis is very small [5]. In future, various kinds of parallel processes can be used by cryptanalyst to attack DES [1][6]. In [3], Bruce Schneier stated that besides uniform XOR distribution, short key length and simple XOR function are responsible for weaknesses in DES. The differential cryptanalysis calculates the differences at start and monitors the difference on the path [4]. At the end of 15 rounds due to this difference monitoring, cryptanalysts can calculate the 48 bits of key. Thus DES should be improved against differential cryptanalysis attack. In order to strengthen the cryptographic security of the DES, this paper proposes the following suggestions.

S.K. Prasad et al. (Eds.): ICISTM 2009, CCIS 31, pp. 325–334, 2009.

First, in order to reduce the probability of N round characteristics against differential cryptanalysis, XOR should be replaced with another set of function and the DES algorithm should be modified structurally so that different function can perform on sub-blocks split from right hand expanded 48 bits. Second, the effective key length of DES should be improved to 112 bits [9]. Third, the cipher should use different keys for different round.

2 Data Encryption Standard

The NBS adopted Data Encryption Standard (DES) as standard on July 15, 1977 [7]. DES was quickly adopted for non-digital media, such as voice grade public telephone lines. Meanwhile, the banking industry, which is the largest user of encryption outside government, adopted DES as a wholesale banking standard. DES is the most widely known symmetric cryptosystem. DES is a block cipher that operates on plaintext blocks of a given size (64 bits) and returns ciphertext blocks of the same size. DES has the same algorithm for the encryption and decryption. DES enciphers a block (64 bits) of data with a 56 bit key and it uses a transformation that alternately applies substitutions and permutations. This process is repeated in 16 rounds. A plaintext block of data is first transposed under an initial permutation (IP) and finally transposed under inverse permutation (IP^{-1}) to give the final result. A block of the 64 bits permuted data is divided into a left sub-block (L_i : i^{th} round) and a right sub-block (R_i) of 32 bits each. Encryption process for the left and right sub-blocks is given in equation (1) and equation (2) respectively.

$$L_i = R_{i-1} \tag{1}$$

$$R_i = L_{i-1} * f(R_{i-1}, K_i) \tag{2}$$

where K_i is the 48 bits round key produced by key scheduling. R_{i-1} is expanded to a 48 bits block $E(R_{i-1})$ using the bit selection table E. Similarly, when using logical characteristic of XOR equation (2), the decryption process is given by equation (3) and equation (4) for the right and left sub-blocks respectively.

$$R_{i-1} = L_i \tag{3}$$

$$L_{i-1} = R_i \oplus f(R_{i-1}, K_i) = R_i \oplus f(L_i, K_i) \tag{4}$$

With the result of the ith round in equation (4), the result of the i-1 round can be derived. The process is explained in fig1. Encryption and Decryption processes are symmetrical in DES as given in equation (5).

$$(M \oplus K) \oplus K = M \tag{5}$$

where M is the Plaintext and K is the key. The sub-block on the left side and a sub-block on the right side are symmetrical. The sub-blocks have the same number of iterative functions during the 16 round performances.

3   Modifications in DES

Various methods have been developed to attack DES. The most famous method is exhaustive attack. Exhaustive attack uses each and every possible key on ciphertext. There are $2^{56} \approx 10^{17}$ possible keys. Assuming that one key is checked every msec. The complete search would take about 10^{11} seconds ≈ 2.2 centuries (worst case) and 1.1 centuries (average case). There are certain methods, which can be applied to DES in parallel. These are known as parallel attacks. Following a model suggested in 1979, if a machine is built with 106 chips and every search takes 1 key/msec, it would take about 10^{5} seconds to search all keys. In an average case, it would take about half a day. The price estimate in 1979 for such a machine was $20,000,000, and each key would cost about $5,000 to crack. This was impractical because such a machine would melt the chips. In 1993, Michael Wiener [10] described a machine consisting of pipelined chips. Each chip was able to test 5×10^{7} keys/sec and the cost of each chip was $10.50. A frame consisting of 5,760 chips could be built for $1,00,000. A machine consisting of 10 frames could be built for about $1,000,000, and the average case would crack a key in 1.5 days. This method was entirely practicable. The next method is differential cryptanalysis, which was introduced, by Eli Biham and Adi Shamir in 1990 [4]. Differential Cryptanalysis is a new method of cryptanalysis. Using this method, Biham and Shamir found a chosen plaintext attack against DES that was more efficient than brute force attack. Differential cryptanalysis looks specifically at ciphertext pairs; pairs of ciphertexts whose plaintexts have particular differences. It analyzes the evolution of these differences as the plaintexts propagate through the rounds of DES when the plaintext is encrypted with the same key. The differential cryptanalysis chooses pairs of plaintexts with a fixed difference. The two plaintexts can be chosen at random, as long as they satisfy particular difference conditions; the cryptanalyst does not even have to know their values (for DES, the term "difference" is defined using XOR). Then, using the differences in the resulting ciphertext, assign different probabilities to different keys. As more and more ciphertexts are analyzed, one key will emerge as the most probable, which is the correct key.

In [6], it was shown that outputs from round 1,2,15 and 16 are very important for cryptanalysts. The outputs from these rounds provide the real picture of difference propagation. Therefore if somehow the output from these could be manipulated, the probability of the characteristic would reduce. The second important aspect of differential cryptanalysis is the difference. The XOR is used to take difference of two ciphertexts and two plaintexts. Imagine a pair of inputs X and X' that have the difference ΔX. The outputs, Y and Y' are known, and therefore so is the difference, ΔY. Both the expansion permutation box and the P-box are known, so ΔA and ΔC are known. The intermediate values, B and B' are not known, but their difference ΔB is known and equal to ΔA. When looking at the difference, the XORing of K_i with A and A' cancels out. A and A' are the values from the substitution box $S(X)$. For any given ΔA, not all values of ΔC are equally likely [3]. The combination of ΔA and ΔC suggests values for bits of A XOR K_i and A' XOR K_i. Since A and A' are known, this gives us information about K_i as shown in fig 1.

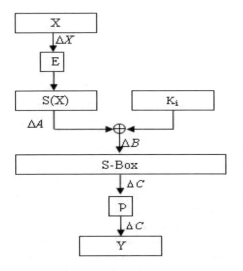

Fig. 1. Data Encryption Standard

The paper proposes two modifications. First modification is of the simple round function F. In [11], a solution of iterative function has been proposed, as round function is weaker. But with iterative function, the algorithm loses symmetry for encryption and decryption. The objective of modification is to maintain the symmetry and strengthen the round function F. To extend DES we must reorganize the structure of DES so that different F functions can be applied on the block. The right hand 32 bits of block is expanded to 48 bits. These 48 bits are divided into two equal blocks of 24 bits. Two different functions performed on left and right blocks are XOR and XOR + Complement respectively as shown in fig 2. The encryption process is given by equation (6) and equation (7) below.

$$L = R_i \qquad (6)$$

$$R = f\left(\left(\left(\frac{R_{i-1}}{2} \oplus K_i\right), \left(\frac{R_{i-1}}{2} \sim \oplus K_i\right)\right) \oplus L_{i-1}\right) \qquad (7)$$

The logical expression shown above in equation (6) and equation (7) shows that difference of plaintext will not be cancelled out with key when XORed with the key. The modified F function produces avalanche effect on 24 bits together i.e. when one input bit is altered certain numbers of output bits get altered. Thus, propagating chosen plaintext difference doesn't produce such avalanche effect. Cryptanalysts have to opt whether chosen plaintexts are to be expanded first and then difference is to be taken or vice versa. But if differential cryptanalysts use this method, the differential cryptanalysis attack loses its definition and is unable to proceed towards the correct key. Due to these all, security of modified DES seems enhancing. The equation pairs satisfy equation (3), (4), (5) and shows symmetrical features for encryption and decryption. Therefore, when logical characteristics of equation (6) and (7) are applied in the equation (3) and (4), decryption is derived for modified DES. The decryption function is same as the encryption.

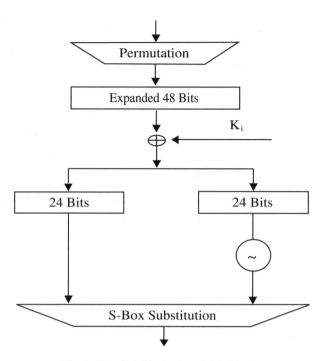

Fig. 2. New Sub-blocks from Right blocks

The decryption process is given in equation (8) and equation (9) below.

$$R_{i-1} = L_i \tag{8}$$

$$L_{i-1} = f\left(\left(\left(\frac{R_i}{2} \oplus K_i\right), \left(\frac{R_i}{2} \sim \oplus K_i\right)\right) \oplus L_i\right) \tag{9}$$

IP and IP^{-1} need not to be improved. The second modification proposed in modified DES is to improve input key length from 64 bits to 128 bits which is divided into 64 bits each, resulting in key 1, key 2 as shown in fig 3. These two keys are XORed. This helps to find out the weak keys easily. If two keys are relevant and produce a pattern repeatedly, then another key can be used. According to the key scheduling of the DES, key will be rotated to left or right side. The key to a particular round is selected with the round number. These two keys are applied to different number of rounds, as for round number 1 and 16 one of two keys is used and for rest of rounds, another key is used. When plaintext is received at round 1, key number 1 is used in feistel structure. At round number 2, key number 2 is used as shown in fig 4.

This arrangement shows that output from round number 2 is entirely different from the output from round number 1 with respect of difference. The outputs from round number 15 and 16 show similar pattern. This change in output causes confusion in monitoring the difference. DES has the same iterative number of F function during the performance from R_0 to L_{16} and from L_0 to R_{16}. It may be attacked easily by

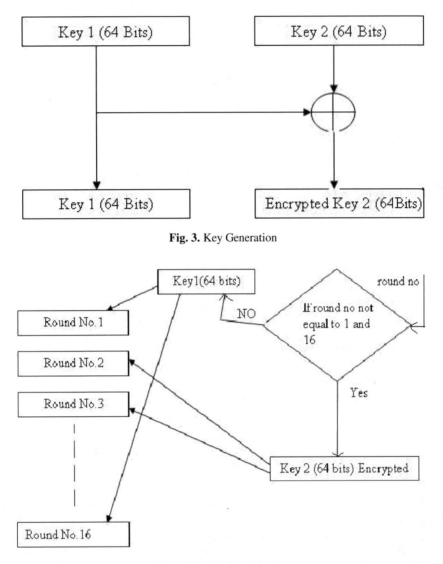

Fig. 3. Key Generation

Fig. 4. Round key Selection

differential cryptanalysis. But this key arrangement increases the difference and so the avalanche effect, therefore modified DES shows more strength against differential cryptanalysis attack. There is no need to design the encryption and decryption separately because the encryption algorithm is same as decryption algorithm.

4   Cryptographic Analysis

There are several analysis methods for cryptosystem. The modified DES is estimated by the differential cryptanalysis, and unicity distance [10]. Differential cryptanalysis under a chosen plaintext attack is a method whereby probability for the effect of particular differences in plaintext pairs and differences in ciphertext pairs corresponding to it can be used to assign probabilities to the possible keys. If the pair of ciphertext is known and we know the output XOR of the F function of the last round, it is easy to calculate the input XOR of the F function of last round. We can find the set of possible keys with the probabilities of N round characteristics. In order to attack with the differential cryptanalysis, the 16 round characteristic of the DES consists of one additional round and the 15 round characteristic that was based on the 2-round iterative characteristic. From fig 5 we can see that if $a' = 0$ then probability of the characteristic will always be 1. But in the next round when $b' = 19600000$, the probability is found as follows. The differential input for second round is 19600000. The expansion operation puts these half bytes into the middle four bits of each S-box in order i.e. 1=0001 goes to S1 and 9=1001 goes to S2 and 6=0110 goes to S3 and 0=0000 goes to S4...S8. Only 1, 9, 6 will provide us with non-zero value, as others are zero. The differential input of S1 is 0 0001 0 = 02, differential input of S2 is 0 1001 0 = 18, differential input of S3 is 0 0110 0 = 0C and other values are zero. Looking at S1's differential distribution table, we find that for $X' = 02$, the highest probability differential output Y' at position C=1100 is 14/64. Highest probability differential output for S2 and S3 at positions 6=0110 and 5=0101 for inputs 9 and 6 is 8/64 and 10/64 respectively All the other S-boxes have $X' = 0$ and $Y' = 0$ with probability 1. Multiplying this probability, value is equal to 0.00427 which is 1/234. The S-box outputs goes through the permutation P before becoming the output of F function. The output of the function F is 0 as shown in fig 5. The output from the second round is zero, giving the differential output as depicted: (00 00 00 00 19 60 00 00x) The differential cryptanalysis attack against the full 16 round DES has the probability of the characteristic as $(1/234)^6 = 2^{-47.2}$. The differential cryptanalysis attack requires the amount of the chosen plaintext messages of $2^{47.2} = 1.6 \times 10^{16}$ (of eight bytes each) [5].

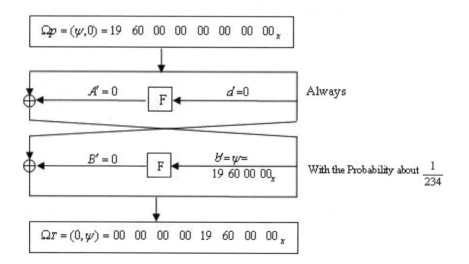

Fig. 5. 2-Round Iterative Characteristics

The 13 round results from iterating this 2 round characteristic six and a half times. In [4], Eli described that if this characteristic is extended to 14 rounds to complete 16 round characteristics, the probability of the characteristic will reduce to $2^{-55.1}$. The probability for characteristics of round 2 to round 14 is calculated as given in previous paragraph. For modified version, the same model is used for cryptanalysis. The output comes from R_1 and R_2 (where R_i is round number) has a difference, which doesn't come in the next round. Suppose that output difference from R_1 is 34 and R_2 is 27 (it can be any value as almost all bits of key are changed). Now from these two values we can see that difference from A' and B' is not related with each other as the probability of characteristic for 34 is multiple of 1x, 2x, 17x and probability of characteristic for 27 is multiple of 3x, 9x. To find more correct case, one more difference is required, as output from first round and the second round doesn't provide us with proper difference. It can be concluded from here that for more difference analysis we have to use one more round and it is round number 14. If the round number is increased by 1 the probability of characteristic reduced to $2^{-55.1}$ as described above. Now round 1 and 16 are encrypted with same key. Exhaustive search will be performed on these two rounds to find out the key. The differential cryptanalysis attack against the full 16 round of the modified DES requires the probability of characteristic of the full 16 round characteristic of $(1/237)^7 = 2^{-55.1}$ at least because the probability of characteristic for the 13 round variant of modified DES is about $(1/237)^7$. Thus, the probability of the characteristic of the modified DES against the differential cryptanalysis is reduced more than the DES.

The unicity distance (UD) [10] describes the amount of cipher text needed in order to break a cipher. More the number of plaintext is required, more secure is cryptosystem. We analyze of UD of the DES and the modified DES for 14 samples. In this analysis, the UD of the modified DES is longer than the UD of the DES for 14 samples. The Unicity Distance is increased to 237.85 per 1000 KB on average. This proves that the cryptographic security of the modified DES is improved as compared to DES.

Table 1and 2 shows the analytical and behavioral comparison of the result with DES.

Table 1. Analytical Comparison

Parameters	Data Encryption Standard	Modified Data Encryption Standard
Probability Characteristics	$2^{-47.2}$	$2^{-55.1}$
Unicity Distance	$2^{47.2}$	$2^{55.1}$

The behavioral comparison in Table 2 shows that modified DES takes 2 sec more than DES for 1000 KB file to execute. This comparison helps in understanding the actual effect of modification. This 50% more time is due to extra machine instructions added in the form of extra function and decision taking capabilities. The analytical comparison shows that number of plaintext pair required for modified DES is $2^{55.1}$. Thus $2^{55.1} - 2^{47.2} \Rightarrow (2^{7.9-1}).2^{47.2}$ more text for modified DES is sign of enhanced security.

Table 2. Behavioral Comparison

Parameters	Data Encryption Standard	Modified Data Encryption Standard
No. of keys	1	2
Running Time	4.21 Sec (1000 KB File)	6.21 Sec (1000 KB File)
No. of Function	1	1
Total Bytes of keys	56 Bits	112 Bits

5 Conclusion

The volume of information exchanged by electronic means such as Internet, wireless phones, Fax etc, is increasing very rapidly. Information transferred through Internet is vulnerable to hackers and privacy of wireless phones without security can be invaded. Thus there is a need to develop improved cryptosystems to provide greater security. This paper designed the modified structure of DES called modified DES. The modified DES divided the right block into sub-blocks of 24 bits each. Then two different functions XOR and XOR + Complement were performed on these two blocks. The proposed architecture extends the DES algorithm so that the number of functions during the full 16 round of right block is different. The objective is to decrease the probability of full 16 round characteristic against differential cryptanalysis. The analytical comparison of the modified DES with DES is given below:

1. We showed that the probability of the characteristic of the modified DES against the differential cryptanalysis is decreased by a factor of $2^{-7.9}$ as compared to probability of DES.
2. To increase time complexity and space complexity against the exhaustive attack and the time memory trade off, the key length is increased to 112 bits (K_1, K_2).
3. The UD is increased to 237.85 per 1000 KB for all of samples.

References

1. Hellman, M.E.: DES will be totally insecure with ten years. IEEE Spectrum 16(7), 32–39 (1979)
2. Davis, R.M.: The Data Encryption Standard in Perspective. IEEE Communications Magazine 16, 5–9 (1978)
3. Schneier, B.: Applied Cryptography. John Wiley & Sons, New York (1996)
4. Biham, E., Shamir, A.: Differential cryptanalysis of the full 16-round DES. In: Brickell, E.F. (ed.) CRYPTO 1992. LNCS, vol. 740, pp. 487–496. Springer, Heidelberg (1993)
5. Biham, E., Shamir, A.: Differential cryptanalysis of DES-like cryptosystems. Journal of Cryptology 4, 3–72 (1991)
6. Heys, H.: A Tutorial on Linear and Differential Cryptanalysis. Technical Report, United States Military Academy, vol. XXVI(3) (2002)

7. Daley, W.M.: Data Encryption Standard. FIPS Pub. 46, U.S, National Bureau of Standards, Washington DC (1977)
8. Coppersmith, D.: The data encryption standard (DES) and its strength against attacks. IBM Research Journal 38(3), 243–250 (1994)
9. Smld, M.E., Branstad, D.K.: The Data Encryption Standard: Past and Future. Proc. of IEEE 76(5), 550–559 (1988)
10. Stinson, D.: Cryptography- Theory and Practice. CRC Press, Boca Raton (2002)
11. Han, S., Oh, H., Park, J.: The improved Data Encryption Standard (DES) Algorithm. In: 4th IEEE International Symposium on Spread Spectrum Techniques and Applications, vol. 3, pp. 1310–1314. IEEE Press, Los Alamitos (1996)

A Novel Framework for Executing Untrusted Programs

Dibyahash Bordoloi and Ashish Kumar

Indian Institute of Technology Kharagpur
Kharagpur, India
{dibyahash,ashish}@cse.iitkgp.ernet.in

Abstract. Lot of studies has been done on tackling untrusted programs. Most of the studies are aimed at preventing an untrusted program from execution. Models have been established to provide proofs to codes so that a program is allowed to execute only when the program passes a test. The test may be in the form of some proof, certificate etc. The basic idea is not to allow an untrusted program to execute if the safety cannot be guaranteed. This paper proposes a totally novel approach to tackle untrusted programs, infact it is diametrically opposite to the existing policies. It starts with the assumption that all untrusted programs are safe until proven otherwise. The best way to prove the safety of a program is to execute it at least once. We have proposed a new model to execute such programs in a virtual environment and then decide on the trustworthiness of the program.

Keywords: Untrusted Programs, Security.

1 Introduction

We come across a lot of untrusted programs these days. While a lot of them have evil intent, many are very useful and highly effective clean programs that come from sources which are not reliable. Lot of programmers are freewheelers who provide no proof of security of their code. The existing security policies will not allow such programs to run easily on most operating systems.

PCC[1] and MCC[2] are the most popular strategies used to tackle untrusted programs. There is always a margin of error in these techniques and a clever determined hacker can manage to bypass all security measures while a clean code may fail to pass the most basic test.

We have developed a novel framework to deal with untrusted programs. At the core of the framework is the idea that all untrusted programs are safe until proven otherwise. To prove whether a program is safe or unsafe we have to execute the program at least once without any interruption within the framework.

When a program is executed and it turns out to be unsafe it becomes necessary to restore system data to the previous safe state. To account for this the framework ensures that the program does not tamper with actual system data at the first place. It performs its run in a virtual environment with the illusion that it is working in the actual system environment. The framework also keeps a unique signature of a unsafe program so that it can detect it if it is executed again.

S.K. Prasad et al. (Eds.): ICISTM 2009, CCIS 31, pp. 335–336, 2009.
© Springer-Verlag Berlin Heidelberg 2009

2 Framework

There are four major constituents in the framework: A Role Assignor, A Virtual Mode, a Tagger and a Tag log. The Virtual Mode which is the heart of the framework is further subdivided into an Image Creator and Verifier, a Tag Reader, an Installer and a User Interaction system.

We assume that the system is following the RBAC[3][4] model. Any untrusted program is assigned a special role by the Role Assignor. An exact image of the present system state is created by the Image Creator which in turn is used to create the virtual environment. A program which has malicious intentions is identified by studying its activities in a complete execution within this environment. The areas of study are the various security parameters like changes made to the registry, opening of files and subsequent changes made in those files, changes in environment variables etc. This execution is done opaquely and all the changes made in the system state actually happen in the virtual environment with the program having no idea that it is running in a mirror image of the system state. A program which turns out to be safe is installed directly by the framework with the changed parameters in the image state being transferred transparently to the actual system state. This default behavior can however be changed according to the local security policy. An interactive stage informs the user about the unsafe program leaving the final decision of its installation or further usage on the user. When the user decides not to continue with the program the image state is unloaded with no harm being done to the actual system. The program is also given a unique tag by the framework so that it can be detected if it is executed in a later stage.

References

1. Sekar, Venkatakrishnan, V.N., Basu, S., Bhatkar, S., DuVarney, D.: Model Carrying Code: A Practical Approach for safe Execution of Untrusted Applications. In: ACM Symposium on Operating Systems Principles (SOSP 2003), Bolton Landing, New York (2003)
2. Necula, G.: Proof Carrying Code. In: ACM Symposium on Principles of Programming Languages (1997)
3. Sandhu, R.S., Coyne, E.J., Feinseein, H.L., Younman, C.E.: Proceedings of the First ACM Workshop on Role-Based-Access Control. ACM, New York (1996)
4. Sandhu, R., Ferraiodo, D., Kulin, R.: The NIST Model for Role Based Access Control. In: 5th ACM Workshop on Role-Based Access Control, Berlin (2000)

Comparative Analysis of Decision Trees with Logistic Regression in Predicting Fault-Prone Classes

Yogesh Singh, Arvinder Kaur Takkar, and Ruchika Malhotra

University School of Information Technology, Guru Gobind Singh Indraprastha University,
Kashmere Gate, Delhi-110403
ys66@rediffmail.com, arvinder70@gmail.com,
ruchikamalhotra2004@yahoo.com

Abstract. There are available metrics for predicting fault prone classes, which may help software organizations for planning and performing testing activities. This may be possible due to proper allocation of resources on fault prone parts of the design and code of the software. Hence, importance and usefulness of such metrics is understandable, but empirical validation of these metrics is always a great challenge. Decision Tree (DT) methods have been successfully applied for solving classification problems in many applications. This paper evaluates the capability of three DT methods and compares its performance with statistical method in predicting fault prone software classes using publicly available NASA data set. The results indicate that the prediction performance of DT is generally better than statistical model. However, similar types of studies are required to be carried out in order to establish the acceptability of the DT models.

Keywords: Software metrics, fault prediction, logistic regression, decision tree, software quality.

1 Introduction

As the complexity and the constraints under which the software is developed are increasing, it is difficult to produce software without faults. Such faulty software classes may increase development and maintenance costs due to software failures, and decrease customer satisfaction. Effective prediction of fault-prone software classes may enable software organizations for planning and performing testing by focusing resources on fault-prone parts of the design and code. This may result in significant improvement in software quality

Identification of fault prone classes is commonly achieved through binary prediction models by classifying a class as fault-prone or not fault-prone. These prediction models can be built using design metrics, which can be related with faults as independent variables.

DT methods are being successfully applied for solving classification problems. It is therefore important to investigate the capabilities of DT methods in predicting software quality. In this work, we investigate the capability of three DT methods in predicting faulty classes and compare their result with Logistic Regression (LR) method.

S.K. Prasad et al. (Eds.): ICISTM 2009, CCIS 31, pp. 337–338, 2009.

We investigate the accuracy of the fault proneness predictions using design metrics. In order to perform the analysis we validate the performance of the DT and LR methods using public domain NASA KC4 data set (NASA data repository). In this work, we analyze three DT methods: Random Forest (RF), J48 and Adtree.

The contributions of the paper are summarized as follows: First, we performed the analysis of public domain NASA data set (NASA data repository), therefore analyzing valuable data in an important area where empirical studies and data are limited. Second, comparative study of three DT methods with LR method has been performed to find which method performs the best to predict the fault proneness of the code. The results showed that DT methods predict faulty classes with better accuracy. However, since our analysis is based on only one data set, this study should be replicated on different data sets to generalize our findings.

2 Goal and Hypothesis

The goal of this study is to compare the performance of DT against the prediction performance of LR method. The performance is evaluated using design metrics for predicting faulty classes using public domain NASA data set.

DT Hypothesis: DT methods (RF, J48, Adtree) outperforms the LR method in predicting fault-prone software classes (Null Hypothesis: DT methods (RF, J48, Adtree) do not outperform the LR method in predicting fault-prone software classes.

3 Conclusion

Among the DT methods, Adtree method predicted faulty classes with highest values of performance measures. The accuracy of Adtree is 80.54 percent, its precision is 86 percent, its recall is 76 percent, and its F-measure is 79 percent. The accuracy and precision of all DT methods outperformed the LR model. However, recall of J48 is outperformed by LR model. F-measure considers the harmonic mean of both precision and recall measures and is used to overcome the tradeoff between precision and recall measures. The F-measure of J48, RF, and Adtree outperforms the LR model. LR model did not significantly outperformed DT methods in any of the performance measures.

This study confirms that construction of DT models is feasible, adaptable to software systems, and useful in predicting fault prone classes. While research continues, practitioners and researchers may apply these methods for constructing the model to predict faulty classes. The prediction performance of Adtree particularly in recall measure, can help in improving software quality in the context of software testing by reducing risks of faulty classes go undetected.

The future work may include conducting similar type of studies with different data sets to give generalized results across different organizations. We plan to replicate our study to predict model based on genetic algorithms. We will also focus on cost benefit analysis of models that will help to determine whether a given fault proneness model would be economically viable.

Software Security Factors in Design Phase
(Extended Abstract)

S. Chandra, R.A. Khan, and A. Agrawal

Department of Information Technology, Babasaheb Bhimrao Ambedkar University,
Lucknow-226025, India
{nupur_madhur,khanraees,alka_csjmu}@yahoo.com

Software security factors are quantifiable measures that provide the base for mathematical foundations. These mathematical results help to estimate software security. Analysis of security factors is vital for software as well as a software dependent society. Security factors play a significant role in software security estimation, such as confidentiality, integrity, availability, authentication, authorization etc. Integration of security features must be part of software development life cycle. It is important for security analysts to know what to look for when they analyze software architecture during software security analysis. Therefore, it is highly required to develop such a framework able to identify a set of security factors including its perspectives, the user and environment. Many of the ideas in this paper are derived from approaches developed in the software security and software quality measurement fields.

A generic framework to identify security factors is depicted in fig 1. Security factors provide the base for quantitative security estimation. In order to achieve the set objective, ten steps are to be followed, including Security Requirements & Validation, Security Factor Perspective, Security Factor Selection Criteria, Behavioral Requirements of Security Factors, Security Factors, Verify Security Factors, Hierarchy of Security Factors, Impact of Security Factors, Phase-wise Security Factors and Security Factors in Design Phase.

Set of security requirements provides the basis for further processing, so it is required to have a set of goals and objectives from security factor perspective. This leads to issues such as what we expect from security factors, what are the requirements for security factors to maintain security. During security factors identification process, it is important to analyze and identify various security factor perspectives such as effort, time, complexity, size etc. Security factor selection criteria may be use to select security factors, which enhance the role of security factors during security estimation. Behavioral requirements of security factors specifies what inputs are expected from the factors, what outputs will be generated by these factors and kinship between those inputs and outputs. Behavior of security factors depends on the data and transformations of data. Security factors are information, other than cryptographic keys, that is needed to establish and describe the protection mechanism that secures the communications [information/data]. No common accepted set of security factors available so it is required to identify a set of security factors. Identified set of security factors are: Confidentiality, Authenticity, Integrity, Utility, Availability, Possession, Availability, Authentication, Authorization, Non-repudiation, Tamper-proofness, Unavoidability, Verifiability, Privacy, Reliability, Safety, Maintainability, Access-control and Resource – Access.

S.K. Prasad et al. (Eds.): ICISTM 2009, CCIS 31, pp. 339–340, 2009.
© Springer-Verlag Berlin Heidelberg 2009

For verification of security factors, verification of the control data values and the data in motion (i.e. the mechanism used to perform the data transformation) both are demanded. Verification of security factors are required in order to disclose security vulnerabilities. Hierarchy theory was developed to deal with the fundamental differences between one level of complexity and another. In order to get the hierarchy of security factors there is need to categorize security factors and identify its sub factors and at last leveling. In building complex systems, one of the major difficulties is to recognize the interfaces between components at the early stage of software development process. Impact of security factors in a qualitative manner facilitates a way to understand the nature of security factors. Security factors may get affected in more than one phase. Set of phase wise security factors provide the base to analyze the effect of a particular security factor. List of security factors encountered in design phase may be inherited from the list of identified security factors. Identified design phase security factors will facilitate during selection of security factors according to their user, environment, and resources. It will provide base to decide which security factors need to be integrated in the software.

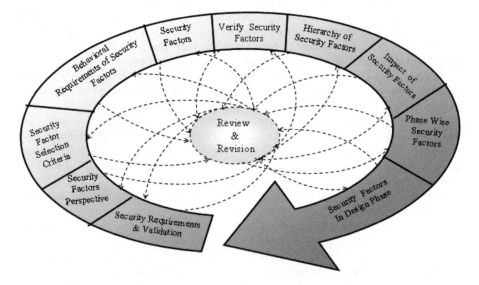

Fig. 1. Framework to identify software security factors

It is important to think about the measures of security of software and how to go about measuring them. To solve the purpose there is a need to figure out what behavior is expected to the software. During security estimation process, it is required to understand the nature of security. The common way is to identify dimensions such as confidentiality, integrity, availability etc. These security factors are encountered in design phase. A structured approach is proposed to model security criteria and perspectives in such a way that security factors are measured and captured. The work deals with the specification of security factors, to identify vulnerability, to mitigate risks and to facilitate the achievement of the secured software systems. Identification of security factors may help to disclose vulnerability and mitigate risks at the early stage of development life cycle.

Mathematical Modelling of SRIRE Approach to Develop Search Engine Friendly Website

Shruti Kohli[1], B.P. Joshi[1], and Ela Kumar[2]

[1] Birla Institute of Technlogy, Noida Center, India
[2] YMCA Institute of Engineering, Faridabad, India
kohli.shruti@gmail.com,
bp_joshi@yahoo.com, ela_kumar@rediffmail.com

Abstract. Endeavor of this paper is to mathematically calculate "Effort of Improvement" required to improve website ranking in search results. This work is the mathematical modeling of the SRIRE approach been developed earlier during the research. A source code has been generated and correctness of algorithm is checked by performing experimental analysis with a popular website.

Keywords: Search Engine ranking, website optimization, SRIRE.

1 Introduction

Search Engine popularity has increased the demand for ranking in its top results. Search Engines have provided guidelines for webpublishers to build quality web pages[1]. In a previous work SRIRE(Search Engine rank improvement using reverse engineering approach)[2] was developed to facilitate webmaster in improving website ranking. SRIRE employs reverse engineering, fig[2] approach to modify basic components of a webpage, fig [1]. In this paper mathematical modelling of SRIRE has been done to calculate "Effort of Improvement" for improved webpage ranking.

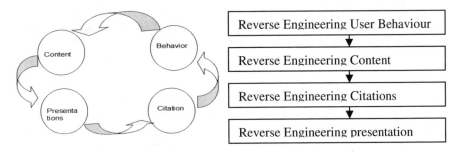

Fig. 1. Components of a Webpage   Fig. 2. Working of SRIRE

Determing "Effort of Improvement"(EOI): "Effort of Improvement" depicts the work need to be done to improve search ranking of the website. Source code for determining Over All Effort(EOI) to improve website and effort required to Improve content of website(IC_EOI) is presented in table [1].

Table 1. Source Code for calculating "Effort of Improvement"

Overall EOI for a webpage(EOI)	EOI for Content Improvement (IC_EOI)
Notations: EOI: Overall Effort of Improvement. I: Total pages indexed in Google M: Number of pages requiring 'EOI' R_s: Current Rank of webpage calculated for the keyword that defines webpage.	N: No. of keywords for webpage. A[M,N]:No. of pages to be improved B[M,N]:pages requiring effort IC_EOI[M]:"IC_EOI" for each webpage T: Number of keywords for i^{th} keyword
``For each x   I
If Rs (x) > 10
{EOI =EOI + 1
A[M]=x;
M=M+1 }
EOI = EOI/I``` Explaination:To find EOI determine number of pages indexed in Google using "Google Analytics". 2. Determine rank of each webpage in search results(Top 10 results is target). Use above source code to find 'EOI'. *EOI is normalized value calculated on a scale of 0-1. It determine efforts required to rank in top 10 search results.*	```For i=1 to M
For j=1 to N
For Each j in B[j] where
(B[i,j]!=NULL)
{ If Rs(A[i]<10 for B[j])
IC_EOI=IC_EOI +1
T=T+1
Next
IC_EOI=IC_EOI/T
I_EOF(i)=IC_EOI}
C_EOI=C_EOI+IC_EOI
Next
Next
C_EOI=C_EOI/N``` Explanation:Determine keyword based ranking.Calculate IC_EOI for low rank page

2 Experiment and Conclusion

This paper quantizes "Effort of Improvement" required to improve rank of a webpage. A news portal having 10 pages indexed in Google was experimented. EOI was calculated as 0.33 which implied 33% pages of website need to be improved. IC_EOI value for different pages showed that two web pages namely Index, Current need improvement. On applying SRIRE their ranking was found to improve.

References

1. Google Webmaster Guidelines (2008), http://www.google.com/support/
2. Kohli, S., Kumar, E.: Application of Reverse Engineering Approach for developing Search Engine friendly e-commerce portals. In: Proc. of CISTM 2008 (December 2008) (accepted in ICFAI)

Personalized Semantic Peer Selection in Peer-to-Peer Networks

Vahid Jalali and Mohammad Reza Matash Borujerdi

Amirkabir University of Technology,
Tehran, Iran
{vjalali, borujerm}@aut.ac.ir

In this poster a personalized approach for semantic peer selection in Peer-to-Peer systems is proposed. For achieving personalized semantic peer selection, an auxiliary ontology, named user ontology is introduced for storing concepts which are of great interest to user. This ontology is then used along with user's query for selecting most promising peers over the network to forward user query to them.

Proposed system is used for sharing academic documents over a Peer-to-Peer network, so potential users of the system would be students and professors. It is clear that students and professors are often experts in specific domains and the field of their research is bounded within a predefined scope. This fact motivates using personalized peer selection in Peer-to-Peer networks with semantic topologies.

The proposed algorithm introduced here is based on Hasse peer selection algorithm [1,2,3] combined by ontosearch personalized conceptual search ideas [4]. In Hasse approach similarity between user's query and resources of a peer is calculated according to Li's similarity measure between concept pairs. It means that the similarity measure between two concepts is determined according to the number of edges between them in an ontology and the depth of their direct common subsumer in the same ontology. There are four major entities in Hasse view of a P2P network which are peers, common ontology, expertise and advertisements. The Peer-to-Peer network consists of a set of peers P. Every peer $p \in P$ has a knowledge base that contains the knowledge that it wants to share. The peers share an ontology O, which provides a common conceptualization of their domain. The ontology is used for describing the expertise of peers and the subject of queries. An expertise description $e \in E$ is an abstract, semantic description of the knowledge base of a peer, based on the common ontology O. This expertise can either be extracted from the knowledge base automatically or specified in some other manner. Advertisements $A \subseteq P \times E$ are used to promote descriptions of the expertise of peers in the network. An advertisement $a \in A$ associates a peer p with an expertise e.

In general matching and peer selection algorithm based on Hasse approach deals with four entities which are: queries, subjects, similarity function and peer selection algorithm. Queries $q \in Q$ are posed by a user and are evaluated against the knowledge bases of the peers. A subject $s \in S$ is an abstraction of a given query q expressed in terms of the common ontology. The similarity function $SF : S \times E \to [0, 1]$ yields the semantic similarity between a subject $s \in S$ and an expertise description $e \in E$. And

S.K. Prasad et al. (Eds.): ICISTM 2009, CCIS 31, pp. 343–344, 2009.

peer selection algorithm returns a ranked set of peers. The rank value is equal to the similarity value provided by the similarity function.

In order to add personalization to this algorithm an auxiliary user ontology is introduced. This ontology consists of concepts which user has shown interest for them over time while working with the system. Each concept in user ontology has its own time label which shows the last time user has chosen a peer with resources related to that concept. In our proposed algorithm, in addition to the concepts in user's query, those concepts in user ontology will also take part in peer selection process. A decay factor will be assigned to concepts of user ontology which diminish their effect in peer selection, according to the time elapsed since user has shown interest in those concepts. Hence in order to select a peer for a specific query, both concepts in the query and in user's ontology (by applying decay factor) are examined against candidate peers' expertise description, and peers with greatest ranks are returned as related to the query.

Proposed algorithm is introduced for academic environments and searching different publications by students or faculties. ACM topic hierarchy with over 1287 topics is selected as the core ontology of the system. Links in ACM topic hierarchy are natural links which convey "is sub class of" and "is super class of". Also the technology for broadcasting messages in the network and getting feed back is JXTA [5]. JXTA provides each peer with various kinds of communication facilities. Depending on the situation, a peer can use JXTA unicast or broadcast channels for communicating with other peers in the network. In addition to these capabilities JXTA provides peer groups which can be used for grouping peers with similar resources together, so that a query can be passed to members of a specific group in the network.

References

1. Hasse, P., Broekstra, J., Ehrig, M., Menken, M., Peter, M., Schnizler, B., Siebes, R.: Bibster – A semantics-based bibliographic peer-to-peer system. In: McIlraith, S.A., Plexousakis, D., van Harmelen, F. (eds.) ISWC 2004. LNCS, vol. 3298, pp. 122–136. Springer, Heidelberg (2004)
2. Haase, P., Siebes, R., van Harmelen, F.: Peer selection in peer-to-peer networks with semantic topologies. In: Bouzeghoub, M., Goble, C.A., Kashyap, V., Spaccapietra, S. (eds.) ICSNW 2004. LNCS, vol. 3226, pp. 108–125. Springer, Heidelberg (2004)
3. Haase, P., Stojanovic, N., Volker, J., Sure, Y.: On Personalized Information Retrieval in Semantics-Based Peer-to-Peer Systems. In: Mueller, W., Schenkel, R. (eds.) Proc. BTW Workshop: WebDB Meets IR. Gesellschaft fuer Informatik (2005)
4. Jiang, X., Tan, A.: Ontosearch: A full-text search engine for the semantic web. In: Proc. of the 21st National Conf. on Artificial Intelligence and the 18th Innovative Applications of Artificial Intelligence Conf. (2006)
5. Li, Y., Bandar, Z., McLean, D.: An Approach for Measuring Semantic Similarity between Words Using Multiple Information Sources. IEEE Transactions on Knowledge and Data Engineering 45 (2003)

An Ant Colony Optimization Approach to Test Sequence Generation for Control Flow Based Software Testing

Praveen Ranjan Srivastava[1] and Vijay Kumar Rai[2]

[1] PhD student, [2] ME student
Computer Science & Information System Group, BITS PILANI – 333031, India
{praveenrsrivastava,vijayrai.19}@gmail.com

Abstract. For locating the defects in software system and reducing the high cost, it's necessary to generate a proper test suite that gives desired automatically generated test sequence. However automatic test sequence generation remains a major problem in software testing. This paper proposes an Ant Colony Optimization approach to automatic test sequence generation for control flow based software testing. The proposed approach can directly use control flow graph to automatically generate test sequences to achieve required test coverage.

1 Software Testing and ACO

Software testing is any activity aimed at evaluating an attribute or capability of a program or system and determining that it meets its required results. The purpose of testing can be quality assurance, verification and validation, or reliability estimation. Correctness testing and reliability testing are two major areas of testing. Reducing the high cost and locating the defects is primary goal. Software testing is a trade-off between budget, time and quality. Three main activities normally associated with software testing are: (1) test data generation, (2) test execution (3) evaluation of test results.

Ant Colony optimization Technique is inspired by observation on real ants. Individually each ant is blind, frail and almost insignificant yet by being able to co-operate with each other the colony of ants demonstrates complex behaviors. One of these is the ability to find the closest route to a food source or some other interesting land mark. This is done by laying down special chemicals called pheromones. As more ants use a particular trail, the pheromone concentration on it increases hence attracting more ants. ACO has been applied now a day frequently in many software testing applications for generating test case sequences. Pre proposed approach addresses the automatic generation of test sequences from the UML State chart diagrams for state-based software testing [5]. The all states test coverage is used as test adequacy Requirement. Specifically, two requirements have been imposed that the generated test suite has to satisfy: (1) *All-state coverage and (2) Feasibility* – Each test sequence in the test.

Evaluation at current vertex α: it will push α into the track set S, evaluate all connections to the current vertex α to determine T, and gathers pheromone levels P from their neighbor.

S.K. Prasad et al. (Eds.): ICISTM 2009, CCIS 31, pp. 345–346, 2009.

1. Move to next vertex: select destination vertex Vi with highest Pheromone level P(Vi) and move to that vertex and update P.
2. In case two destination vertices having same pheromone level, randomly select one vertex destination β and move to that vertex and update P accordingly.
3. Update Pheromone: Update the pheromone level for the current vertex α to P (α) = max (P (α), P (β) +1)
4. In the end move to the destination vertex β, set α: = β, and return to step 1.

After generating Test cases next task is to find feasible path traversed by ant. For this, we proposing a technique where, each vertex will be assigning some weight; it will be on the basis of number of nodes. if between two node i, and j, number of nodes are same, the 50:50 weight ratio will be applied, as in our example, number of node are same between source and destination, so 50:50 rules have been applied.

Using that weight, ant will decide which paths have to be chosen. But in a case where, no of nodes are not same between source and destination, then we will assign, according to number of nodes, it may be 80:20 ratio or 70:30. The path having maximum number of node will assign maximum weight, and ant will choose that path which has maximum number of weight. It is shown in below figure.

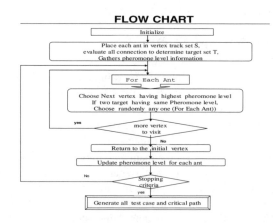

2 Conclusion

This paper presented an ant colony optimization approach to test sequence generation for Control Flow Graph based software testing. Using the developed algorithm, a group of ants can effectively explore the CFG diagrams and automatically generate test sequences Our approach has the following advantages: (1) CFG can be directly used to generate test sequences; (2) the whole generation process is fully automated; (3) redundant exploration of the CFG diagrams is avoided due to the use of ants, resulting in efficient generation of test sequences.

Reference

Li, H.: Optimization of State-based Test Suites for Software Systems: An Evolutionary Approach. International Journal of Computer & Information Science 5(3), 212–223

Component Prioritization Schema for Achieving Maximum Time and Cost Benefits from Software Testing

Praveen Ranjan Srivastava[1] and Deepak Pareek[2]

[1] PhD student, [2] ME student
Computer Science & Information System Group, Bits Pilani – 333031 India
praveenrsrivastava@gmail.com, pareekdeep@gmail.com

Abstract. Software testing is any activity aimed at evaluating an attribute or capability of a program or system and determining that it meets its required results. Defining the end of software testing represents crucial features of any software development project. A premature release will involve risks like undetected bugs, cost of fixing faults later, and discontented customers. Any software organization would want to achieve maximum possible benefits from software testing with minimum resources. Testing time and cost need to be optimized for achieving a competitive edge in the market. In this paper, we propose a schema, called the Component Prioritization Schema (CPS), to achieve an effective and uniform prioritization of the software components. This schema serves as an extension to the Non Homogenous Poisson Process based Cumulative Priority Model. We also introduce an approach for handling time-intensive versus cost-intensive projects.

1 Non-homogenous Poisson Process Based Cumulative Priority Model

The model has a to priority scale with priorities ranging from very high to very low. Each priority level is called a 'priority weight' (from 1 to 5). Each component of the software is assigned a priority weight by as already discussed above. The next step is to attach percentage stringency to each priority weight. Percentage stringency refers to the maximum allowable deviation from the optimum. For example, a percentage stringency of 25% would mean that the actual value may exceed the optimal by at most 25%. The maximum allowable deviation may vary from organization to organization, based on the organizational policies. Every component will have an optimal testing time T^* (calculated from Goel and Okumoto model), and the actual testing time T_a (i.e. the time taken to test the whole component). Similarly, there is an expected testing cost C_0, and the actual cost of testing C_a associated with every component. The Cumulative Priority Model represents the deviation from optimal in terms of three variables: α, β and δ. α is the deviation from optimal testing time, given by the equation $\alpha = (T_a - T^*)/T^*$, and β is the deviation from expected cost of testing. It is calculated by the equation $\beta = (C_a - C_0)/C_0$, where δ, the Limiting Factor, is the sum of deviations from optimal time and deviation from expected cost. It is represented by

S.K. Prasad et al. (Eds.): ICISTM 2009, CCIS 31, pp. 347–349, 2009.
© Springer-Verlag Berlin Heidelberg 2009

the equation $\delta = \alpha + \beta$. According to this model, all the components with very high priority are tested first; followed by all the components with lesser priority and so on. After testing all components of the same priority weight, δ is calculated. An optimal testing strategy can be derived from the proposed Priority based model. If the calculated $\delta \leq \delta_{max}$, it implies that the deviation from optimal time and cost is within acceptable limits. For components with very high, high, or medium priority, If $\delta > \delta_{max}$, the code must be refined before testing any further. But, if $\delta > \delta_{max}$ for components with less or very less priority, we release the software instead of modifying or improvising the code any further, because it is not possible to test the software exhaustively.

2 Component Prioritization Schema

Every software product is divided into components during the design phase. According to the Cumulative Priority Model, these components can be prioritized by a committee of experts which would consist of managers from within the organization, representatives from the client, project developers and testers.

But the perception of code complexity may vary from individual to individual. This varied perception might lead to confusion and render the priority weights ineffective. Therefore it becomes imperative to introduce a schema which ensures that the component prioritization is uniform and more effective.

We propose a schema for component prioritization i.e. assigning priority weights to the software components. While designing the CPS, few factors were considered more critical than others. These factors are found to be common for any kind of software project, and their exact quantification is quite feasible. The factors that have been selected for the prioritization schema are:

Customer-assigned priority, Code Complexity and Risk Factor, Function Usage, Requirements Volatility, Code Reusability.

3 Priority Weight Calculation

For calculating the priority weight of a component, we need to combine all the five factors mentioned above. Priority Wight for a component is calculated using the following formula:

- Priority Weight = Ceiling (Customer Priority Factor + Code Complexity Factor + Requirements Volatility Factor + Reusability Factor + Function Usage Factor)

Note here that we take the ceiling of the sum of all the factors for computing priority weight. This step is essential to ensure that components with a sum of factors < 1 are not assigned a priority weight of 0. Another reason for using the ceiling function is to avoid fractional priority weights, because the Cumulative Priority Model   does not deal with fractional weights.

4 Conclusion

The proposed extension to the Non Homogenous Poisson Process based Cumulative Priority Model provides a prioritization schema to ensure that software components are prioritized uniformly and efficiently. The paper also introduces a schema to introduce higher weight factors with time or cost, thereby decreasing the time or cost stringency of the project.

Reference

Ranjan, P., Srivastava, G., et al.: Non -Homogenous Poisson Process Based Cumulative Priority Model for Determining Optimal Software Testing Period. ACM SIGSOFT Software Engineering Notes 33(2) (March 2008)

Optimization of Software Testing Using Genetic Algorithm

Praveen Ranjan Srivastava

Computer Science and Information System Group
Bits Pilani - 333031 India
praveenrsrivastava@gmail.com

Abstract. Test data generation is one of the key issues in software testing. A properly generated test suite may not only locate the errors in a
software system, but also help in reducing the high cost associated with
software testing. It is often desired that test data in the form of test
sequences within a test suite can be automatically generated to achieve
required test coverage. This paper proposes Genetic Algorithm to test
data generation for the optimizing software testing.

Keywords: Software testing, Genetic Algorithm, Test-data Generation,
Test Cases, Sorting.

1   Test Data Generation Using GA

The use of GA in the generation of test data is one of the main milestones of
this project. The overall goal of the project is to provide a methodology in which
strategies inspired by nature play a major role on the testing (data) process while
eliminating the bias that arise from experts' decisions. The code generates random numbers for sorting them using QuickSort. The numbers generated are
expected to be integers within 0 and 100. A research using genetic algorithms
is done to automate the generation of data such that it follows the constraint.
GA generates chromosomes for the initial generation which is the binary representation of every random number generated. With the initial generation as the
current generation, the fitness of every chromosome is evaluated using the fitness
function. Once the fitness is evaluated, the individuals are sorted according to
their fitness. After which GA proceeds to the iterative phase, which has several
steps:

2   Algorithm Implemented

The algorithm is described below:

1. Let N be the number elements to be sorted which is a user input.
2. GATestDataGeneration;(generates random numbers which are then
     tested and then sorted using Quick sort)

S.K. Prasad et al. (Eds.): ICISTM 2009, CCIS 31, pp. 350–351, 2009.

3. GAGenerationGenerator;(generates bit string corresponding to
   each random number)
4. GA Fitness Evaluation;(evaluates fitness for every member of
   the population)
5. Assigning the bit string and the fitness to Individual of
   the population;
6. GAFitnessSort;(sorts the individual as per its fitness)
7. While (individual's fitness >= N)
       Find probability;(Random number generated between 0.0 and 1.0)
       if (probability > 0.8)
         CrossOver();
       else
         Mutation();
   Repeat until all individuals are fit to be in the population.
8. Convert the bit string of every fit individual into integer.
9. This test data is then passed to Quick Sort for sorting.
10. Finally the time taken for the entire implementation is found.

GAFitnessEval:
Input: array of the generated binary string
Output: array of the evaluated fitness
FitnessEvaluation(): generates fitness for every bit string using the fitness
function mentioned below.

```
for (every bit string generated) {
  if (integer value of str[i] > 1100100(binary value of 100))
  then
    fitness[i] = N + i
  else
    fitness[i] = i
```

3   Conclusion

This paper described how genetic algorithms can be used to generate test data
for achieving limitation and constraints. I have found the time taken for the
execution of GA to test the data generated which is found to be really negligible
measure to describe the types of programs where GA testing would outperform
random testing. GA approach is very useful for generation of test data. The
theme of this paper is generation of valid test data only.

Reference

Goldberg, D.E.: Genetic Algorithms: In Search, Optimization and Machine Learning.
    Addison Wesley, MA (1989)

Study and Analysis of Software Development and User Satisfaction Level

Deepshikha Jamwal, Pawanesh Abrol, and Devanand Padha

Dept. of Computer Sci and Inf. Tech., Univ. of Jammu, India
jamwal.shivani@gmail.com, pawanesh_a@yahoo.com, dpadha@rediffmail.com

Abstract. It has been observed that the design and development of the software is usually not carried out keeping in mind the satisfaction level of user, software cost and problems encountered during usage. The user satisfaction level may depend upon parameters including software exploration SE, software quality satisfaction level SQ, software cost SC, problems encountered during work SPE and time taken to rectify the problems ST. However, mostly the satisfaction level of the software is not as high as the cost involved. An analysis of the level of satisfaction of the software vis-a-vis the cost can help understand and further improve the software development process [1].

1 Introduction

The most important aspect of a software application is its development, as it undergoes a number of development stages to reach to its final shape. The development steps, which need to be followed for developing software project, are project planning, feasibility study, requirement analysis, design etc. However the most important aspect of software development is the system design. At this stage, the features and their operations are described, which need to be incorporated in the software. In this phase one also needs to work out various screen layouts, business rules, and process diagrams, with proper documentation. Many a times, it has also been reported that due to the improper design of the software it is not explored to its full potential may be due many reasons [2]. One more aspect of the software application is its maintenance. The software must be designed and developed in such a fashion that it should be explicitly maintenance free or should involve the least maintenance [3].

2 Methodology

In general there are different categories of software available for different purposes. Due to wide application it was very difficult to select the proper software to study. The different parameters like software usage, software cost, software designing, software application areas etc. were identified. Primary data has been collected using questionnaire methods, which comprised of questions related to above parameters. These questionnaires were floated to thirty-five different chosen organizations and five questionnaires were collected from each organization.

S.K. Prasad et al. (Eds.): ICISTM 2009, CCIS 31, pp. 352–353, 2009.

In addition to the questionnaire various people associated with the computer related activities in these organizations were interviewed personally or contacted through telephone. The data collected was analyzed using certain statistical tools for further investigations [4,5].

3   Observations

Software Quality Satisfaction Level vs Software Cost
Software quality satisfaction level (Sq) parameter defines the extent of user's satisfaction after investing money in buying the software, i.e, the software cost (Sc).

Software Quality vs Problem Encountered
The parameter problem encountered (Spe) shows how much problems does the user have to face while exploring the software. As non-computer professionals do not use the full capability of software because of lack of proper training as educated people are exposed to. Another reason for non-exhaustive software usage is that they have no requirement for all features of the software.

Users Satisfaction Level wrt to Software Cost
The parameter users satisfaction level (Ss) defines the extent of satisfaction of user while doing work on the software, as he has to encounter many problems on exploring the software.

4   Conclusion

From the study it can be concluded that the satisfaction level of the user depends upon all the above-mentioned parameters. So it can be inferred that these parameters must be considered while designing different software. However we are doing further research work with additional software parameters like the software complexity, help desk response time, relative available with the open source software etc. Satisfaction level can be attributed to the lack of proper training, help and complexity of software.

References

1. Jamwal, D., et al.: Critical Analysis of Software Usage vs. Cost (February 2008) ISSN 0973-7529 ISBN No 978-81-904526-2-5
2. Maxwell, K., Van Wassenhove, L., Dutta, S.: Performance Evaluation of General Specific Models in Software Development. Effort Estimation 45(6), 787–803 (1999)
3. Estublier, J., et al.: Impact of software engineering research on the practice of software configuration management 14(4), 383–430 (October 2005) ISSN: 1049-331X
4. Adelson, B., Soloway, E.: The Role of Domain Experience in Software Design. IEEE Trans. Software Engineering (11), 1351–1360 (1985)
5. Gotterbarn, D., et al.: Enhancing risk analysis using software development impact statements. In: SEW 2001, pp. 43–51. IEEE Computer Society, Los Alamitos (2001)

Influence of Hot Spots in CDMA Macro Cellular Systems

Rekha Agarwal[1], E. Bindu[1], Pinki Nayak[1], and B.V.R. Reddy[2]

[1] Amity School of Engineering and Technology, New Delhi
{rarun96@yahoo.com, bindusugathan@gmail.com, pinki_dua@yahoo.com}
[2] Professor, USIT, GGSIP University, Delhi
bvrreddy64@rediffmail.com

Cell sectorization improves the performance of Code Division Multiple Access (CDMA) cellular system by decreasing the intracell interference [4], but the presence of Hot Spot (HS) in neighbouring cell increases the intercell interference in the home cell. This paper analyses the intercell and intra cell interferences from hot spots and their influence on overall system.

Three-sector hexagonal cell configuration with N uniformly distributed users each receiving a signal of strength S due to perfect power control is considered (Fig. 1). Intra-sector Interference with N_s users per sector can be evaluated as [2],

$$I_{intra} = (N_s - 1).S, \tag{1}$$

and inter-sector interference at home sector Base Station (BS) is evaluated as [4],

$$I_{inter} = S.\rho \int_r \int_\theta \left(\frac{r}{r_0}\right)^4 .10^{(\xi_0-\xi)/10}.rd\theta.dr. \tag{2}$$

The inter-sector interference power received by home sector BS due to presence of a Mobile Station (MS) in the neighbouring sector significantly dependents on the position (r, θ) of MS with respect to its BS (Fig. 1). The analysis (Fig. 2) show that interference is very less (in the order of -40dB) till the distance r is approximately 0.2 of cell radius and independent of the orientation θ of MS with respect to its BS. When r increases beyond 0.35 of cell radius, the interference starts depending on the angular position θ of the MS and increases as θ decreases.

Taking a HS with M uniformly distributed users at r from neighboring BS give interference of $M \cdot I(r, \theta)$ at home sector BS. The SIR at the home cell BS is,

$$\left(\frac{S}{I}\right) = \left(\frac{S}{(N-1).S + I(r,\theta).M}\right) = \left(\frac{1}{(N-1) + \frac{M}{S}.I(r,\theta)}\right) \tag{3}$$

The optimum number of users that can be accommodated in the home sector, due to presence of HS in a neighbouring sector at location (r, θ), for an acceptable threshold SIR of -13.8 dB [1] at the home sector BS can be evaluated based on (3). It is observed that, in order to maintain a threshold level of -13.8dB, the number of users decrease from approx. 23 to 16 in the home cell as user density in HS increases from 10 to 30 (Fig. 3). Keeping user density and angle of HS constant, if location is varied, the SIR in home cell will decrease when HS moves far from the neighbouring BS

S.K. Prasad et al. (Eds.): ICISTM 2009, CCIS 31, pp. 354–355, 2009.
© Springer-Verlag Berlin Heidelberg 2009

(Fig. 4). With HS located at cell boundary ($r=1.0R$), the number of users that can be supported reduces to nearly 5. Analysis shows the effect of location and user concentration of HS on system capacity.

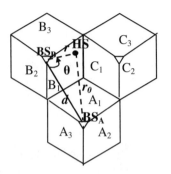

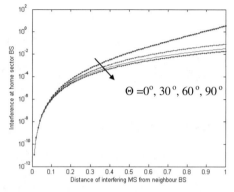

Fig. 1. The HS is in neighbouring sector B_1 at (r, θ). Cell radius is unity[3].

Fig. 2. Interference on home sector BS due to the MS in the neighbouring sector

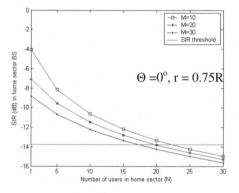

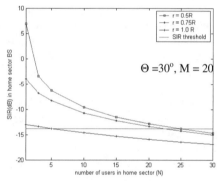

Fig. 3. SIR against users in home sector BS, with different HS concentrations

Fig. 4. SIR against users in home sector, with different hot spot locations

References

1. Wu, J.S., Chung, J.K., Wen, C.C.: Hot-spot traffic Relief with a Tilted Antenna in CDMA Cellular Networks. IEEE Transaction on Vehicular Technology 47, 1–9 (1998)
2. Gilhousen, K.S., Jaccobs, I.M., Padovani, R., Viterbi, A.J., Weaver, L.A., Wheatley, C.E.: On Capacity of a Cellular CDMA System. IEEE Transaction on Vehicular Technology 40, 303–312 (1991)
3. Nguyen, Trung, V., Dassanayake, P.: Estimation of Intercell Interference in CDMA. In: Australian Telecommunications Networks and Applications Conference (2003)
4. Nguyen, T., Dassanayake, P., Faulkner, M.: Capacity of Cellular Systems with Adaptive Sectorisation and non-uniform Traffic. In: IEEE Vehicular Technology Conf. (2001)

An Overview of a Hybrid Fraud Scoring and Spike Detection Technique for Fraud Detection in Streaming Data

Naeimeh Laleh and Mohammad Abdollahi Azgomi

Department of Computer Engineering, Iran University of Science and Technology, Tehran, Iran
naimeh_laleh@comp.iust.ac.ir, azgomi@iust.ac.ir

1 Extended Abstract

Credit card and personal loan applications have increased significantly. Application fraud is present when the application forms contain plausible and synthetic identity information or real stolen identity information. The monetary cost of application fraud is often estimated to be in the billions of dollars [1].

In this paper we propose a hybrid fraud scoring and spike detection technique for fraud detection in streaming data over time and space. The algorithm itself differentiates normal, fraud and anomalous links, and increases the suspicion of fraud links with a dynamic global black list. Also, it mitigates the suspicion of normal links with a dynamic global white list. In addition, this technique uses spike detection [3], because it highlights sudden and sharp rises in intensity, relative to the current identity attribute value, which can be indicative of abuse. In the most subsequent mining methods on data sets the assumption is that the data is already labeled. In credit card fraud detection, we usually do not know if a particular transaction is a fraud until at least one month later. Due to this fact, most current applications obtain class labels and update the existing models in preset frequency, which are usually synchronized with data refresh [2, 4]. Our work foils fraudsters' attempts to commit identity fraud with credit applications with our unique type of anomaly detection, spike detection and known fraud detection strategy. As identity fraudsters continuously morph their styles to avoid detection, our approach is rapid as it is not completely dependent on flagging of known fraudulent applications.

For each incoming identity example, this technique creates one of the three types of single links (black, white, or anomalous) against any previous example within a set window [5]. Subsequently, it integrates possible multiple links to produce a smoothed numeric suspicion score [9]. Also, this technique is able to detect both known frauds and new fraud patterns. This technique uses spike detection. For each identity example, it detects spikes in single attribute values and integrates multiple spikes from different attributes to produce a numeric suspicion score.

The input data is personal identity applications arrive in streams [6]. Each stream is continuously extracted from the entire database. These extracted applications are either previously scored (previous stream which exists within the window), or the ones to be scored in arrival order (the current stream). Each application is made up of a mixture of both dense and sparse attributes. Sparse attributes usually consist of string occurrences and identifiers with an enormous number of possible values which

S.K. Prasad et al. (Eds.): ICISTM 2009, CCIS 31, pp. 356–357, 2009.

can occur at widely spaced intervals (for example, personal names and telephone numbers). Dense attributes are usually numerical and have a finite number of values and therefore occur more frequently (for example, street numbers and postcodes) [7, 10].

The purpose is to derive two accurate suspicion scores for all incoming new applications in real-time. To reach this goal, each current attribute value a_{ik} is compared to the previous ones $a_{i-1k}, a_{i-2k}, ..., a_{i-wk}$ to detect sparse and dense attributes and give the first score to each application for spike detection [11, 12]. At the same time, every current non-scored application v_i, will be matched against all previous scored applications v_j within a window W to generate the second numeric suspicion scores on credit applications based on the links to each other, where W is the window size of the previously extracted applications in the user-specified temporal representation [8]. An alarm is raised on an application if the Second suspicion score is higher than T_{fraud} or higher than zero and first suspicion score is higher than previous application's suspicion score.

References

1. Kou, Y., Lu, C., Sinvongwattana, S., Huang, Y.P.: Survey of Fraud Detection Techniques. In: Proc. of IEEE Networking, Taiwan, March 21-23 (2004)
2. Bolton, R.J., Hand, D.J.: Statistical Fraud Detection: A Review (2002)
3. Phua, C., Gayler, R., Lee, V., Smith-Miles, K.: Adaptive Spike Detection for Resilient Data Stream Mining. In: Proc. of 6th Australasian Data Mining Conf. (2007)
4. Fan, W., Huang, Y.-a., Wang, H., Yu, P.S.: Active Mining of Data Streams. IBM T. J. Watson Research, Hawthorne, NY 10532 (2006)
5. Phua, C., Lee, V., Smith-Miles, K.: The Personal Name Problem and a Recommended Data Mining Solution. In: Encyclopedia of Data Warehousing and Mining, Australian (2007)
6. Christen, P.: Probabilistic Data Generation for Deduplication and Data Linkage. Australian National University (Visited: 20-10-2008), http://datamining.anu.edu.au/linkage.html
7. Witten, I., Frank, E.: Data Mining: Practical Attribute Ranking Machine Learning Tools and Techniques. Morgan Kaufmann, Elsevier, San Francisco (2005)
8. Phua, C., Gayler, R., Lee, V., Smith-Miles, K.: On the Communal Analysis Suspicion Scoring for Identity Crime in Streaming Credit Applications. European J. of Operational Research (2006)
9. Macskassy, S., Provost, F.: Suspicion Scoring based on Guilt-by-Association, Collective Inference. In: Proc. of Conf. on Intelligence Analysis (2005)
10. Phua, C., Gayler, R., Lee, V., Smith, K.: Temporal representation in spike detection of sparse personal identity streams. In: Chen, H., Wang, F.-Y., Yang, C.C., Zeng, D., Chau, M., Chang, K. (eds.) WISI 2006. LNCS, vol. 3917, pp. 115–126. Springer, Heidelberg (2006)
11. Li, T., Li, Q., Zhu, S., Ogihara, M.: A Survey on Wavelet Applications in Data Mining 4(2), 49–69 (2002)
12. Steiner, S.H.: Exponentially Weighted Moving Average Control Charts with Time Varying Control Limits and Fast Initial Respons. Journal of Quality Technology 31, 75–86 (1999)

PCNN Based Hybrid Approach for Suppression of High Density of Impulsive Noise

Kumar Dhiraj, E. Ashwani Kumar, Rameshwar Baliar Singh,
and Santanu Kumar Rath

National Institute of Technology Rourkela
Rourkela-769008, Orissa, India
kumardhiraj.nit.rourkela@gmail.com

Abstract. Many image processing applications requires Impulsive noise elimination. Windyga's peak-and-valley filter used to remove impulsive noise, its
main disadvantage is that it works only for low density of noises. In this Paper,
a variation of the two dimensional peak-and-valley filters is proposed to overcome this problem. It is based on minimum/maximum values present in the
noisy image, which replaces the noisy pixel with a value based on neighborhood information based on the outcomes of PCNN (Pulse Coupled Neural
Network). This method preserves constant and edge areas even under high impulsive noise probability. Extensive Computer simulations show that the proposed approach outperforms other filters in the noise reduction and the image
details preservation.

Keywords: PCNN, Peak-and-valley Filter, k-NN, Impulsive noise.

1 Introduction

The structure of PCNN was proposed by Eckhorn et al. The standard PCNN [1]
model works as iteration by the following equations:

$$F_{ij}[n] = S_{ij} \tag{1}$$

$$L_{ij} = \sum (W_{ijkl} * Y[n-1])_{ij} \tag{2}$$

$$U_{ij}[n] = F_{ij} \cdot (1 + \beta \cdot L_{ij}[n]) \tag{3}$$

$$Y_{ij}[n] = \begin{cases} 1 & U_{ij} if\ (n) > \theta_{ij}[n-1] \\ 0 & otherwise \end{cases} \tag{4}$$

$$\theta_{ij}[n] = \theta_{ij}[n-1] \cdot e^{-\alpha_\theta} + V_\theta \cdot Y_{ij}[n-1] \tag{5}$$

2 Improved Peak-and-Valley

Intuition Behind Modification of Standard Peak-and-Valley Filter:
This is based on k-NN based classification model. According to this, a pixel is said to
be a true pixel when majority of k-nearest neighbor falls in the same category otherwise it is used to be said as a noisy pixel and vice versa.

S.K. Prasad et al. (Eds.): ICISTM 2009, CCIS 31, pp. 358–359, 2009.

3 Proposed Variant of Peak-and-Valley Filter Based on PCNN

The characteristics of pulse noise and the feature of similar neurons firing synchro-nously and bursting pulse of PCNN can realize the detecting of noisy pixels. The pulse noised pixels, either bright (1) or dark (0), are quite different from others. In PCNN, the neurons corresponding to bright (dark) pixels are inspired firstly (finally) [2]. The PCNN outputs can be used to control the peak-and-valley filtering so that image details and other geometrical information are well kept while de-noising. The initial thresh-olds of PCNN were set as 0. From Eqs. (1)~ (5), we knew that the threshold reached to the maximum in the second iteration. If the parameters were selected appropriately, with iteration and decay of thresholds, the neuron with high brightness (noisy points) would output 1 firstly, while others would output) 0. Similarly we repeat this process for high darkness (noisy points) and would give output 1, while others would output) 0. After detecting the noisy pixels (1 or 0), the improved peak-and-valley filtering was performed in the 3X3 (5X5 or 7X7) neighboring area of noisy points, and the bright (1) and dark (0) were removed effectively. Since our proposed approach only processed the noise pixels, it was able to keep the edges and details as well.

Table 1. PCNN Based Improved Peak-and-Valley Filtering

Noise	Original	Median	Peak & Valley	PCNN With Median	PCNN with improved P& V
10	15.45	28.61	19.78	**35.45**	31.88
15	13.64	27.31	16.74	**31.73**	30.41
20	12.40	25.03	14.21	28.13	**29.24**
25	11.52	22.88	13.26	25.57	**27.96**
30	10.64	20.63	11.91	22.86	**26.03**

4 Conclusion

Proposed PCNN based improved peak-and-valley works better in following conditions:

1) It performs better in compare to a) Standard median filter and b) standard peak-and-valley filter; for low texture information and intensity.

2) Also, it performs better compare to a) standard median filter, b) standard 2-dimensional peak-and-valley filter, c) improved peak and valley and d) PCNN with median filter; for high density of impulsive noises.

References

[1] Johnson, J.L.: PCNN Models and Applications. IEEE Trans. Neural Networks 10(3), 480–498 (1999)
[2] Chan, R.H., Hu, C., Nikolova, M.: An iterative procedure for removing random valued impulse noise. IEEE Signal Process. Lett. 11(12), 921–924 (2004)

An Overview of a New Multi-formalism Modeling Framework

Hamid Mohammad Gholizadeh and Mohammad Abdollahi Azgomi

Performance and Dependability Engineering Lab., Department of Computer Engineering, Iran
University of Science and Technology, Tehran, Iran
`hgholizadeh@comp.iust.ac.ir, azgomi@iust.ac.ir`

1 Extended Abstract

Despite the fact that several techniques are proposed for adding compositionality to
formalisms have appeared in the literature [1, 2], composition of models of different
formalisms is still a concern. A very recent trend in multi-formalism is to take advantage of meta-modeling approaches [3, 4]. We have exploited this technique in our
framework and blended it with object-orientation to add flexibility and compositionality to our framework.

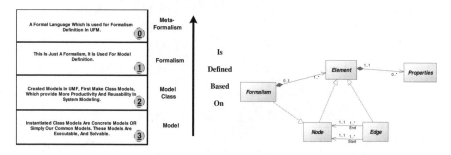

Fig. 1. Framework model definition data structure and Formalism metadata in the framework

Fig. 1. depicts 4-layers and their reinterpretation which is used in framework for its
structure definition. Layer 0 is formal description of a formalism, which is called
meta-formalism. In our framework Formalism is collection of *ELEMENTS* and each
ELEMENT has type of {*Node, Edge, Model*}. For example, in SAN formalism all
places, activities, input and output gates are considered as ELEMENTS of Node type
and arcs are considered as Edge Type. It is possible that a model contains other models which are called sub-models. Those sub-models are considered as type of *Model*.
ELEMENTS might have some *properties* too. These properties make each element
distinct. For example, a token is property of place, and gate function is property of
gates in SAN formalism. These corresponding properties can be defined as *int* type
and *function* type, respectively.

We can define heterogeneous composed models in framework. Each Sub-Model is
connected by an arc to at least one container Node or Sub-Model .Semantic of this
connection is defined by *Relation Function*. It is a simple programming function and

S.K. Prasad et al. (Eds.): ICISTM 2009, CCIS 31, pp. 360–361, 2009.
© Springer-Verlag Berlin Heidelberg 2009

theatrically can be written in any procedural programming language. It defines how the values in sub model can be affected by the values in container model and vise versa. By that we can exploit strengthen of procedural programming language in defining relationship between models and sub models. *Relation Function* is executed by solvers while communication is needed between corresponding models.

Perfect description of formalism rules is not possible, using only concepts like *Properties* or *ELEMENTS*. To address this issue OCL expressions [5, 6] are used. they are based on meta-data diagram in Fig.1. OCL is standard formal language for defining constraints in UML diagrams, and it is easier for formalism definers to use it, rather than using a mathematical language.

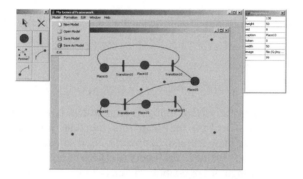

Fig. 2. Screenshots of a tool for framework showing the producer/consumer model

We implemented a framework tool with Java™. We designed it based on Model-View-Controller (MVC) model [7], since it should be interactive and event oriented application (Fig. 2). The tool is so customizable that you can easily switch between different defined formalism for making a composed model.

References

1. Peccoud, J., Courtney, T., Sanders, W.H.: Möbius: An Integrated Discrete-Event Modeling Environment. Bioinformatics 23(24), 3412–3414 (2007)
2. CPN Tools (Visited: 15-04-2008), http://www.daimi.au.dk/CPNtools/
3. de Lara, J., Vangheluwe, H.: AToM3: A Tool for Multi-Formalism and Metamodelling. In: Kutsche, R.-D., Weber, H. (eds.) FASE 2002. LNCS, vol. 2306, p. 174. Springer, Heidelberg (2002)
4. Vittorini, V., et al.: The OsMoSys Approach to Multi-formalism Modeling of Systems. J. Software and Systems Modeling (SoSyM) 3(1), 68–81 (2004)
5. The Object Management Group (OMG): OCL 2.0 Specification. Version 2 (2005)
6. Warmer, J., Kleppe, A.: The Object Constraint Language: Getting your Models Ready for MDA, 2nd edn. Addison-Wesley, Reading (2003)
7. Burbeck, S.: Applications Programming in Smalltalk-80 (TM): How to Use Model-View-Controller (MVC)

Association Rule Based Feature Extraction for Character Recognition

Sumeet Dua and Harpreet Singh

Department of Computer Science, Louisiana Tech University,
Ruston, LA 71272, USA
sdua@coes.latech.edu, hsi001@latech.edu

Keywords: Association rules, feature extraction, supervised classification.

1 Introduction

Association rules that represent isomorphisms among data have gained importance in exploratory data analysis because they can find inherent, implicit, and interesting relationships among data. They are also commonly used in data mining to extract the conditions among attribute values that occur together frequently in a dataset [1]. These rules have wide range of applications, namely in the financial and retail sectors of marketing, sales, and medicine.

2 Proposed Methodology

In this paper we present the novel idea of representing association rules in a multidimensional format so that they can be used as input for classifiers in place of raw features. Association rules (AR) are based on the strong rule interestingness measures (*support* and *confidence*) for discovering relationships present in large-scale databases, typically "market-basket" datasets, and have gained increasing importance in recent years. Assuming a Dataset (D) of items in market basket format a typical AR is represented as $A \Rightarrow B$, where both A and B are combination of *itemsets* for e.g. $i_1, i_2 \Rightarrow i_3, i_4$ where $i_1, i_2, i_3, i_4 \in D$. This rule is associated with two measures called the support and confidence defined as

$$\text{Support}(i_2, i_3 \Rightarrow i_5) = P((i_5 \cup i_2) \cup i_3) = \frac{numtrans(i_2, i_3, i_5)}{N} \quad conf\,idence(i_2, i_3 \Rightarrow i_5) = P(i_5 \mid i_2, i_3) = \frac{numtrans(i_2, i_3, i_5)}{numtrans(i_2, i_3)}$$

where *numtrans(x)* is the number of transactions in database D having x items and N is the total number of transactions. Our methodology consists of three major parts namely data preprocessing, association rule generation, and classifier training and classification. The outline of the proposed methodology can be seen in Figure 1. During data preprocessing the first four central moments of data (expectation, variance, skewness and kurtosis) are generated, and raw feature values are replaced with these moments. Data is normalized and discretized to have a subtle representation. Association rules [1, 3] are then generated from this dataset, which satisfy

S.K. Prasad et al. (Eds.): ICISTM 2009, CCIS 31, pp. 362–364, 2009.
© Springer-Verlag Berlin Heidelberg 2009

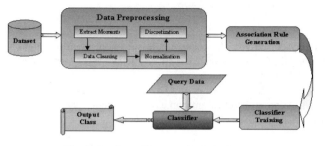

Fig. 1. Outline of Proposed Methodology

minimum support and confidence values. The confidence is fixed high at 90% and support is varied from 15% to 30%. The final step of the process is to aggregate the association rules from instances of same class into a class rule set and represent it in a structured format keeping only support and confidence for classification.

3 Results

We ran several sets of experiments where 100 instances in each class (16%) were used for training, and 500 instances were used for testing the Artificial Character dataset [2]. The first set of experiments showed the comparison of raw feature values as input to classifier versus (four) moments-based AR as an input to the classifier. The minimum support was set at 30%, and the minimum confidence was set at 90% for association rule mining. Four different classifiers were used for comparison: F-KNN, Naïve Bayes, SVM, and Adaboost [3]. Extensive experimentation was performed. It was observed that when raw features were used as input, the highest accuracy was only 26%, and when AR were used for classification, the number jumped to 45%. The jump in accuracy was consistent over all the classifiers although the increase varied from classifier to classifier. In the next set of experiments we evaluated the effect of using only support values or only confidence values for the classification of instances. Using only support values gave us better accuracy than using only confidence values. This variation in accuracy was consistent over both four and five sets of feature moments. The FKNN classifier [1] was used with two support and confidence pairs for four moments (Sup= 30% and 25%, and Conf = 90%) and three support and confidence measures for five moment features (Sup= 30%, 25%, and 20%, and Conf = 90%). The final set of experiments was performed to see the effect of using only a selected set of rules as input to FKNN. Initially, all the rules from a class were combined into a rule set and then only the top 50 or top 100 unique rules were selected. The accuracy increased with the number of rules.

4 Conclusion

In this paper we have demonstrated that employing association rules based on moments of features is superior to the raw feature values of data for classification. Association rules contain discriminatory properties, which when carefully exploited can enhance the efficacy of supervised classification algorithms.

References

1. Dua, S., Singh, H., Thompson, H.W.: Associative Classification of Mammograms using Weighted Rules based Classification, Expert Systems With Applications, doi:10.1016/j.eswa.2008.12.050
2. http://archive.ics.uci.edu/ml/datasets/Artificial+Characters
3. Witten, I., Frank, E.: Data Mining: Practical machine learning tools and techniques, 2nd edn. Morgan Kaufmann, San Francisco (2005)

Author Index